HUMAN BIOLOGY

Writing Team:
James Torrance
James Fullarton
Clare Marsh
James Simms
Caroline Stevenson

Diagrams by James Torrance

HODDER
GIBSON
AN HACHETTE UK COMPANY

The Publishers would like to thank the following for permission to reproduce copyright material:

Photo credits: p.1 (background) and Unit 1 running head image © Alexandr Mitiuc – Fotolia.com, (inset left) © STEVE GSCHMEISSNER/SCIENCE PHOTO LIBRARY, (inset centre) © 3D4MEDICAL.COM/SCIENCE PHOTO LIBRARY, (inset right) © Michael J. Gregory, Ph.D./Clinton Community College; p.3 © Deco Images II / Alamy; p.5 © STEVE GSCHMEISSNER/SCIENCE PHOTO LIBRARY; p.6 © 3D4MEDICAL.COM/SCIENCE PHOTO LIBRARY; p.13 © James King-Holmes/Science Photo Library; p.14 © James Torrance; p.15 © SCIENCE PHOTO LIBRARY; p.22 © Science Source/Science Photo Library; p.45 © Eric Martz Image prepared by Eric Martz from 7tim using Jmol (MolviZ.Org); p.49 © PHOTOTAKE Inc. / Alamy; p.58 © MAURO FERMARIELLO/SCIENCE PHOTO LIBRARY; p.91 © Michael J. Gregory, Ph.D./Clinton Community College; p.115 © Orlando Florin Rosu – Fotolia.com; p.125 (background) and Unit 2 running head image © Imagestate Media (John Foxx), (inset left) © Dan Marschka/AP/Press Association Images, (inset centre) © DR NAJEEB LAYYOUS/SCIENCE PHOTO LIBRARY, (inset right) © DR P. MARAZZI/SCIENCE PHOTO LIBRARY; p.135 © Dan Marschka/AP/Press Association Images; p.143 © DR NAJEEB LAYYOUS/SCIENCE PHOTO LIBRARY; p.146 © BSIP SA / Alamy; p.166 © ANDY CRUMP, TDR, WHO/SCIENCE PHOTO LIBRARY; p.183 © DR P. MARAZZI/SCIENCE PHOTO LIBRARY; p.205 (background) and Unit 3 running head image © Sebastian Kaulitzki – Fotolia.com, (inset left) © DR P. MARAZZI/SCIENCE PHOTO LIBRARY, (inset centre) © STEVE GSCHMEISSNER/SCIENCE PHOTO LIBRARY, (inset right) © MEDICAL BODY SCANS/JESSICA WILSON/SCIENCE PHOTO LIBRARY; p.219 © DR P. MARAZZI/SCIENCE PHOTO LIBRARY; p.221 © Cindee Madison and Susan Landau, UC Berkeley; p.222 © PHILIPPE PSAILA/SCIENCE PHOTO LIBRARY; p.226 © C.J Bucher, Verlag, Munchen, Germany; p.227 © Tim Graham/Getty Images; p.235 (left) © James Torrance, (right) © From a study by Bugelski and Alampay, 1961; p.236 (left) © Ely William Hill in 1915, published in *Puck*, an American humour magazine, on 6 November 1915; p.249 © TRACY DOMINEY/SCIENCE PHOTO LIBRARY; p.252 © Sebastian Kaulitzki – Fotolia.com; p.254 © STEVE GSCHMEISSNER/SCIENCE PHOTO LIBRARY; p.256 © DON FAWCETT/SCIENCE PHOTO LIBRARY; p.259 © Paul Kingsley / Alamy; p.265 © Clare Marsh; p.269 © MEDICAL BODY SCANS/JESSICA WILSON/SCIENCE PHOTO LIBRARY; p.274 (top) © MAURO FERMARIELLO/SCIENCE PHOTO LIBRARY, (bottom) © James Torrance; p.285 (left) © S Harris / www.cartoonstock.com, (right) © NiDerLander – Fotolia.com; p.286 (top) © Norman Thelwell, (bottom) © auremar – Fotolia.com; p.288 © David Cole / Alamy; p.289 © TEMISTOCLE LUCARELLI – Fotolia.com; p.291 © Print Collector / HIP / TopFoto; p.302 © Norman Thelwell; p.311 (background) and Unit 4 running head image © Sebastian Kaulitzki – Fotolia.com, (inset left) © DAVID SCHARF/SCIENCE PHOTO LIBRARY, (inset centre) © DAVID SCHARF/SCIENCE PHOTO LIBRARY, (inset right) © JOHN BAVOSI/SCIENCE PHOTO LIBRARY; p.314 (left) © SCIENCE PHOTO LIBRARY, (right) © EYE OF SCIENCE/SCIENCE PHOTO LIBRARY; p.323 © FLPA / Alamy; p.324 © Dey Pharma; p.327 © STEVE GSCHMEISSNER/SCIENCE PHOTO LIBRARY; p.332 (top right) © DAVID SCHARF/SCIENCE PHOTO LIBRARY, (middle left) © NIBSC/SCIENCE PHOTO LIBRARY, (middle right) © DAVID SCHARF/SCIENCE PHOTO LIBRARY, (bottom left) © AMI IMAGES/SCIENCE PHOTO LIBRARY, (bottom right) © LONDON SCHOOL OF HYGIENE & TROPICAL MEDICINE/SCIENCE PHOTO LIBRARY; p.333 © Custom Medical Stock Photo / Alamy; p.334 © Elizabeth Fletcher. Product images courtesy of Omega Pharma (TCP), Novartis Consumer Health UK Limited (Savlon), Reckitt Benckiser Group plc (Dettol); p.335 (top left) © Copyright 2006 Carolina K. Smith – Fotolia.com, (top right) © demarfa – Fotolia.com, (bottom left) © CDC/SCIENCE PHOTO LIBRARY; p.345 (top) © JOHN BAVOSI/SCIENCE PHOTO LIBRARY, (bottom) © WIM VAN CAPPELLEN/REPORTERS/SCIENCE PHOTO LIBRARY; p.346 © PROF. S.H.E. KAUFMANN & DR J.R GOLECKI/ SCIENCE PHOTO LIBRARY; p.348 © BARBARA RIOS/SCIENCE PHOTO LIBRARY.

Acknowledgements The authors and publisher would like to extend grateful thanks to Jim Stafford for assistance offered at manuscript stage of this book, as well as for further guidance and editorial advice during the production process.

Every effort has been made to trace all copyright holders, but if any have been inadvertently overlooked the Publishers will be pleased to make the necessary arrangements at the first opportunity.

Although every effort has been made to ensure that website addresses are correct at time of going to press, Hodder Gibson cannot be held responsible for the content of any website mentioned in this book. It is sometimes possible to find a relocated web page by typing in the address of the home page for a website in the URL window of your browser.

Hachette UK's policy is to use papers that are natural, renewable and recyclable products and made from wood grown in sustainable forests. The logging and manufacturing processes are expected to conform to the environmental regulations of the country of origin.

Whilst every effort has been made to check the instructions of practical work in this book, it is still the duty and legal obligation of schools to carry out their own risk assessments.

Orders: please contact Bookpoint Ltd, 130 Milton Park, Abingdon, Oxon OX14 4SB. Telephone: (+44) 01235 827720. Fax: (+44) 01235 400454. Lines are open 9.00–5.00, Monday to Saturday, with a 24-hour message answering service. Visit our website at www.hoddereducation.co.uk. Hodder Gibson can be contacted direct on: Tel: 0141 848 1609; Fax: 0141 889 6315; email: hoddergibson@hodder.co.uk

© James Torrance, James Fullarton, Clare Marsh, James Simms, Caroline Stevenson 2013

First published in 2013 by
Hodder Gibson, an imprint of Hodder Education,
An Hachette UK Company
2a Christie Street
Paisley PA1 1NB

Without Answers
Impression number   5 4
Year   2016

ISBN: 978 1444 182156

With Answers
Impression number   5 4
Year   2016
ISBN: 978 1444 182132

Cover photo © V. Yakobchuk – Fotolia.com
Illustrations by James Torrance
Typeset in Minion Pro 11pt by Fakenham Prepress Solutions, Fakenham, Norfolk NR21 8NN
Printed in Italy

A catalogue record for this title is available from the British Library

Contents

The book you are holding is from a second (or subsequent) printing of this title. In this version the Chapter/Section names and page numbers have been amended from the first printing. These changes have been made to be in line with amendments that were made to the ordering of the Higher syllabus in summer 2014.

Preface

This book has been written to act as a valuable resource for students studying Higher Grade Human Biology. It provides a **core text** which adheres closely to the SQA syllabus for *Higher Human Biology* introduced in 2014. Each section of the book matches a unit of the revised syllabus; each chapter corresponds to a content area. In addition to the core text, the book contains a variety of special features:

Learning Activities

Each chapter contains an appropriate selection of learning activities in the form of *Case Studies, Case Histories, Related Topics, Research Topics, Related Information, Related Activities* and *Investigations*, as laid down in the SQA Course Support Notes. These non-essential activities are highlighted throughout in yellow for easy identification. They do not form part of the basic mandatory course content needed when preparing for the final exam but are intended to aid understanding and to support research tasks during course work.

Practical Activity and Report

An assignment designed to match the required performance criteria and provide students with the opportunity to write a *Scientific Report*, which includes description of procedure, recording of results, drawing of conclusions and evaluation of procedure. This report satisfies the requirements of a mandatory part of SQA assessment.

Testing Your Knowledge

Key questions incorporated into the text of every chapter and designed to continuously assess *Knowledge and Understanding*. These are especially useful as homework and as instruments of diagnostic assessment to check that full understanding of course content has been achieved.

What You Should Know

Summaries of key facts and concepts as *'Cloze' Tests* accompanied by appropriate word banks. These feature at regular intervals throughout the book and provide an excellent source of material for consolidation and revision prior to the SQA examination.

Applying Your Knowledge and Skills

A variety of questions at the end of each unit designed to give students practice in exam questions and foster the development of *Skills of Scientific Experimentation, Investigation and Enquiry* (i.e. selection of relevant information, presentation of information, processing of information, planning experimental procedure, evaluating, drawing valid conclusions and making predictions and generalisations). These questions are especially useful as extensions to class work and as homework.

Updates and syllabus changes: important note to teachers and students from the publisher

This book covers all course arrangements for Revised Higher and CfE Higher, but does not attempt to give advice on any 'added value assessments' or 'open assignments' that may form part of a final grade in the CfE version of Higher Human Biology (2015 onwards).

Please remember that syllabus arrangements change from time to time. We make every effort to update our textbooks as soon as possible when this happens, but – especially if you are using an old copy of this book – it is always advisable to check whether there have been any alterations to the arrangements since this book was printed. You can check the latest arrangements at the SQA website (www.sqa.org.uk), and you can also check for any specific updates to this book at www.hoddereducation.co.uk/HigherScience.

We make every effort to ensure accuracy of content, but if you discover any mistakes please let us know as soon as possible – see our contact details on the back cover.

Human Cells

Division and differentiation in human cells

A multicellular organism such as a human being consists of a large number of cells. Rather than each cell carrying out every function for the maintenance of life, a **division of labour** occurs and most of the cells become differentiated. **Differentiation** is the process by which an unspecialised cell becomes altered and adapted to perform a specialised function as part of a permanent tissue.

Differentiation in human cells

A human being begins life as a fertilised egg (zygote), as shown in Figure 1.1. The zygote divides repeatedly by mitosis and cell division to form an embryo. Like the cells in an adult, each **embryonic cell** possesses all the genes for constructing the whole organism. However, unlike those in adult cells, all the genes in cells at this early stage are switched on or have the potential to become switched on.

As embryological development proceeds, the unspecialised cells of the early embryo undergo **differentiation** and become **specialised** in structure and biochemical properties, making them perfectly adapted for carrying out a particular function. For example, a motor neuron (see Figure 1.1) is a type of nerve cell that possesses an axon (a long, insulated cytoplasmic extension). This structure is ideally suited to the transmission of nerve impulses. Similarly the cells of the epithelial lining of the windpipe are perfectly suited to their job of sweeping dirty mucus up and away from the lungs.

Selective gene expression

Once a cell becomes differentiated, it only expresses the genes that code (see chapter 3) for the proteins **specific** to the workings of that particular type of cell. For example, in a nerve cell, genes that code for the formation of neurotransmitter substances are switched on and continue to operate but those for mucus are switched off. The reverse is true of the genes in a goblet cell in the lining of the windpipe. Only a fraction of the genes in a specialised cell are expressed (e.g. 3–5% in a typical human cell).

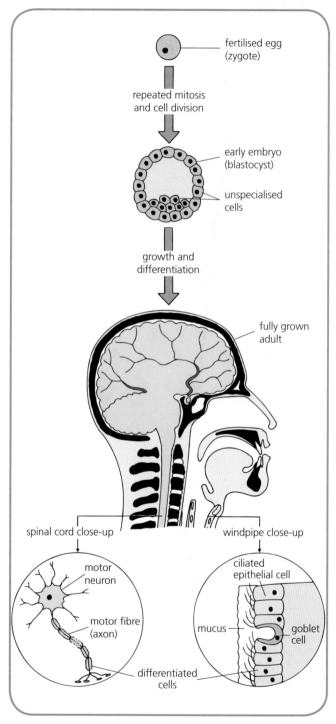

Figure 1.1 Differentiation

Stem cells

Stem cells are unspecialised cells that can:

- **reproduce** ('self-renew') themselves by repeated mitosis and cell division while remaining undifferentiated
- **differentiate** into specialised cells when required to do so by the multicellular organism that possesses them.

Embryonic stem cells

A human **blastocyst** is an early embryo consisting of a ball of **embryonic stem cells** (see Figures 1.2 and 1.3).

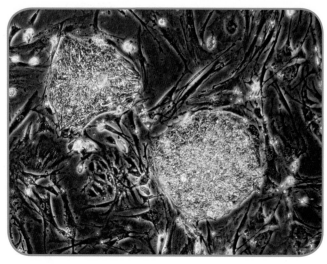

Figure 1.3 Embryonic stem cells

All of the genes in an embryonic cell have the potential to be switched on, so the cell is capable of differentiating into almost all of the cell types (more than 200) found in the human body. Because of this ability, embryonic stem cells are described as being **pluripotent** (see also page 12).

Tissue (adult) stem cells

Tissue (adult) stem cells are found in locations such as skin and red bone marrow (see Figure 1.2). They have a much **narrower differentiation potential** than embryonic stem cells because many of their genes are already switched off. However, they are able to replenish continuously the supply of certain differentiated cells needed by the organism such as skin, intestinal lining and blood.

Tissue (adult) stem cells can only give rise to a limited range of cell types closely related to the tissue in which they are normally located. However they are able to make all the cell types found in that particular tissue and are therefore described as being **multipotent**. For example, blood (haematopoietic) stem cells in red bone marrow give rise to red blood cells, white blood cells (various forms of phagocytes and lymphocytes) and platelets. (Also see the Related Topic on page 4.)

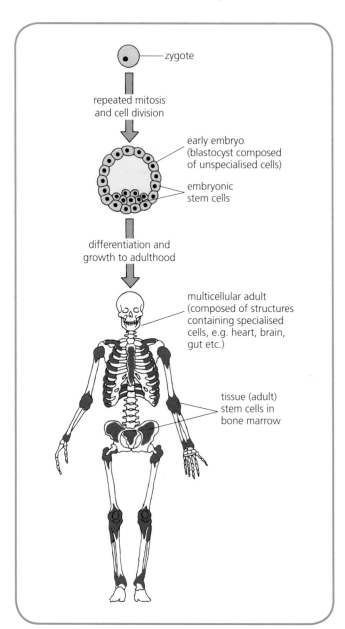

zygote

repeated mitosis
and cell division

early embryo
(blastocyst composed
of unspecialised cells)

embryonic
stem cells

differentiation and
growth to adulthood

multicellular adult
(composed of structures
containing specialised
cells, e.g. heart, brain,
gut etc.)

tissue (adult)
stem cells in
bone marrow

Figure 1.2 Two types of stem cell

Origin of blood cells

Tissue (adult) stem cells in red bone marrow (see Figure 1.4) give rise to red blood cells, platelets and specialised white blood cells including phagocytes, natural killer cells and B and T lymphocytes – the cells of the immune system.

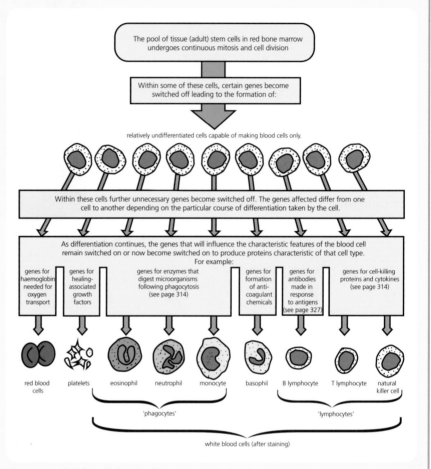

Figure 1.4 Origin of blood cells

Differentiation in somatic cells

All differentiated cells (except reproductive cells) derived from stem cells are called **somatic** cells. Somatic cells form several different types of body **tissue**.

Epithelium

Epithelial tissue is composed of cells that unite to form membranes. These can be made of a single layer of cells or a number of layers. Epithelium provides the body surface with a protective, multilayered covering – the skin. Epithelium also lines body cavities and tubular structures such as the oesophagus (see Figure 1.5) and the blood vessels. Materials such as food and respiratory gases are able to move easily through the thin layer of epithelial cells that lines blood vessels.

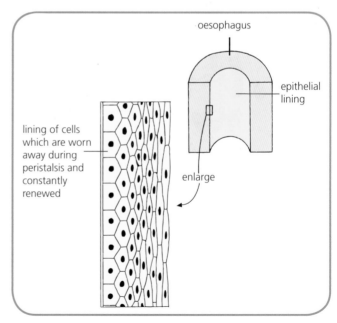

Figure 1.5 Epithelium lining the oesophagus

Connective tissue

Bone, cartilage and blood are all examples of **connective tissues**. Connective tissue is characterised by the large quantity of extracellular material present in the spaces between its cells. This matrix may be solid (e.g. in bone), fibrous or gelatinous (e.g. in different types of cartilage) or liquid (e.g. as plasma in blood).

Bone

Bone consists of concentric layers of calcified material laid down around blood vessels. Live bone cells deep within the calcified material receive oxygen and essential nutrients via tiny canals in contact with blood vessels (see Figures 1.6 and 1.7).

Figure 1.6 Bone cell

Cartilage

There are several forms of **cartilage**. They differ in the composition of their extracellular material. It might, for example, be solid and smooth, as in the type of cartilage on the ends of long bones at joints, or it might consist of dense fibres embedded in a solid background, as in the slightly flexible type of cartilage found at the knee joint (see Figure 1.8).

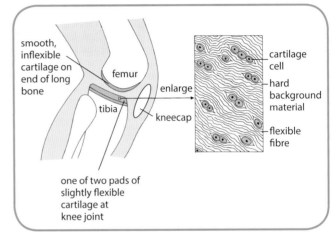

Figure 1.8 Cartilage

Blood

Blood is classified as a connective tissue because its extracellular space (about half of its volume) is composed of plasma (see page 163).

Muscle tissue

Muscle tissue consists of cells capable of contraction. The three types of **muscle** tissue are skeletal, smooth and cardiac. In **skeletal** muscle (see Figure 1.9) the cells

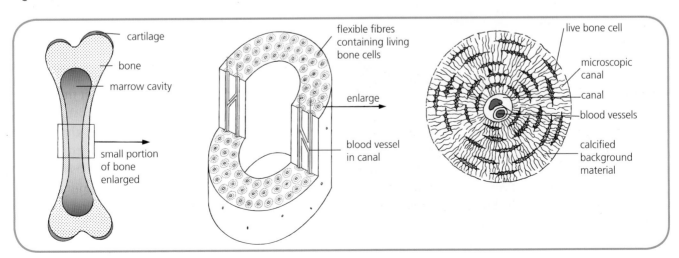

Figure 1.7 Structure of bone

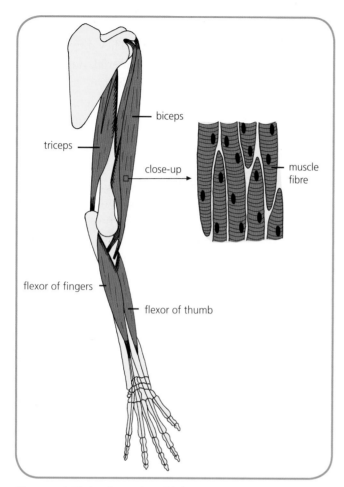

Figure 1.9 Skeletal muscles of left arm

take the form of striped fibres. In **smooth** (involuntary) muscle, the cells are spindle-shaped (see Figure 1.10) and arranged in sheets, which form part of the walls of large blood vessels and of the wall of the alimentary canal. In **cardiac** muscle (see Figure 1.11) each cell has one or more branches in contact with adjacent cells.

Figure 1.10 Smooth muscle cells

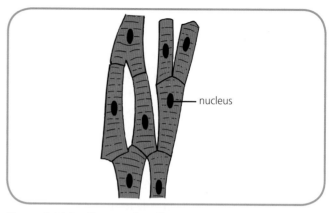

Figure 1.11 Cardiac muscle cells

Nervous tissue

This tissue is composed of a network of nerve cells called **neurons**, which receive and transmit nerve impulses, and **glial** cells, which support and maintain the neurons (see chapter 17).

Increasing levels of complexity

Somatic **cells** make up the different **tissues** in the human body. A group of tissues makes up each of the body's **organs**. The skin, for example, is an organ composed of a variety of tissues including epithelium, blood, muscle and nerves. A collection of tissues and organs make up a **system**. For example, the circulatory system is composed of blood, heart and blood vessels. All the systems make up the whole, integrated human **organism**.

Maintenance of chromosome number

Certain somatic cells divide during growth and tissue repair. In humans, each somatic cell contains **46 chromosomes** and is **diploid** (i.e. contains 23 pairs of homologous chromosomes). Prior to nuclear division (**mitosis**), the genetic material undergoes replication and becomes doubled in quantity for a brief time. Nuclear division quickly follows and the genetic material is divided equally between two daughter nuclei. Therefore each somatic cell formed receives an identical copy of the full set of 46 chromosomes and the diploid chromosome number is **maintained**. This process ensures that each somatic cell receives a complete set of the genetic information.

Differentiation in germline cells

A **germline cell** is one that eventually leads to the formation of sex cells (gametes). Germline cells are set aside from somatic cells early in the course of human development. A germline cell, like a somatic cell, is **diploid**. Its nucleus contains 23 pairs of homologous chromosomes and it is able to undergo mitosis and form more germline cells.

Meiosis

These germline cells are then able to undergo a second form of nuclear division called **meiosis**. During this process the genetic material is doubled by replication as before. However, this time, during nuclear division, it is divided between **four** nuclei. Each receives a **single set of 23 chromosomes**, resulting in the formation of four **haploid** gametes. The human life cycle alternates between haploid and diploid cells as summarised in Figure 1.12.

Mutation in germline and somatic cells

If a **mutation** occurs in a **germline** cell, it is passed on to offspring (see Figure 1.13). For example, a certain gene on chromosome 7 may mutate to the recessive form that results in the formation of abnormally thick and sticky mucus. This mutant allele is passed on during meiosis to gametes and then to offspring in the next generation. If a gamete with this recessive allele fuses with a gamete carrying a copy of the same allele (perhaps the result of a mutation in a germline cell many generations ago), the zygote formed will develop into a sufferer of **cystic fibrosis** (see page 58).

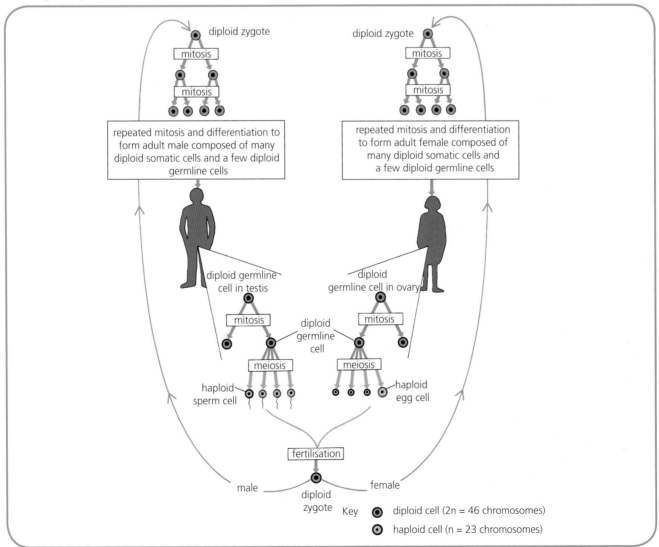

Figure 1.12 Human life cycle

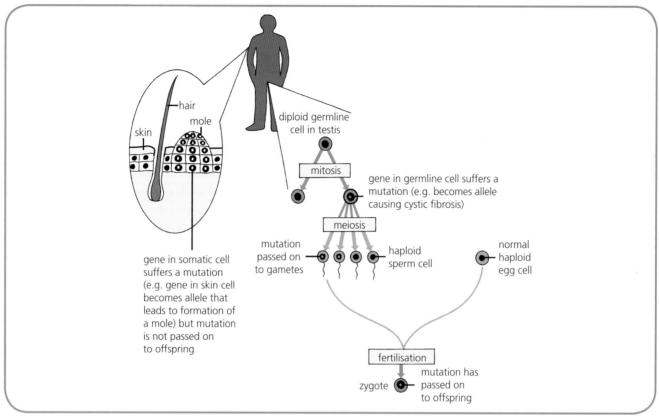

Figure 1.13 Transmission of mutation in germline cell

A mutation can occur in a **somatic** cell. It might even lead to a localised change in phenotype such as the development of a mole from a mutated skin cell.

However, this type of somatic cell mutation is not passed on to the next generation.

Testing Your Knowledge 1

1 a) Define the term *differentiation*. (2)
  b) In what way is a ciliated epithelial cell a good example of a specialised cell? (1)
  c) A goblet cell in the lining of the windpipe produces mucus but not insulin. Explain briefly how this specialisation is brought about with reference to genes. (2)

2 a) Give TWO characteristics of stem cells. (2)
  b) i) Name TWO types of stem cell found in humans.
    ii) For each type, identify ONE location where these cells could be found. (4)
  c) Which type of stem cell is capable of differentiating into all the types of cell that make up the organism to which it belongs? (1)

3 a) Name THREE different types of differentiated tissue derived from somatic cells in addition to connective tissue. (3)
  b) Arrange the following terms into order of increasing complexity: *body, cell, organ, system, tissue*. (1)
  c) Briefly describe how chromosome number is maintained in a somatic cell at cell division. (2)

4 a) What is the difference between the terms *haploid* and *diploid*? (2)
  b) What is a *germline* cell? (1)
  c) i) Briefly describe how a diploid germline cell produces haploid gametes.
    ii) In which type of cell (somatic or germline) is a mutation *not* passed on to the members of the next generation? (3)

Research value of stem cells

Much of the research to date has been carried out using stem cells from mice and humans. Human stem cells can be grown in optimal culture conditions provided that certain growth factors are present. In the absence of these growth factors, the stem cells differentiate rapidly.

By investigating why stem cells continue to multiply in the presence of a certain chemical yet undergo differentiation in its absence, scientists are attempting to obtain a fuller understanding of cell processes such as growth and differentiation. It is hoped that this will lead in turn to a better understanding of gene regulation (including the molecular biology of cancer).

Models

A **model organsim** is one that is suitable for laboratory research because its biological characteristics are similar to those of a group of related (but often unavailable) organisms. For example, mice are used in research as model organisms for humans since they possess many genes equivalent to the genes in humans that are responsible for certain inherited diseases. Research work on the models helps to provide an understanding of the malfunctioning of these genes and may lead the way to the development of new treatments.

Similarly **stem cells**, which are genetically identical to differentiated somatic cells, can be used in research as **model cells** to investigate:

- the means by which certain diseases and disorders develop
- the responses of cells to new pharmaceutical drugs.

Therapeutic value of stem cells

In recent times the therapeutic value of stem cells has been appreciated and they have been successfully put to use in bone marrow transplantation, skin grafts for burns and repair of damaged corneas. (Also see Case Study below.)

Case Study | **Therapeutic use of stem cells**

Bone marrow transplantation

Leukaemia

Cancers of the blood such as **leukaemia** and **lymphoma** result from the uncontrolled proliferation of white blood cells. Treatment involves the destruction of the patient's own cancerous bone marrow cells by radiation or chemotherapy and their replacement with a **'bone marrow' transplant** of normal, blood-forming stem cells.

HSCs

Haematopoietic stem cells (**HSCs**) are tissue stem cells that can multiply and differentiate into a variety of specialised blood cells. HSCs are present in bone marrow, peripheral (circulating) blood and umbilical cord blood.

In the past, the most common source of HSCs was **bone marrow** which was drawn from the donor using a syringe. Most transplants now use HSCs obtained by harvesting them from the donor's **peripheral blood**. A few days before the harvest, the donor is injected with a chemical that coaxes additional HSCs to migrate from the marrow to the bloodstream. As a result, up to 20% of the white cells collected in the sample are HSCs. This is at least double the number present in a bone marrow sample and its use normally leads to a speedier recovery by the patient.

Although **umbilical cord** blood is rich in HSCs, the volume of blood obtained is relatively small. Therefore its use in transplantation tends to be restricted to very small adults and children.

Figure 1.14 shows a simplified version of the procedure carried out during bone marrow transplantation.

Skin graft

In a traditional skin graft, a relatively large section of skin is removed from a region of the person's body and grafted to the site of injury. This means that the person has two bodily areas that need careful treatment and time to heal.

A skin graft using stem cells only requires a **small sample** of skin to be taken to obtain stem cells. Therefore the site needs much less healing time and suffers minimum scarring. The sample is normally taken from an area close to and similar in structure to the site of injury. Enzymes are used to isolate and loosen the stem cells, which are then cultured. Once a **suspension of new stem cells** has developed, they are sprayed over the damaged area to bring about regeneration of missing skin.

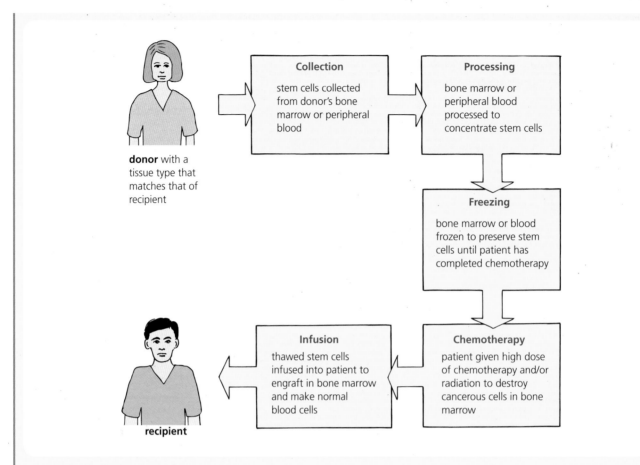

Figure 1.14 Bone marrow transplantation

Cornea repair

In recent years scientists have shown that **corneal damage** by chemical burning can be successfully treated using stem cell tissue. This can be grown from the patient's own stem cells located at the edge of the cornea. In many cases the person's eyesight can be restored following grafting of the stem cell tissue from the healthy eye to the surface of the damaged eye. Since the skin graft and cornea repair techniques use the affected individual's own cells, there is no risk of the transplanted tissue being rejected.

Future therapeutic potential of stem cells

Recently human embryonic stem cells grown on synthetic scaffolds have been successfully used to treat burn victims. The stem cell tissue provides a source of **temporary skin** while the patient is waiting for grafts of their own skin to develop.

Embryonic stem cells are able to differentiate into any type of cell in the body. Therefore they are believed to have the potential to provide treatments in the future for a wide range of disorders and degenerative conditions, such as **diabetes**, **Parkinson's disease** and **Alzheimer's disease**, that traditional medicine has been unable to cure. Already scientists have managed to generate nerve cells from embryonic stem cells in culture. It is hoped that this work will eventually be translated into effective therapies to treat neurological disorders such as multiple sclerosis. However, the use of embryonic stem cells raises questions of **ethics**.

Ethical issues

Ethics refers to the moral values and rules that ought to govern human conduct. The use of stem cells raises several **ethical issues**. For example the extraction of human embryonic stem cells to create a stem cell line (a continuous culture) for research purposes results in the destruction of the human embryo. Many people believe strongly that this practice is unethical (see Case Study – Embryonic stem cell debate).

Ethical issues are also raised by the use of **induced pluripotent stem cells** (see Case Study – Sources of stem cells) and by the use of **nuclear transfer technique** (see Case Study – Nuclear transfer technique).

Case Study | Embryonic stem cell debate

At present the creation of a human embryonic stem cell line using cells from a human embryo (of no more than 14 days) results in the destruction of the embryo. The ethical debate about the use of embryonic stem cells most commonly rests on the controversial question: 'Is a human embryo of less than 2 weeks a human *person*?'

People on one side of the debate believe that the embryo is definitely a human person and argue that fatally extracting stem cells from it constitutes murder. People on the other side of the debate feel certain that the embryo is not yet a person and believe that removing stem cells from it is morally acceptable.

The people who are against stem cell research using human embryos often support their case with the following claims:

- A human life begins when a sperm cell fuses with an egg cell and it is inviolable (i.e. it is sacred and must not be harmed).
- A unique version of human DNA is created at conception.
- A fertilised egg is a human being with a soul.
- Stem cell research violates the sanctity of life.

The people who are in favour of stem cell research using human embryos often support their case with the following arguments:

- An embryo is not a person although it has the potential to develop into a person.
- At 14 days or less an embryo is not sentient (i.e. it does not have a brain, a nervous system, consciousness or powers of sensation).
- The death of a very young embryo is not of serious moral concern when it has the potential to benefit humanity (particularly people whose daily lives are compromised by debilitating medical conditions).
- Abortion is legal in many countries including the UK. Destroying a 14-day-old embryo is far less objectionable to most people than terminating a fetus at 20 weeks.
- Stem cell research uses embryos that were generated for IV fertilisation but were not used and would be destroyed as a matter of course.

Possible solutions to the problem
In the future the ethical issues raised by the use of embryonic stem cells may become less heated if advances in **induced pluripotent stem cell** technology (see page 12) and increased use of stem cells from **amniotic fluid** (see page 12) reduce the need for their use.

Case Study | Sources of stem cells

Donated embryos

At present patients undergoing infertility treatment may agree to donate any **extra embryos** that are not required for their treatment to medical science. These very early embryos provide an immediate source of embryonic stem cells for research. In addition **long-term cultures** originally set up using cells isolated from donated embryos and continued for many years provide a further source of embryonic stem cells. However, the number of human embryonic cells available for research remains limited and this restricts the ability of scientists to carry out research work in this important area.

Amniotic fluid

Scientists continue to search for new sources of stem cells. One of these is **amniotic fluid**. The stem cells that it contains can be harvested from the fluid removed from pregnant women for amniocentesis tests (see Figure 1.15). The stem cells obtained are capable of differentiating into many types of specialised cell such as bone, muscle, nerve and liver. One advantage of using stem cells from amniotic fluid is that it does not involve the destruction of a human embryo.

Induced pluripotent stem cells

A **totipotent** stem cell is one that is able to differentiate into any cell type and is capable of giving rise to the complete organism. A **pluripotent** stem cell is a descendent of a totipotent stem cell and is capable of differentiating into many different types of cell.

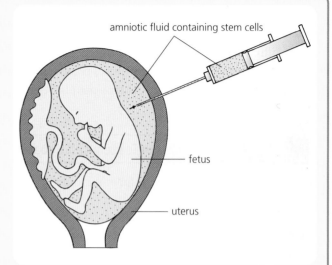

Figure 1.15 Stem cells in amniotic fluid

Induced pluripotent stem cells are, strictly speaking, not true stem cells. They are differentiated cells (e.g. from human skin) that have been genetically reprogrammed using transcription factors (see chapter 3) to switch some of their turned-off genes back on again. As a result they act as stem cells and can be used for research. However, the viruses used as vectors to deliver the transcription factors have been shown to cause cancers in mouse models. Therefore a significant amount of research is essential in this area before induced pluripotent cells can be considered for use as part of a routine medical procedure.

Case Study | Nuclear transfer technique

This technique involves removing the nucleus from an egg (see Figure 1.16) and then replacing it with a nucleus from a donor cell. Some cells constructed in this way divide normally, producing undifferentiated stem cells.

Using this technique, a nucleus from a human cell (e.g. skin) can be introduced into an **enucleated** animal cell (e.g. an egg cell from a cow), as shown in Figure 1.17. The cell formed is called a **cytoplasmic hybrid cell**. Once it begins to divide, stem cells can be extracted after 5 days and used for research. However,

they are not 100% human and must not be used for therapeutic procedures.

Some people feel that it is unethical to mix materials from human cells with those of another species even if the hybrid cells formed are used strictly for research purposes only. Other people support the production of cytoplasmic hybrid cells because it helps to relieve the shortage of human embryonic stem cells available for research. In addition, they point out that the practice allows the nucleus from a diseased human cell (e.g. from a patient suffering a

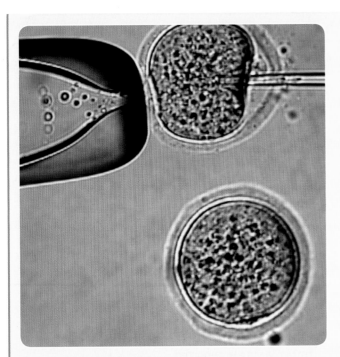

Figure 1.16 Removing the nucleus from an egg

degenerative disease or cancer) to be introduced into the enucleated animal egg. This may allow scientists to study the gene expression in these cells, observe how the disease develops and eventually develop new treatments that disrupt the disease process.

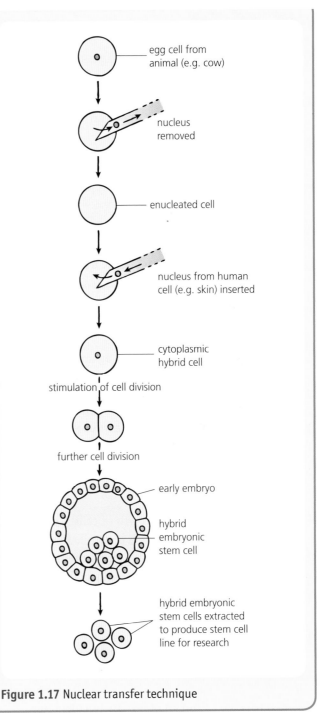

egg cell from animal (e.g. cow)

nucleus removed

enucleated cell

nucleus from human cell (e.g. skin) inserted

cytoplasmic hybrid cell

stimulation of cell division

further cell division

early embryo

hybrid embryonic stem cell

hybrid embryonic stem cells extracted to produce stem cell line for research

Figure 1.17 Nuclear transfer technique

Regulation

In the UK, the use of stem cells in research is carefully **regulated** by laws such as the *Human Fertilisation and Embryology Act* and the *Human Reproductive Cloning Act*. A licence from the *Human Fertilisation and Embryology Authority* must be obtained before research

involving stem cells may be carried out (see Case Study – Regulation of stem cell research).

The aim of this strict regulation is to ensure that the quality of the stem cells used and the safety of the procedures carried out are of the highest order and that abuses of the system are prevented.

Case Study | Regulation of stem cell research

The Human Fertilisation and Embryology Authority only grants a licence if it is satisfied that:

- the use of human embryos is necessary and that the research work could not be carried out in another way
- the purpose of the work is to increase knowledge about serious disease
- the knowledge obtained will be applied in the development of treatments for the serious disease.

Time limit

An **embryo** may be used up to 14 days after conception. Similarly a **cytoplasmic hybrid cell** resulting from nuclear transfer (see page 12) may be grown in the laboratory for up to 14 days. During this time stem cells may be removed for research purposes.

Use of embryos beyond 14 days is against the law. They must be destroyed. Experts have chosen this time limit because it is the stage by which the embryo (as a blastocyst) normally becomes **implanted** in the uterus and begins to develop a nervous system.

Inter-species ban

It is illegal to place a human embryo in an animal such as a cow or to place an animal's embryo inside a human. Similarly it is against the law to allow a blastocyst from a cytoplasmic hybrid cell to develop within a human or an animal's body to enable it to grow.

Safeguards

A recent European directive entitled *Tissues and Cells* ensures that the following further safeguards apply to the use of stem cells now and in the future:

- Safety and quality of stem cells is ensured.
- Donors are selected carefully.
- Transfer of stem cells from donor to recipient is tracked.
- Adverse effects (e.g. illness) following stem cell transplants are reported.
- Sources of all materials used are able to be traced.

Cancer cells

A **cancer** is an uncontrolled growth of cells. Cell division in normal healthy cells is controlled by factors such as cell cycle regulators and external chemical signals. Cancerous cells do not respond to these **regulatory signals**.

Tumours

Cancer cells divide uncontrollably to produce a mass of abnormal cells called a **tumour**. A tumour is described as **benign** if it remains as a discrete group of abnormal cells in one place within an otherwise normal tissue. Most benign tumours (e.g. warts – see Figure 1.18) do not cause problems and can be successfully removed.

A tumour is said to be **malignant** if some of its cells lose the surface molecules that keep them attached to the original cell group, enter the circulatory system and spread through the body. This enables them to invade other tissues and do harm by 'seeding' new tumours in other parts of the body (see Figure 1.19).

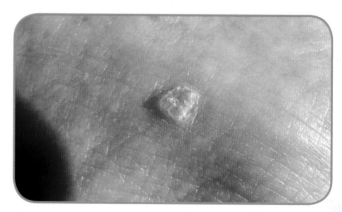

Figure 1.18 Wart

Genetic errors

Most cancers originate from a cell that has undergone a succession of **mutations** to the genes involved in the control of cell division. As these genetic errors accumulate, a point is reached where control of cell division is lost. The cell can now divide excessively, unhindered by any regulatory signal or control mechanism.

Normally it takes a very long time for a cell to

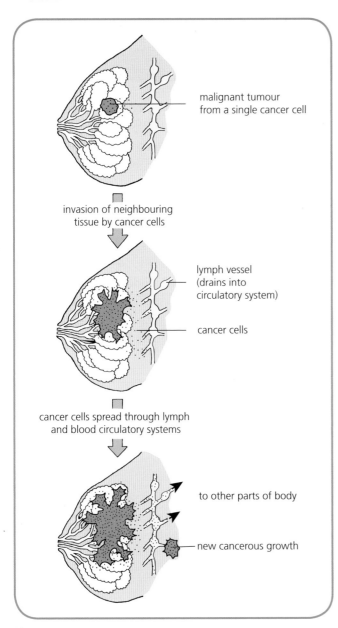

Figure 1.19 Breast cancer

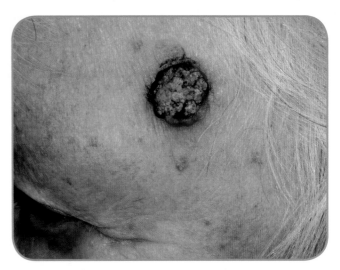

Figure 1.20 Skin cancer

Testing Your Knowledge 2

1 a) i) Name ONE medical condition that is routinely treated using tissue stem cells.
       ii) From where in the human body are these cells obtained? (2)
   b) Give an example of a medical condition that might be treated in the future using stem cells. (1)
   c) Why can the stem cells used to treat the medical condition you gave as your answer to **a)** not be used to treat patients suffering the condition you gave as your answer to **b)**? (2)
2 a) One definition of the word *ethical* is 'in accordance with principles that are morally correct'. Briefly explain why stem cell research using human embryos raises ethical issues. (2)
   b) Why is it important that stem cell research is carefully regulated? (2)
3 Identify TWO characteristics of cancer cells. (2)

accumulate, and be affected by, several different mutations. It is for this reason that the risk of cancer increases with age. However, the risk is also increased by exposure to **agents that cause genetic damage** such as smoking, pollution and excessive exposure of skin to ultraviolet radiation (see Figure 1.20). Cancer is particularly common in skin, lung and bowel tissue because the mutation rate is higher than normal in these cells that have a high frequency of division.

2 Structure and replication of DNA

DNA (deoxyribonucleic acid) is a complex molecule present in all living cells. It is the molecule of inheritance and it stores genetic information in its sequence of bases. This sequence determines the organism's genotype and the structure of its proteins.

Structure of DNA

A molecule of DNA consists of two strands each composed of repeating units called **nucleotides**. Each DNA nucleotide consists of a molecule of **deoxyribose** sugar joined to a **phosphate** group and an organic **base**. Figure 2.1 shows the carbon skeleton of a molecule of deoxyribose. Figure 2.2 shows the four types of base present in DNA.

note
³C = 3′ carbon atom
⁵C = 5′ carbon atom

Figure 2.1 Deoxyribose

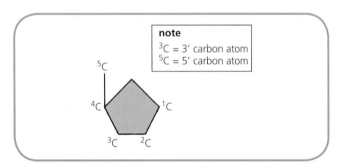

adenine (A)   thymine (T)   guanine (G)   cytosine (C)

Figure 2.2 Four types of organic base

Figure 2.3 shows how the deoxyribose molecule in a nucleotide has a base attached to its carbon 1 and a phosphate attached to its carbon 5. Since there are four types of base, there are four types of nucleotide.

Sugar–phosphate backbone

A strong **chemical bond** forms between the phosphate group of one nucleotide and the carbon 3 of the deoxyribose on another nucleotide (see Figure 2.4). By this means neighbouring nucleotides become joined together into a long, permanent strand in which sugar molecules alternate with phosphate groups, forming the DNA molecule's **sugar–phosphate backbone**.

Base-pairing

Two of these strands of nucleotides become joined together by weak **hydrogen bonds** forming between their bases (see Figure 2.5). However, the hydrogen bonds can be broken when it becomes necessary for the two strands to separate.

Each base can only join up with one other type of base: adenine (A) always bonds with thymine (T) and guanine (G) always bonds with cytosine (C). A–T and G–C are called **base pairs**.

Antiparallel strands

A DNA strand's **3′ end** on deoxyribose is distinct from its **5′ end** at a phosphate group. The chain is only able to grow by adding nucleotides to its 3′ end. In Figure 2.6 the DNA strand on the left has its 3′ growing end at the bottom of the diagram and its 5′ end at the top. The reverse is true of its complementary strand on the right. This arrangement of the two strands with their sugar–phosphate backbones running in **opposite directions** is

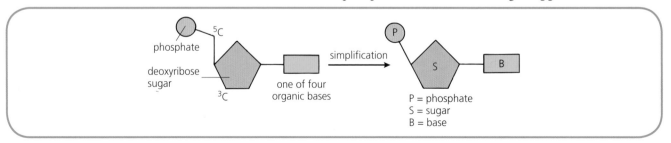

phosphate

deoxyribose sugar

⁵C

³C

one of four organic bases

simplification

P

S

B

P = phosphate
S = sugar
B = base

Figure 2.3 Structure of a DNA molecule

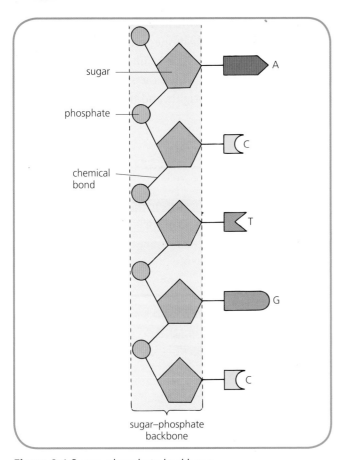

Figure 2.4 Sugar–phosphate backbone

Figure 2.6 Antiparallel strands

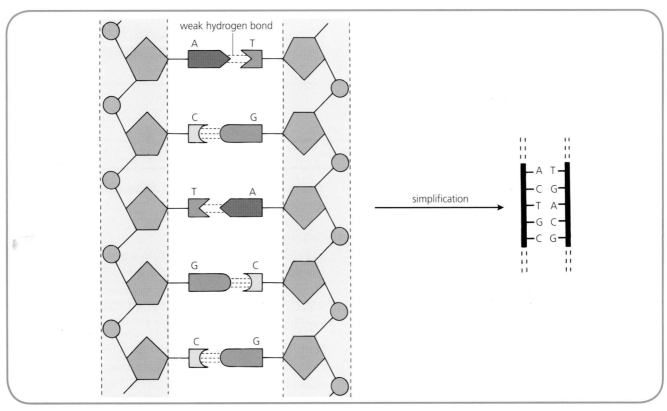

Figure 2.5 Base-pairing

described as **antiparallel**. (For the sake of simplicity the letters in a diagram are normally all written the same way up.)

Double helix

In order for the base pairs to align with each other, the two strands in a DNA molecule take the form of a twisted coil called a **double helix** (see Figure 2.7) with the sugar–phosphate backbones on the outside and the base pairs on the inside. As a result, a DNA molecule is like a spiral ladder in which the sugar–phosphate backbones form the uprights and the base pairs form the rungs.

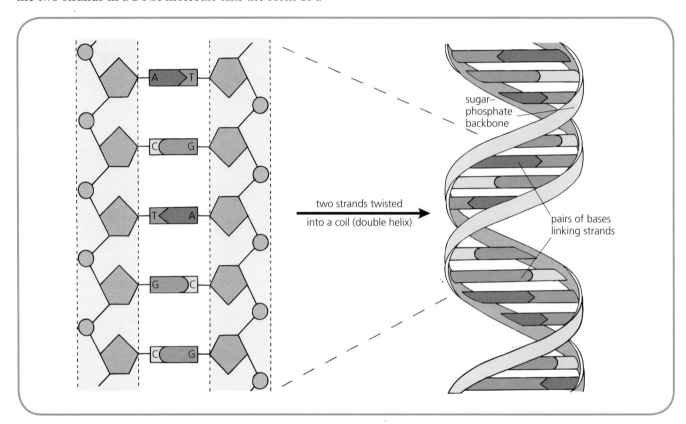

Figure 2.7 Double helix

Case Study   Identity of genetic material

Proteins or DNA?

During the early part of the twentieth century, genes were known to be located on chromosomes and chromosomes to be composed of protein and DNA. However, scientists did not know for certain which of these two classes of molecule made up the genetic material.

Proteins vary greatly in structure and therefore it was widely believed at the time that they were the hereditary material. DNA on the other hand seemed to lack diversity. Therefore it was thought unlikely that DNA would be able to carry the vast quantity of genetic information needed for the transmission and expression of all the traits inherited by every species.

Bacterial transformation

In 1928 a scientist called Griffith was working with two strains of the bacterium *Streptococcus pneumoniae*. Strain S caused pneumonia in mammals; strain R was harmless. Griffith used these two strains in the experiment shown in Figure 2.8.

From this experiment Griffith concluded that some chemical component had passed from dead S cells to live R cells. These live R cells had become **transformed** into live S cells. He called the chemical

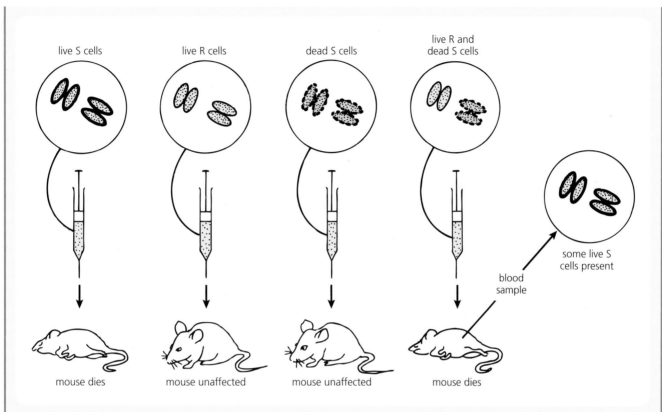

Figure 2.8 Bacterial transformation experiment

substance the **transforming principle** but he did not know which constituent of the bacterial cell was responsible.

Positive identification

The R cell/S cell line of enquiry was pursued by a team of scientists led by Avery. They repeated Griffith's experiment and systematically isolated each constituent of the disease-causing strain of the bacterium. Over a period of 14 years they tested each substance for its ability to transform R cells to S cells. In 1944 they were able to report that without doubt only **DNA fragments** (and not proteins) were able to bring about this transformation. However, many scientists remained sceptical.

Compelling evidence

Background

A bacteriophage (phage for short) is a type of virus that attacks a bacterium and multiplies inside it. Figure 2.9 shows a bacteriophage that attacks *Escherichia coli* bacteria. The virus's DNA strand is contained inside a protein coat from which a tail projects. The virus launches an attack by injecting its DNA into the bacterial host cell, which then becomes a virus-producing 'factory' (see Figure 2.10).

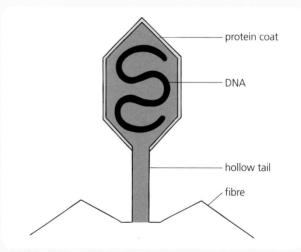

Figure 2.9 Bacteriophage virus

Phage experiments

In 1952 Hershey and Chase devised experiments to investigate whether the information needed to alter the host cell's biochemical machinery and make new viral particles resided in the original virus's DNA or in its protein coat.

Sulphur is found in protein but not in DNA. Therefore they labelled the protein coats of a group

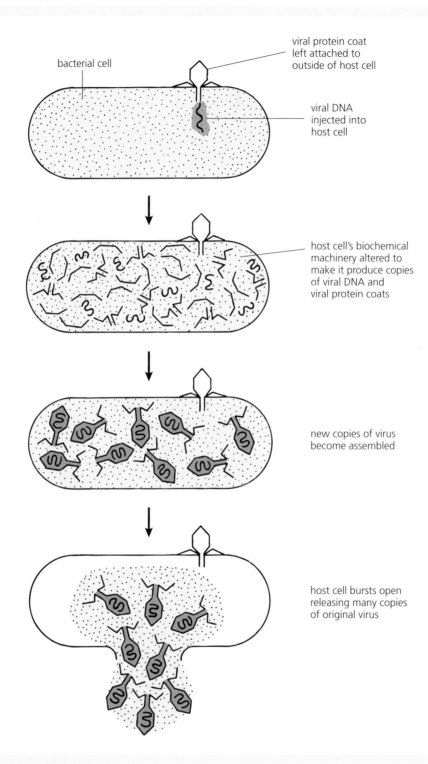

bacterial cell

viral protein coat left attached to outside of host cell

viral DNA injected into host cell

host cell's biochemical machinery altered to make it produce copies of viral DNA and viral protein coats

new copies of virus become assembled

host cell bursts open releasing many copies of original virus

Figure 2.10 Multiplication of bacteriophage virus

of phage particles by growing them in *E. coli* cells cultured in the presence of ^{35}**S** (a radioactive isotope of sulphur).

Phosphorus is found in DNA but not in protein. They labelled the DNA strands of a different group of phage particles by growing them in *E. coli* cells cultured in the presence of ^{32}**P** (a radioactive isotope of phosphorus).

Their experiments are shown in Figure 2.11. From these results it was concluded that the viral protein (labelled with ^{35}S) did not enter the *E. coli* cells but that the viral DNA (labelled with ^{32}P) did enter the

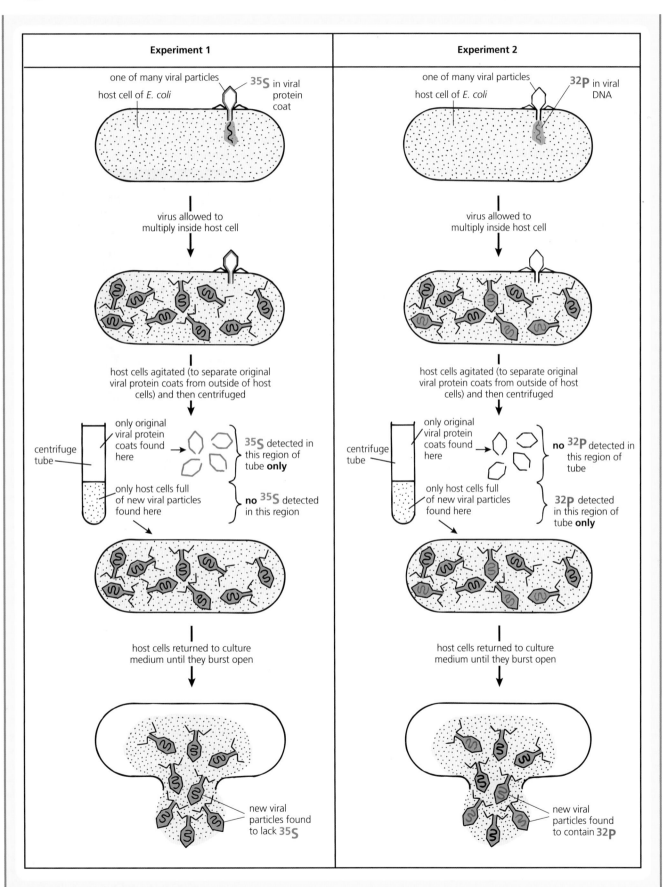

Figure 2.11 Phage experiments of Hershey and Chase

cells. Since the viral DNA was responsible for causing the host cell to produce new copies of the virus, this experiment provided compelling evidence to support the theory that DNA (and not protein) is the genetic material responsible for passing on hereditary information.

Establishing the structure of DNA

Once DNA was known for certain to be the genetic material, scientists became keen to establish the details of the molecule's three-dimensional structure.

Chemical analysis

In the late 1940s, Chargaff analysed the base composition of DNA extracted from a number of different species. He found that the quantities of the four bases were not all equal but that they always occurred in a **characteristic** ratio regardless of the source of the DNA. These findings, called Chargaff's rules, are summarised as follows:

- The number of adenine bases = the number of thymine bases (i.e. A:T = 1:1).
- The number of guanine bases = the number of cytosine bases (i.e. G:C = 1:1).

However, Chargaff's rules remained unexplained until the double helix was discovered.

X-ray crystallography

At around the time that Chargaff was carrying out chemical analysis of DNA, Wilkins and Franklin were employing **X-ray crystallography**. When X-rays are passed through a crystal of DNA, they become deflected (diffracted) into a **scatter pattern**, which is determined by the arrangement of the atoms in the DNA molecule (see Figure 2.12).

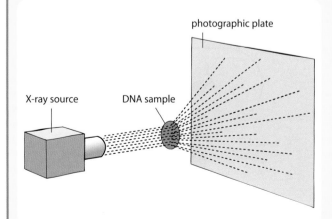

Figure 2.12 X-ray crystallography

When the scatter pattern of X-rays is recorded using a photographic plate, a **diffraction pattern** of spots is produced (see Figure 2.13). This reveals information that can be used to build up a **three-dimensional picture** of the molecules in the crystal. Wilkins and Franklin found that the X-ray diffraction patterns of DNA from different species (e.g. bull, trout and bacteria) were identical.

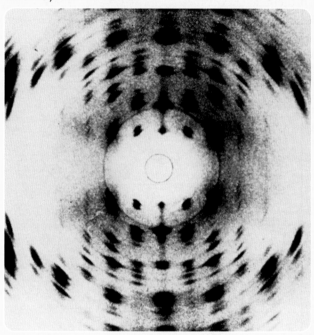

Figure 2.13 X-ray diffraction pattern of DNA

Formation of an evidence-based conclusion

From the X-ray diffraction patterns of DNA (produced by Wilkins and Franklin) Watson and Crick figured out that the DNA must be a long, thin molecule of constant diameter coiled in the form of a **helix**. In addition, the density of the arrangement of the atoms indicated to them that the DNA must be composed of **two strands**.

From Chargaff's rules they deduced that base A must be paired with T and base G with C. They figured that this could only be possible if DNA consisted of two strands held together by specific **pairing of bases**. Taking into account further information about distances between atoms and angles of bonds, Watson and Crick set about building a wire model of DNA and in 1953 were first to establish the three-dimensional **double helix** structure of DNA.

Arrangement of DNA in chromosomes

A strand of DNA is several thousand times longer than the length of the cell to which it belongs. Therefore it is essential that DNA molecules are organised in such a way that a chaotic tangle of strands in the nucleus is prevented. This is achieved by the molecules of DNA becoming **tightly coiled** and packaged around bundles of protein like beads on a string (see Figure 2.14). They are able to unwind again when required to do so.

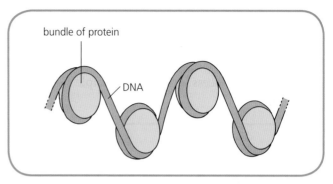

Figure 2.14 Structure of a chromosome

Testing Your Knowledge 1

1 a) i) How many different types of base molecule are found in DNA?
   ii) Name each type. (3)
   b) Which type of bond forms between the bases of adjacent strands of a DNA molecule? (1)
   c) Describe the base-pairing rule. (1)
2 a) Figure 2.15 shows part of one strand of a DNA molecule.

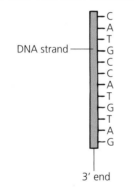

DNA strand —

C
A
T
G
C
C
A
T
G
T
A
G

3′ end

Figure 2.15

   i) Redraw the strand and then draw the complementary strand alongside it.
   ii) Label the 3′ end and the 5′ end on each strand. (2)
   b) DNA consists of two strands whose backbones run in opposite directions. What term is used to describe this arrangement? (1)
3 a) What name is given to the twisted coil arrangement typical of a DNA molecule? (1)
   b) If DNA is like a spiral ladder, which part of it corresponds to the ladder's:
   i) rungs
   ii) uprights? (2)

Replication of DNA

DNA is a unique molecule because it is able to direct its own **replication** and reproduce itself exactly. The replication process is shown in a simple way in Figure 2.16.

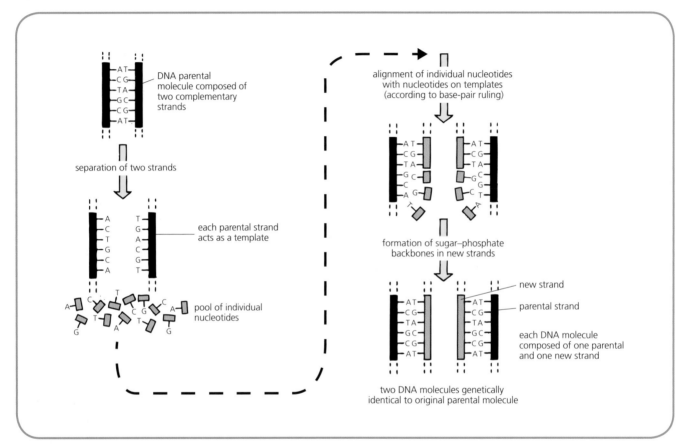

Figure 2.16 DNA replication

Case Study | Establishing which theory of DNA replication is correct

Watson and Crick accompanied their model of DNA with a theory for the way in which it could replicate. They predicted that the two strands would unwind and each act as a template for the new complementary strand. This would produce two identical DNA molecules each containing one 'parental' strand and one newly synthesised strand. This so-called **semi-conservative** replication remained a theory until put to the test by Meselson and Stahl.

Hypotheses
Figure 2.17 shows three different hypotheses, each of which could explain DNA replication.

Testing the hypotheses

Background
When E. coli bacteria are cultured, they take up nitrogen from the surrounding medium and build it into DNA. They can be cultured for several generations in medium containing the common isotope of nitrogen (^{14}N) or the heavy isotope (^{15}N) as their only source of nitrogen. When DNA is extracted from each type of culture and centrifuged, the results shown in Figure 2.18 are obtained.

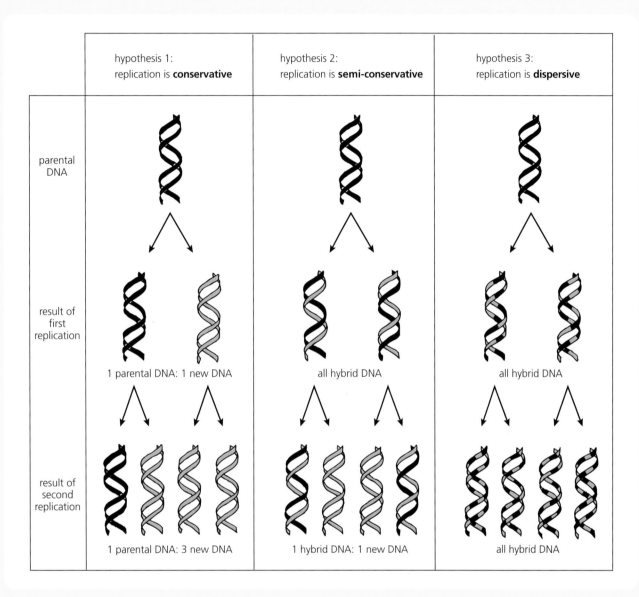

hypothesis 1: replication is **conservative**	hypothesis 2: replication is **semi-conservative**	hypothesis 3: replication is **dispersive**

parental DNA

result of first replication

1 parental DNA: 1 new DNA | all hybrid DNA | all hybrid DNA

result of second replication

1 parental DNA: 3 new DNA | 1 hybrid DNA: 1 new DNA | all hybrid DNA

Figure 2.17 DNA replication hypotheses

Putting the hypotheses to the test

Meselson and Stahl began with *E. coli* that had been grown for many generations in medium containing ^{15}N. They then cultured these bacteria in medium containing ^{14}N and sampled the culture after 20 minutes (the time needed by the bacteria to replicate DNA once). Figure 2.19 predicts the outcome of the experiment for each of the three hypotheses. Figure 2.20 shows the actual results of the experiment. From these results it is concluded that hypothesis 2 is supported and that the replication of DNA is **semi-conservative**.

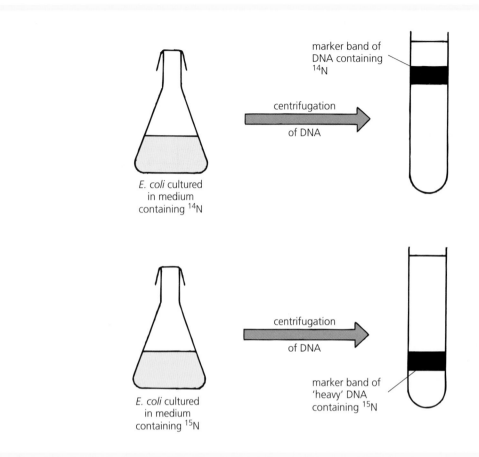

Figure 2.18 Labelling DNA with ¹⁴N or ¹⁵N

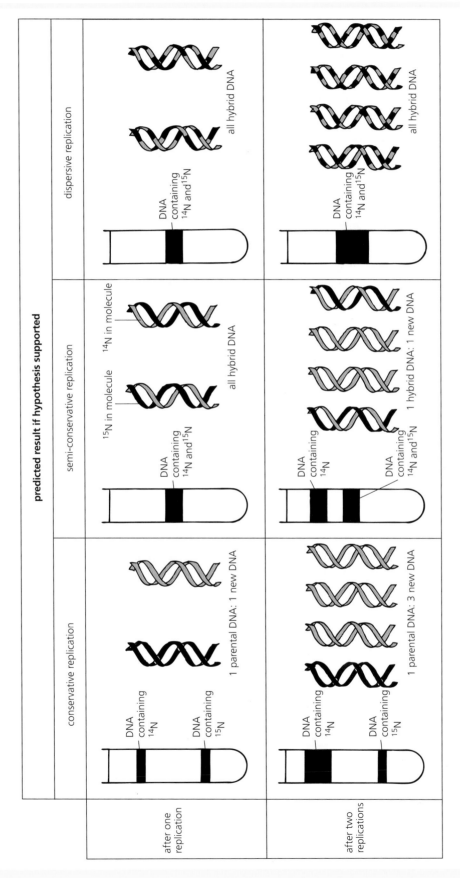

Figure 2.19 Predictions

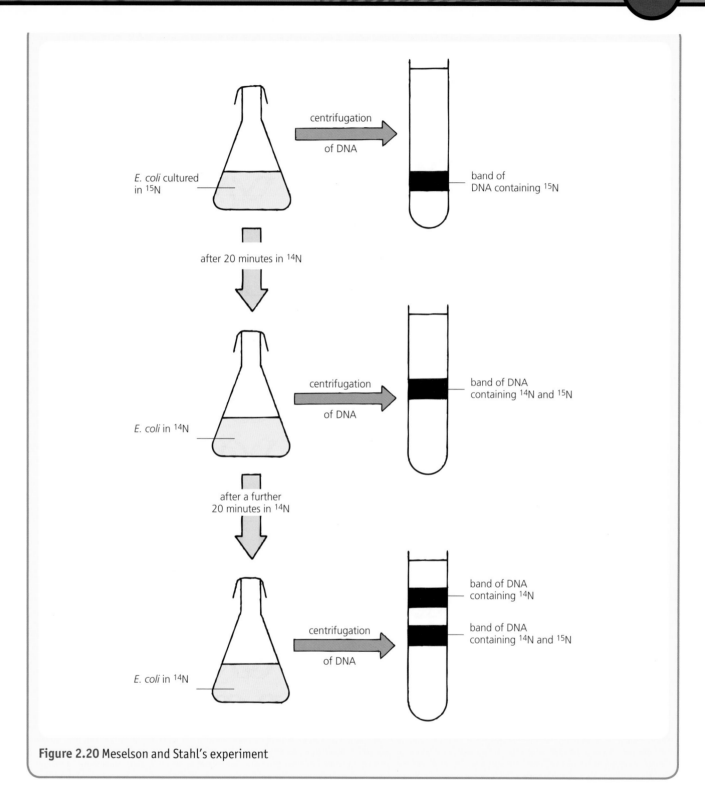

Figure 2.20 Meselson and Stahl's experiment

Enzyme control of DNA replication

DNA replication is a complex process involving many enzymes. It begins when a starting point on DNA is recognised. The DNA molecule unwinds and weak hydrogen bonds between base pairs break, allowing the two strands to separate ('unzip'). These template

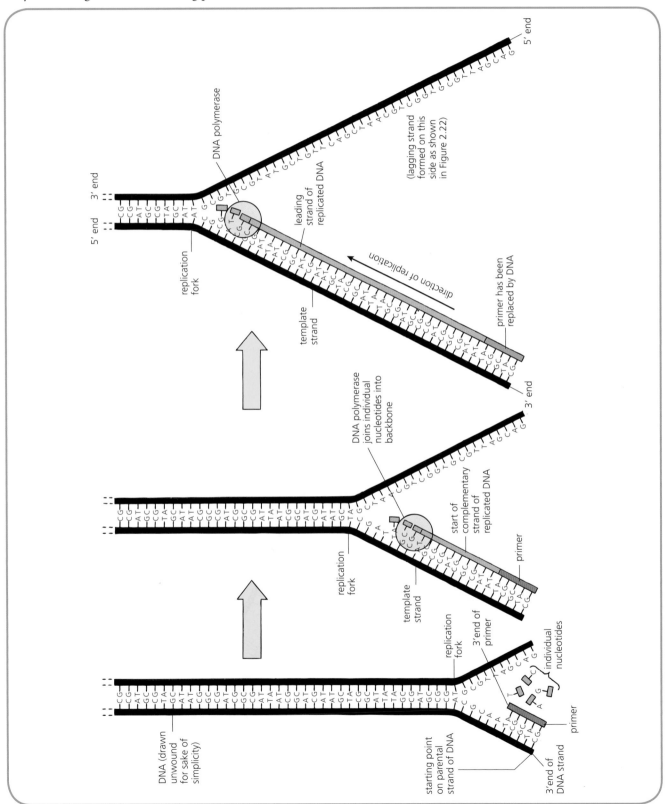

Figure 2.21 Formation of leading strand of replicated DNA

strands become stabilised and expose their bases at a Y-shaped **replication fork** (see Figure 2.21).

Formation of the leading DNA strand

The enzyme that controls the sugar–phosphate bonding of individual nucleotides into the new DNA strand is called **DNA polymerase**. This enzyme can only add nucleotides to a pre-existing chain. For it to begin to function, a **primer** must be present. This is a short sequence of nucleotides formed at the 3′ end of the parental DNA strand about to be replicated, as shown in Figure 2.21.

Once individual nucleotides have become aligned with their complementary partners on the template strand (by their bases following the base-pairing rules), they become bound to the 3′ end of the primer and formation of the complementary DNA strand begins.

Formation of sugar–phosphate bonding between the primer and an individual nucleotide and between the individual nucleotides themselves is brought about by DNA polymerase. Replication of the parental DNA strand that has the 3′ end is **continuous** and forms the **leading** strand of the replicated DNA.

Formation of the lagging DNA strand

DNA polymerase is only able to add nucleotides to the free 3′ end of a growing strand. Therefore the DNA parental template strand that has the 5′ end has to be replicated in fragments, each starting at the 3′ end of a primer, as shown in Figure 2.22.

Each fragment must be primed as before to enable the DNA polymerase to bind individual nucleotides together. Once replication of a fragment is complete, its primer is replaced by DNA. Finally an enzyme called

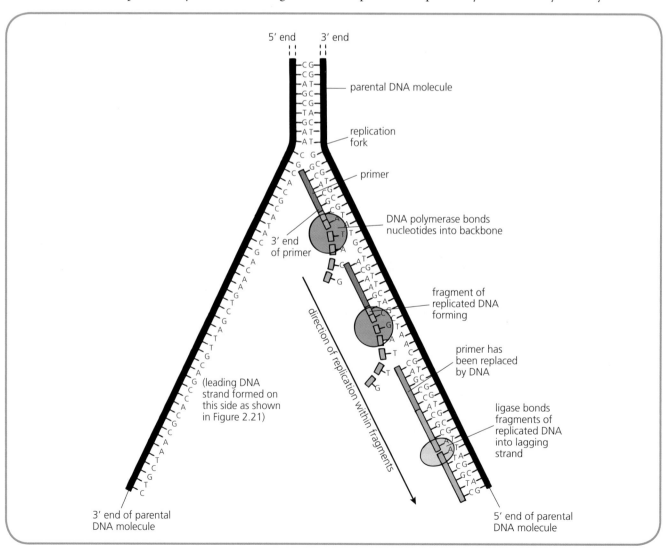

Figure 2.22 Formation of lagging strand of replicated DNA

ligase joins the fragments together. The strand formed is called the **lagging** strand of replicated DNA and its formation is described as **discontinuous**.

Many replication forks

When a long chromosome (e.g. one from a mammalian cell) is being replicated, many replication forks operate simultaneously to ensure speedy copying of the lengthy DNA molecule (see Figure 2.23).

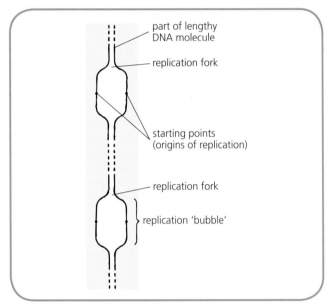

Figure 2.23 Replication forks

Requirements for DNA replication

For DNA replication to occur, the nucleus must contain:

- **DNA** (to act as a template)
- **primers**
- a supply of the four types of DNA **nucleotide**
- the appropriate **enzymes** (e.g. DNA polymerase and ligase)
- a supply of **ATP** (for energy – see page 97).

Importance of DNA

DNA is the molecule of inheritance and it encodes the hereditary information in a **chemical language**. This takes the form of a sequence of organic bases, unique to each species, which makes up its **genotype**. DNA replication ensures that an exact copy of a species' genetic information is passed on from cell to cell during growth and from generation to generation during reproduction. Therefore DNA is essential for the continuation of life.

Testing Your Knowledge 2

1 Decide whether each of the following statements is true or false and then use T or F to indicate your choice. Where a statement is false, give the word that should have been used in place of the word in bold print. (5)
   a) During DNA replication, each DNA parental strand acts as a **template**.
   b) A guanine base can only pair up with a **thymine** base.
   c) An adenine base can only pair with a **cytosine** base.
   d) Complementary base pairs are held together by weak **antiparallel** bonds.
   e) Each new DNA molecule formed by replication contains one **parental** and one new strand.

2 Figure 2.24 shows part of a DNA molecule undergoing replication. Match numbers 1–6 with the following statements. (5)
   a) DNA polymerase promotes formation of a fragment of the lagging strand of replicating DNA.
   b) The parental double helix unwinds.
   c) DNA polymerase bonds nucleotide to primer.
   d) Ligase joins fragments onto the lagging strand of replicating DNA.
   e) DNA polymerase promotes formation of the leading strand of replicating DNA.
   f) The DNA molecule becomes stabilised as two template strands.

3 a) Name FOUR substances that must be present in a nucleus for DNA replication to occur. (4)
   b) Briefly explain why DNA replication is important. (2)

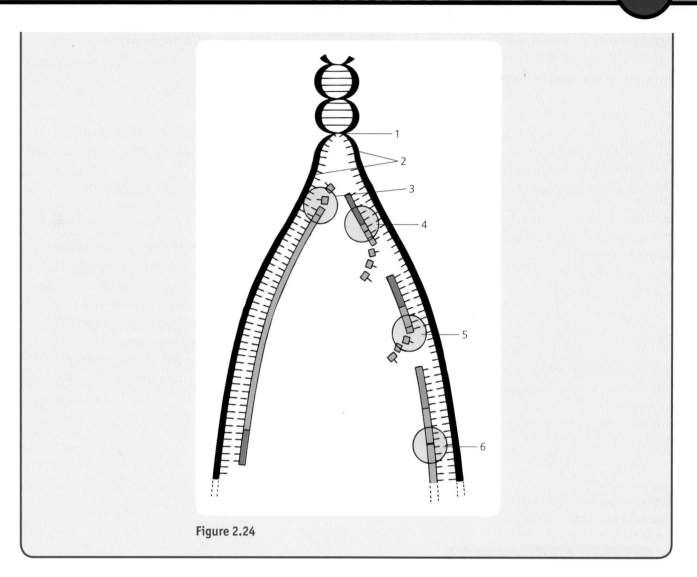

Figure 2.24

3 Gene expression

A cell's **genotype** (its genetic constitution) is determined by the sequence of the DNA bases in its genes (the genetic code). A cell's **phenotype** (its physical and chemical state) is determined by the proteins that are synthesised when the genes are expressed. Gene expression involves the processes of **transcription** and **translation** (discussed in this chapter). It is affected by environmental factors acting inside and outside the cell. Only a fraction of the genes in a cell are expressed.

Structure of RNA

The second type of nucleic acid is called **RNA** (ribonucleic acid). Each nucleotide in an RNA molecule is composed of a molecule of **ribose** sugar, an organic base and a phosphate group (see Figure 3.1). In RNA, the base **uracil** (**U**) replaces thymine found in DNA.

Characteristic	RNA	DNA
Number of nucleotide strands present in one molecule	One	Two
Complementary base partner of adenine	Uracil	Thymine
Sugar present in a nucleotide	Ribose	Deoxyribose

Table 3.1 Differences between RNA and DNA

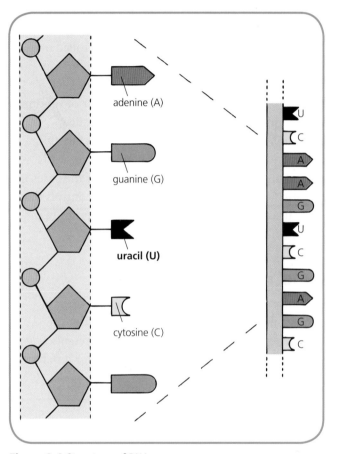

adenine (A)

guanine (G)

uracil (U)

cytosine (C)

Figure 3.2 Structure of RNA

Unlike DNA, which consists of two strands, a molecule of RNA is a **single strand**, as shown in Figure 3.2. The differences between RNA and DNA are summarised in Table 3.1.

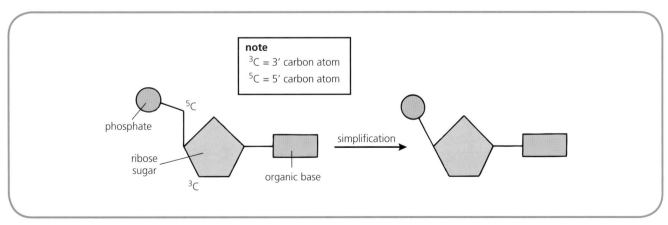

note
³C = 3' carbon atom
⁵C = 5' carbon atom

phosphate

⁵C

ribose sugar

³C

organic base

simplification

Figure 3.1 Structure of an RNA molecule

Control of inherited characteristics

The sequence of bases along the DNA strands contains the **genetic instructions** that control an organism's inherited characteristics. These characteristics are the result of many biochemical processes controlled by enzymes, which are made of **protein**. Each protein is made of one or more **polypeptide** chains. Each polypeptide is composed of a large number of subunits called **amino acids**. A protein's exact molecular structure, shape and ability to carry out its function all depend on the **sequence** of its amino acids. This critical order is determined by the order of the bases in the organism's DNA. By this means DNA controls the structure of enzymes and, in doing so, determines the organism's inherited characteristics.

Genetic code

The information present in DNA takes the form of a molecular language called the **genetic code**. The sequence of bases along a DNA strand represents a sequence of 'codewords'. DNA possesses four different types of base. Proteins contain 20 different types of amino acid. If the bases are taken in groups of three then this gives 64 (4^3) different combinations (see Appendix 1). It is now known that each amino acid is coded for by one or more of these 64 **triplets** of bases. Thus an individual's genetic information is encoded in its DNA with each strand bearing a series of base triplets arranged in a specific order for coding the particular proteins needed by that individual.

Gene expression through protein synthesis

The genetic information for a particular polypeptide is carried on a section of DNA in the nucleus. However, assembly of amino acids into a genetically determined sequence takes place in the cell's cytoplasm in tiny structures called **ribosomes**. Figure 3.3 gives an overview of gene expression through protein synthesis. A molecule of **mRNA** (messenger RNA) is formed (**transcribed**) from the appropriate section of the DNA strand and carries that information to ribosomes. There the mRNA meets **tRNA** (transfer RNA) and the genetic information is **translated** into protein.

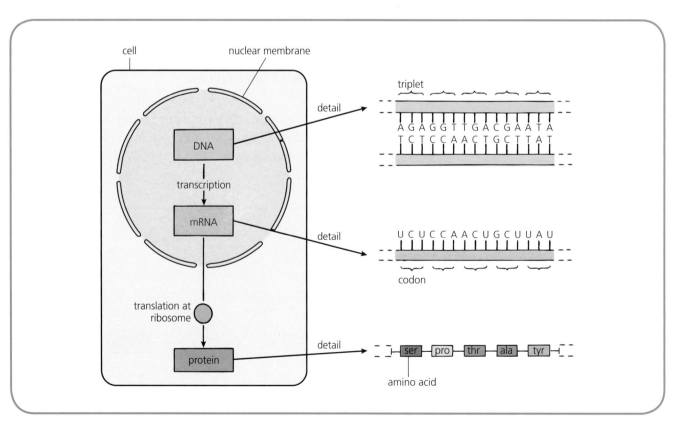

Figure 3.3 Overview of gene expression

Transcription

Transcription is the synthesis of mRNA from a section of DNA. A promoter is a region of DNA in a gene where transcription is initiated, as shown in Figure 3.4 (where the DNA strand has been drawn uncoiled for the sake of simplicity).

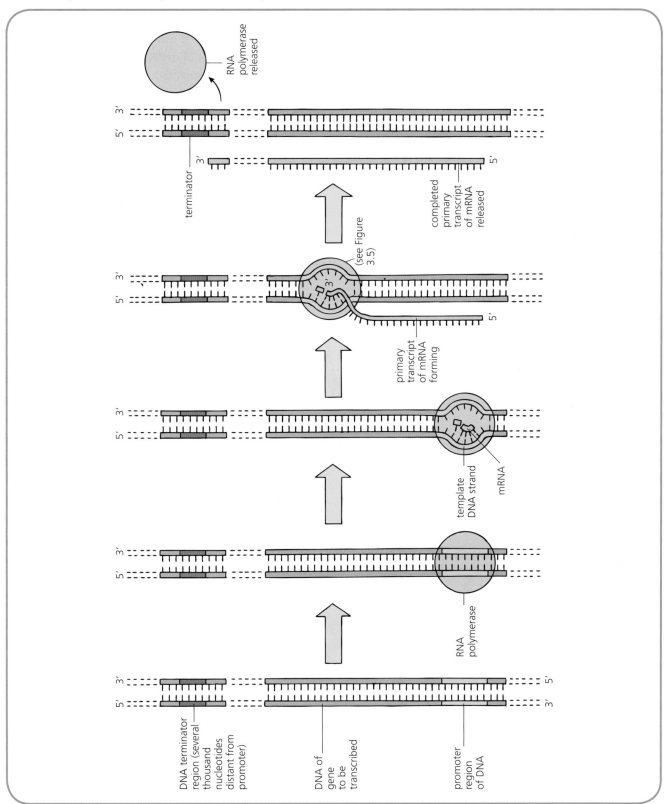

Figure 3.4 Transcription of mRNA

RNA polymerase is the enzyme responsible for transcription. As it moves along the gene from the promoter, unwinding and 'unzipping' the DNA strand, it brings about the synthesis of an **mRNA** molecule. As a result of the base-pairing rule, the mRNA gets a nucleotide sequence complementary to one of the two DNA strands (the template strand), as shown in Figure 3.5.

RNA polymerase can only add nucleotides to the 3′ end of the growing mRNA molecule. The molecule elongates until a terminator sequence of nucleotides is reached on the DNA strand. The resultant mRNA strand that becomes separated from its DNA template is called a **primary transcript** of mRNA.

Modification of primary transcript

Normally the region of DNA transcribed to mRNA is about 8000 nucleotides long yet only about 1200 nucleotides are needed to code for an average-sized polypeptide chain. This is explained by the fact that long stretches of DNA that exist within a gene do not play a part in the coding of the polypeptide. These non-coding regions, called **introns**, are interspersed between the coding regions, called **exons**. Therefore the region in the primary transcript of mRNA responsible for coding the polypeptide is fragmented.

Splicing

Figure 3.6 shows how the introns are cut out and removed from the primary transcript of mRNA and the exons are **spliced** together to form mRNA with a continuous sequence of nucleotides. This modified mRNA called the **mature transcript** of mRNA passes out of the nucleus into the cytoplasm (see Figure 3.7) and moves on to the next stage of protein synthesis where it becomes translated into a sequence of amino acids.

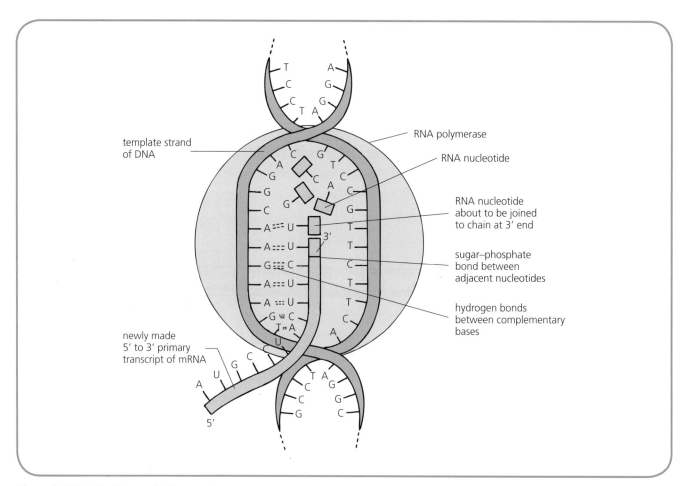

Figure 3.5 Detail of transcription

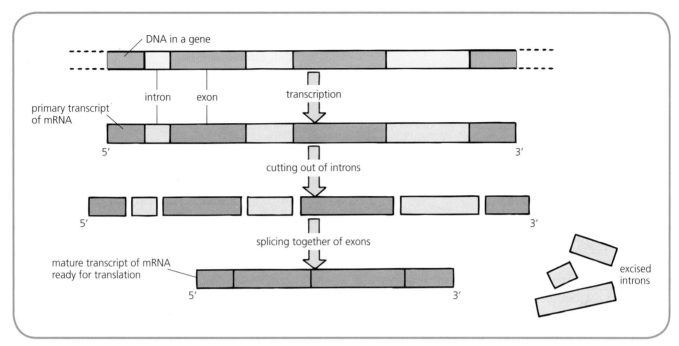

Figure 3.6 Modification of primary mRNA transcript

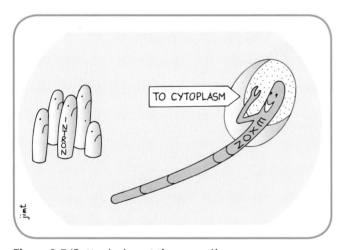

Figure 3.7 'Better luck next time, guys!'

Testing Your Knowledge 1

1 State THREE ways in which RNA and DNA differ in structure and chemical composition. (3)

2 a) In what way does the DNA of one species differ from that of another, making each species unique? (1)

b) How many bases in the genetic code correspond to one amino acid? (1)

3 a) Draw a diagram of the mRNA strand that would be transcribed from section X of the DNA molecule shown in Figure 3.8. (2)

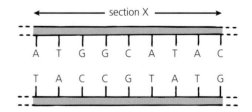

Figure 3.8

b) Name the enzyme that would direct this process. (1)

4 a) What is the difference between an *exon* and an *intron*? (1)

b) Which of these must be removed from the primary transcript of mRNA? (1)

c) By what process are they removed? (1)

Translation

Translation is the synthesis of **protein** as a polypeptide chain under the direction of **mRNA**. The genetic message carried by a molecule of mRNA is made up of a series of base triplets called **codons**. The codon is the **basic unit** of the genetic code. Each codon is complementary to a triplet of bases on the original template DNA strand.

Transfer RNA

A further type of RNA is found in the cell's cytoplasm. This is called **tRNA** (transfer RNA) and it is composed of a single strand of nucleotides. However, a molecule of tRNA has a three-dimensional structure because it is folded back on itself in such a way that hydrogen bonds form between many of its nucleotide bases, as shown in Figure 3.9. Each molecule of tRNA has only one particular triplet of bases exposed. This triplet is called an **anticodon**. It is complementary to an mRNA codon and corresponds to a specific amino acid carried by that tRNA at its **attachment site**.

Table 3.2 shows the relationship between mRNA's codons, tRNA's anticodons and the amino acids coded. Many different types of tRNA are present in a cell – one or more for each type of amino acid. Each tRNA picks up its appropriate amino acid molecule from the cytoplasm's amino acid pool at its site of attachment. The amino acid is then carried by the tRNA to a ribosome and added to the growing end of a polypeptide chain. By this means, the genetic code is translated into a sequence of amino acids.

The mRNA codon AUG (complementary to tRNA anticodon UAC) is unusual in that it codes for methionine (met) *and* acts as the **start codon**. mRNA codons UAA, UAG and UGA do not code for amino acids but instead act as **stop codons**.

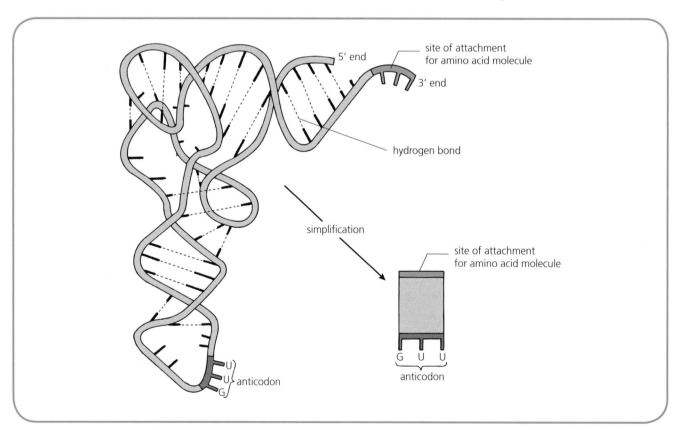

Figure 3.9 Structure of transfer RNA (tRNA)

Codon (mRNA)	Anticodon (tRNA)	Amino acid	Codon (mRNA)	Anticodon (tRNA)	Amino acid	Codon (mRNA)	Anticodon (tRNA)	Amino acid	Codon (mRNA)	Anticodon (tRNA)	Amino acid
UUU	AAA	phe	UCU	AGA	ser	UAU	AUA	tyr	UGU	ACA	cys
UUC	AAG	phe	UCC	AGG	ser	UAC	AUG	tyr	UGC	ACG	cys
UUA	AAU	leu	UCA	AGU	ser	UAA	AUU	STOP	UGA	ACU	STOP
UUG	AAC	leu	UCG	AGC	ser	UAG	AUC	STOP	UGG	ACC	trp
CUU	GAA	leu	CCU	GGA	pro	CAU	GUA	his	CGU	GCA	arg
CUC	GAG	leu	CCC	GGG	pro	CAC	GUG	his	CGC	GCG	arg
CUA	GAU	leu	CCA	GGU	pro	CAA	GUU	gln	CGA	GCU	arg
CUG	GAC	leu	CCG	GGC	pro	CAG	GUC	gln	CGG	GCC	arg
AUU	UAA	ile	ACU	UGA	thr	AAU	UUA	asn	AGU	UCA	ser
AUC	UAG	ile	ACC	UGG	thr	AAC	UUG	asn	AGC	UCG	ser
AUA	UAU	ile	ACA	UGU	thr	AAA	UUU	lys	AGA	UCU	arg
AUG	UAC	met or START	ACG	UGC	thr	AAG	UUC	lys	AGG	UCC	arg
GUU	CAA	val	GCU	CGA	ala	GAU	CUA	asp	GGU	CCA	gly
GUC	CAG	val	GCC	CGG	ala	GAC	CUG	asp	GGC	CCG	gly
GUA	CAU	val	GCA	CGU	ala	GAA	CUU	glu	GGA	CCU	gly
GUG	CAC	val	GCG	CGC	ala	GAG	CUC	glu	GGG	CCC	gly

Table 3.2 mRNA codons, tRNA anticodons and the amino acids coded
(See Appendix 1 for full names of amino acids.)

Ribosomes

Ribosomes are small, roughly spherical structures found in all cells. They contain ribosome RNA (rRNA) and enzymes essential for protein synthesis. Many ribosomes are present in growing cells which need to produce large quantities of protein.

Binding sites

A ribosome's function is to bring tRNA molecules (bearing amino acids) into contact with mRNA. A ribosome has one binding site for mRNA and three binding sites for tRNA, as shown in Figure 3.10.

Of the tRNA binding sites:

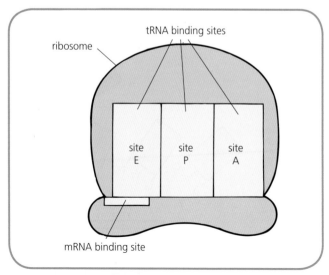

Figure 3.10 Binding sites on a ribosome

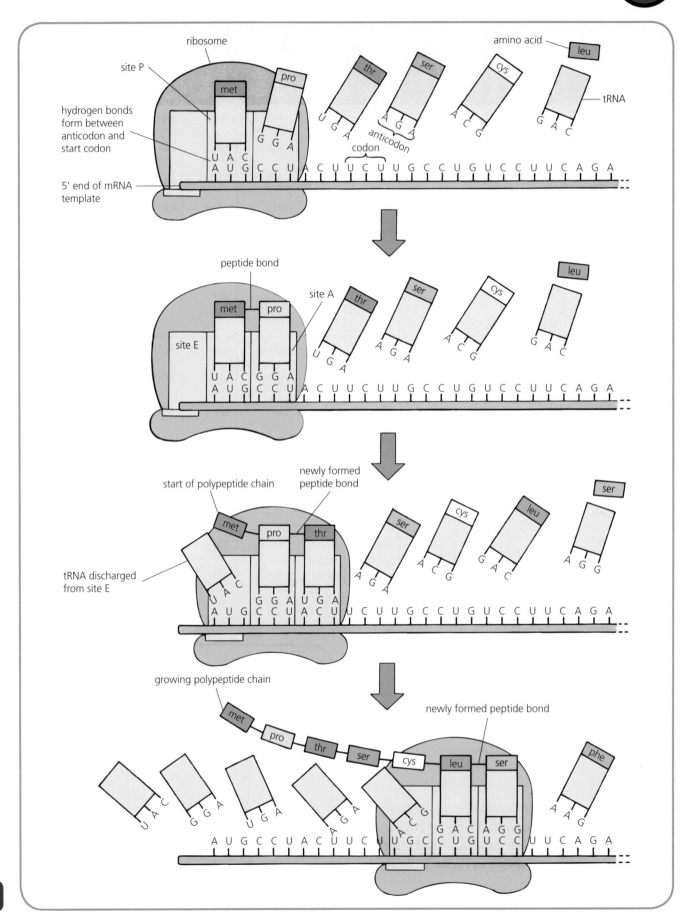

Figure 3.11 Translation of mRNA into polypeptide

- site P holds the tRNA carrying the growing polypeptide chain
- site A holds the tRNA carrying the next amino acid to be joined to the growing chain by a peptide bond
- site E discharges a tRNA from the ribosome once its amino acid has become part of the polypeptide chain.

Start and stop codons in action

Before translation can begin, a ribosome must bind to the 5′ end of the mRNA template so that the mRNA's **start codon** (AUG) is in position at binding site P. Next a molecule of tRNA carrying its amino acid (methionine) becomes attached at site P by hydrogen bonds between its anticodon (UAC) and the start codon (see Figure 3.11).

The mRNA codon at site A recognises and then forms hydrogen bonds with the complementary anticodon on an appropriate tRNA molecule bearing its amino acid. When the first two amino acid molecules are adjacent to one another, they become joined by a **peptide bond**.

As the ribosome moves along one codon, the tRNA that was at site P is moved to site E and discharges from the ribosome to be reused. At the same time the tRNA that was at site A is moved to site P. The vacated site A becomes occupied by the next tRNA bearing its amino acid, which becomes bonded to the growing peptide chain. The process is repeated many times allowing the mRNA to be translated into a complete **polypeptide chain**.

Eventually a **stop codon** (see Table 3.2) on the mRNA is reached. At this point, site A on the ribosome becomes occupied by a release factor, which frees the polypeptide

from the ribosome. The whole process needs energy from ATP (see chapter 7).

Polyribosome

A single molecule of mRNA is normally used to make many copies of the polypeptide. This **multiple translation** is achieved by several ribosomes becoming attached to the mRNA and translating its message at the same time. Such a string of ribosomes on the same mRNA molecule is called a **polyribosome** (see Figure 3.12).

One gene, many proteins

Alternative RNA splicing

Figure 3.6 on page 37 shows a primary transcript of mRNA being cut up and its exons being spliced together to form a molecule of mRNA ready for translation. This molecule of mRNA is not the only one that can be produced from that primary transcript. Depending on circumstances, **alternative segments of RNA** may be treated as the exons and introns. Therefore the same primary transcript has the potential to produce several mRNA molecules each with a different sequence of base triplets and each coding for a different polypeptide. In other words, one gene can code for several different proteins and a limited number of genes can give rise to a wide variety of proteins.

One gene, two antibodies – an example of alternative splicing

An antibody is a Y-shaped protein molecule. The two antibody molecules (P and Q) shown in Figure 3.13 are coded for by the same gene yet they are different in structure. P possesses a membrane-anchoring unit coded for by an exon present in its mRNA. However,

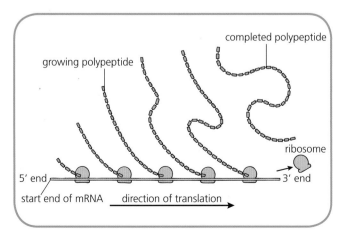

Figure 3.12 Polyribosome

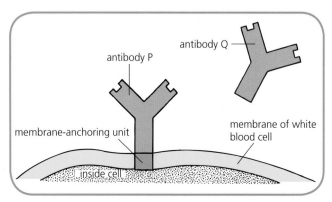

Figure 3.13 Products of alternative RNA splicing

this membrane-anchoring unit is absent from Q because its mRNA lacks the necessary exon (discarded as an intron at the splicing stage). As a result, antibody P functions as a membrane-bound protein on the outer surface of a white blood cell whereas antibody Q operates freely in the bloodstream.

Post-translational modifications

Once translation is complete, further modification (in addition to the folding and coiling described on page 45) may be required to enable a protein to perform its specific function.

Cleavage

A single polypeptide chain may need to be cut (**cleaved**) by enzymes to become active. The protein **insulin**, for example, begins as a single polypeptide chain but requires its central section to be cut out by protease enzymes. This results in the formation of an active protein consisting of two polypeptide chains held together by sulphur bridges, as shown in Figure 3.14.

Molecular addition

A protein's structure can be modified by adding a **carbohydrate** component or a **phosphate** group to it. **Mucus**, for example, is a glycoprotein consisting of protein to which carbohydrate has been added. **Regulatory proteins** often require the addition of a phosphate group to make them functional. **p53**, for example, is a regulatory protein that is normally inactive. However, in situations where a cell's DNA has become damaged, phosphate is added to p53. This process of phosphorylation (see page 99) makes inactive p53 change in structure and become **active p53 tumour-suppressor protein**. It then brings about an appropriate outcome such as repair of DNA or, in extreme cases, programmed cell death.

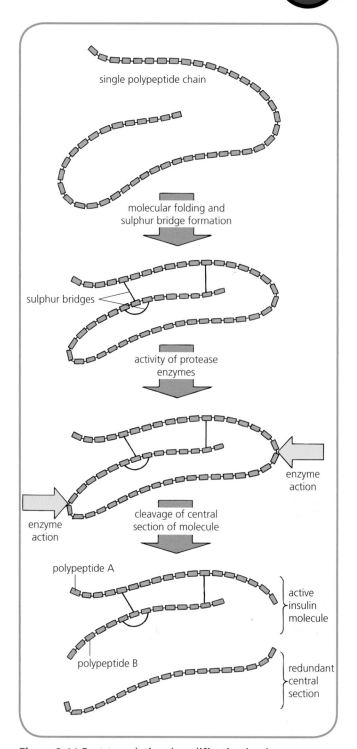

single polypeptide chain

molecular folding and sulphur bridge formation

sulphur bridges

activity of protease enzymes

enzyme action

enzyme action

cleavage of central section of molecule

polypeptide A

active insulin molecule

polypeptide B

redundant central section

Figure 3.14 Post-translational modification by cleavage

Testing Your Knowledge 2

1 a) How many anticodons in a molecule of tRNA are exposed? (1)
   b) Each molecule of tRNA has a site of attachment at one end. What becomes attached to this site? (1)

2 a) What is a *ribosome*? (1)
   b) i) How many tRNA binding sites are present on a ribosome?
      ii) To what does a tRNA's anticodon become bound at one of these sites? (2)
   c) What type of bond forms between adjacent amino acids attached to tRNA molecules? (1)
   d) What is the fate of a tRNA molecule once its amino acid has been joined to the polypeptide chain? (2)

3 Copy and complete Table 3.3. (2)

Stage of synthesis	Site in cell
Formation of primary transcript of mRNA	
Modification of primary transcript of mRNA	
Collection of amino acid by tRNA	
Formation of codon–anticodon links	

Table 3.3

4 Choose the correct answer from the underlined choice for each of the following statements. (6)
   a) The basic units of the genetic code present on mRNA are called anticodons/codons.
   b) The synthesis of mRNA from DNA is called transcription/translation.
   c) A non-coding region of mRNA is called an intron/exon.
   d) Protein synthesis occurs in a cell's nucleus/cytoplasm.
   e) Cleavage of the insulin polypeptide is an example of pre-/post-translational modification.
   f) To become functional, some molecules of regulatory proteins need the addition of phosphate/ribose groups.

What You Should Know Chapters 1–3

adenine	fragments	research
anticodons	genetic	ribose
antiparallel	guanine	ribosomes
backbone	helix	RNA polymerase
bonds	introns	splicing
characteristic	ligase	starting
codons	modification	therapy
complementary	nucleotides	thymine
cytosine	polypeptide	tissue
deoxyribose	primary	transcription
differentiate	primer	translation
DNA polymerase	protein	tumour
embryonic	regulators	unspecialised
exons	replication	uracil

Table 3.4 Word bank for chapters 1–3

1 Stem cells are _____ cells that can reproduce themselves and can _____ into specialised cells.

2 _____ stem cells are able to differentiate into all the cell types that make up the human body. _____ (adult) stem cells are only able to regenerate a limited range of cell types.

3 A differentiated cell only expresses the genes for the proteins _____ of that cell type.

4 Stem cells are used in _____ to gain a better understanding of cell growth and gene regulation. In the future, several debilitating conditions may be treated successfully using stem cell _____.

5 Cancerous cells fail to respond to cell cycle _____ and divide excessively to form a _____.

6 DNA consists of two strands twisted into a double _____. Each strand is composed of _____. Each nucleotide consists of _____ sugar, phosphate and one of four types of base (_____, thymine, _____ and cytosine).

7 Adenine always pairs with _____; guanine always pairs with _____.

8 Within each DNA strand neighbouring nucleotides are joined by chemical _____ into a sugar–phosphate _____. The backbones of complementary strands are _____ because they run in opposite directions.

9 DNA is unique because it can direct its own _____. This begins by DNA unwinding and its two strands separating at a _____ point. A _____ forms beside the DNA strand with the 3′ end. Individual nucleotides aligned with _____ nucleotides on the DNA strand become joined into a new DNA strand by the enzyme _____.

10 The DNA strand with the 5′ end is replicated in _____ that are joined together by the enzyme _____.

11 RNA differs from DNA in that it is single-stranded, contains _____ (not deoxyribose) and the base _____ in place of thymine.

12 DNA contains an individual's _____ information as a coded language determined by the sequence of its bases arranged in triplets called _____. Expression of this information through _____ synthesis occurs in two stages when a gene is switched on.

13 The first stage, _____, begins when the enzyme _____ becomes attached to, and moves along, the DNA, bringing about the synthesis of a _____ transcript of mRNA from individual RNA nucleotides. Primary RNA is cut and spliced to remove non-coding regions called _____ and to bind together coding regions called _____.

14 The second stage, _____, occurs at _____ where codons on the mRNA strand match up with the _____ on tRNA molecules carrying amino acids. These become joined together by peptide bonds to form a _____ chain whose amino acid sequence reflects the code on the mRNA.

15 Alternative _____ of primary mRNA and post-translational _____ of protein structure enable a gene to be expressed as several proteins.

4 Genes and proteins in health and disease

Structure of proteins

All **proteins** contain the chemical elements carbon (C), hydrogen (H), oxygen (O) and nitrogen (N). Often they contain sulphur (S). Each protein is built up from a large number of subunits called **amino acids** of which there are 20 different types. The length of a protein molecule varies from many thousands of amino acids to just a few. Insulin, for example, contains only 51.

Polypeptides

Amino acids become joined together into chains by chemical links called **peptide bonds**. Each chain is called a **polypeptide** and it normally consists of hundreds of amino acid molecules linked together. During the process of protein synthesis (see page 34), amino acids are joined together in a **specific order**, which is determined by the sequence of bases on a portion of DNA. This sequence of amino acids determines the protein's ultimate structure and function.

Hydrogen bonds

Chemical links known as **hydrogen bonds** form between certain amino acids in a polypeptide chain, causing the chain to become coiled or folded as shown in Figure 4.1 on page 46.

Further linkages

During the folding process, different regions of the chain(s) come into contact with one another. This allows interaction between individual amino acids in one or more chains. It results in the formation of various types of cross-connection including **bridges** between **sulphur** atoms, attraction between positive and negative charges and further hydrogen bonding. These cross-connections occur between amino acids in the same polypeptide chain and those on adjacent chains. They are important because they cause the molecule to adopt the final **three-dimensional** structure that it needs to carry out its specific function.

Some types of protein molecule are formed by several spiral-shaped polypeptide molecules becoming linked together in parallel when bonds form between them. This gives the protein molecule a rope-like structure (see Figure 4.1). Other types of protein molecule consist of one or more polypeptide chains folded together into a roughly spherical shape like a tangled ball of string (see Figure 4.1). The exact form that the folding takes depends on the types of further linkage that form between amino acids on the same and adjacent chains.

A computer-generated representation of a protein molecule's three-dimensional structure is shown in Figure 4.2.

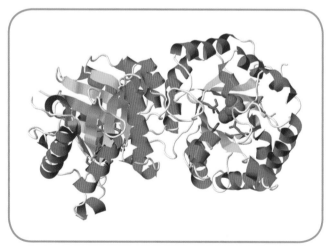

Figure 4.2 Protein molecule as visualised by RasMol software

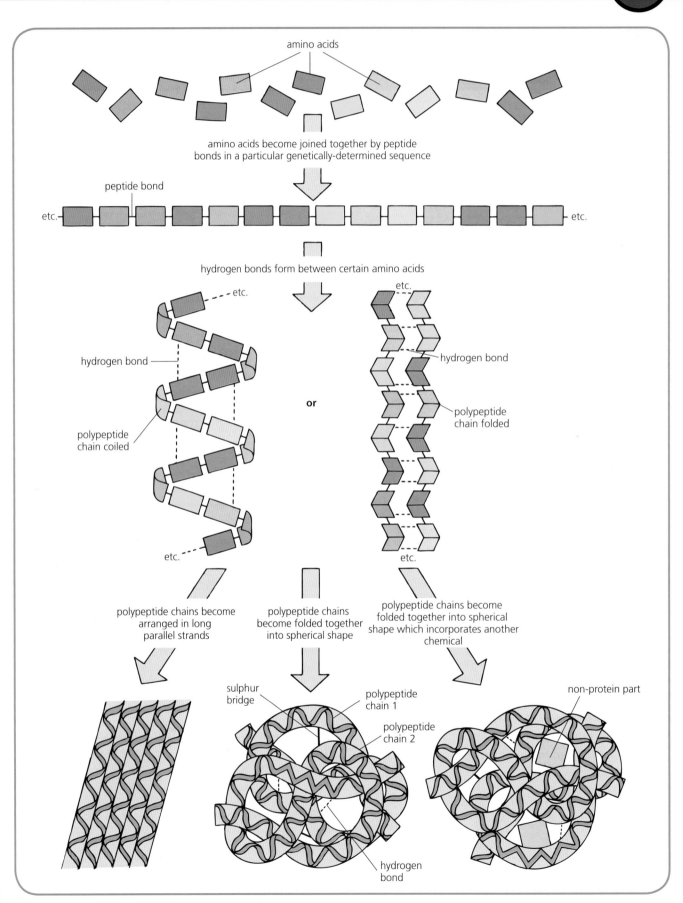

Figure 4.1 Structure of proteins

Separation of fish proteins by agarose gel electrophoresis

Gel electrophoresis is a technique used to separate electrically charged molecules by subjecting them to an electric current, which forces them to move through a sheet of gel.

Eliminating two variable factors

The behaviour of a protein molecule during gel electrophoresis is affected by three variable factors: its shape, its size and its net electric charge. In this experiment variation in shape of the protein molecule and variation in electric charge (e.g. positive or negative) are eliminated by subjecting each protein sample to a negatively charged detergent, a session of heat treatment and a reducing agent. These processes give the molecule a uniformly negative charge and disrupt its hydrogen bonding and sulphur bridges. All the protein molecules become converted to one or more negatively charged linear polypeptides and therefore only vary in **size** (molecular weight).

How the process works

When negatively charged molecules of protein are placed at one end of a sheet of gel and subjected to an electric current, they move towards the opposite, positively charged end of the gel. However, the molecules do not all move through the gel at the same rate. Smaller molecules move at a faster rate and are therefore found to have **moved further** than larger molecules in a given period of time.

The technique illustrated in Figure 4.3 on page 48 is used to separate the proteins present in extracts from four species of fish (W, X, Y and Z) and those in a standard sample of known proteins.

Identification of fish proteins

Protein	Molecular weight (kDa)
a	210
b	107
c	90
d	42
e	35
f	30
g	5

Table 4.1 Proteins in standard sample

Figure 4.4 shows a gel with five lanes resulting from the electrophoresis process. The standard sample is known to contain the seven fish proteins listed in Table 4.1. The band nearest to the well in lane 1 represents the largest protein molecule, which has moved the shortest distance (i.e. protein a). Similarly, band b represents the next largest and so on to band g, which has moved the furthest distance and must be the smallest. Comparison of each of lanes 2–5 with the standard reveals the identity of the proteins present in the extract from a particular species of fish.

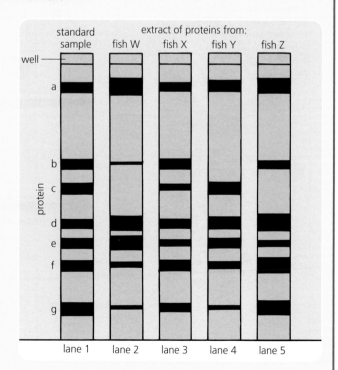

Figure 4.4 Banding results of gel electrophoresis

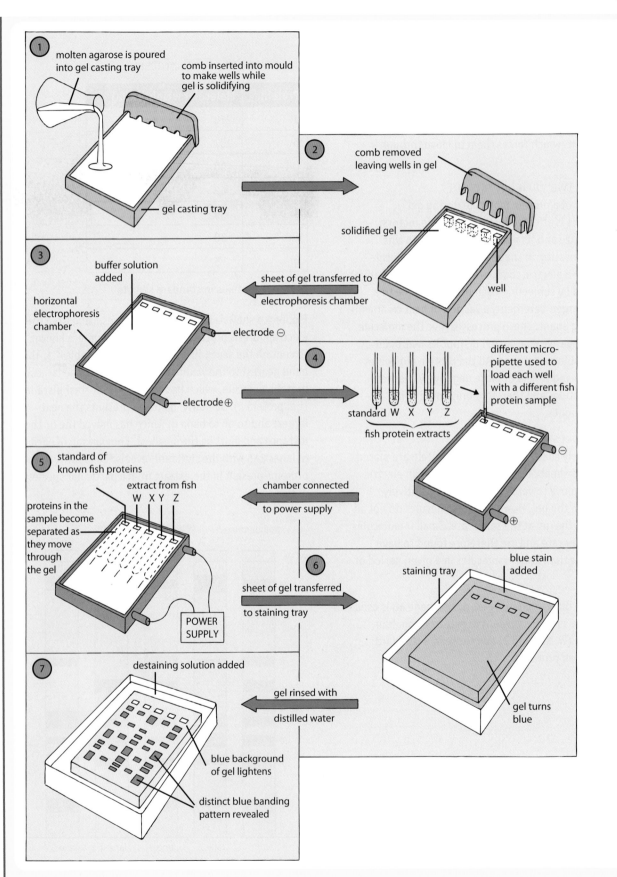

Figure 4.3 Separation of fish proteins by gel electrophoresis

Functions of proteins

A vast variety of structures and shapes exists among proteins and as a result they are able to perform a wider range of functions than any other type of molecule in the body. Some are found in bone and muscle, where their strong fibres provide support and allow movement. Others are vital components of all living cells and play a variety of roles, as follows.

Enzymes

Each molecule of **enzyme** is made of protein and is folded in a particular way to expose an active surface that readily combines with a specific substrate (see chapter 6). Since intracellular enzymes speed up the rate of biochemical processes such as respiration and protein synthesis, they are essential for the maintenance of life.

Structural proteins

Protein is one of two main components that make up the **membrane** surrounding a living cell. Similarly it forms an essential part of all membranes possessed by a cell's organelles. Therefore this type of protein plays a vital structural role in every living cell.

Hormones

These are **chemical messengers** transported in an animal's blood to 'target' tissues where they exert a specific effect. Some hormones are made of protein and exert a regulatory effect on the animal's growth and metabolism. A few examples are given in Table 4.2.

Antibodies

Antibodies are also made of protein. They have a characteristic Y-shape (see Figure 4.5). They are produced by white blood cells to defend the body against antigens (see chapter 22).

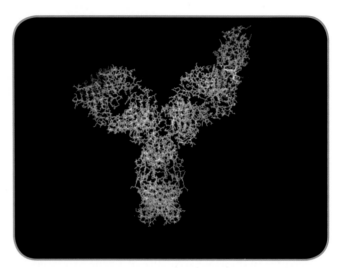

Figure 4.5 Antibody molecule

Associations with other chemicals

Some types of protein molecule are associated with non-protein chemicals (see Figure 4.1). A **glycoprotein** is composed of protein and carbohydrate. An example is mucus, the slimy viscous substance secreted by epithelial cells for protection and lubrication. **Haemoglobin** is the oxygen-transporting pigment in blood. It consists of protein associated with non-protein structures containing iron.

Hormone	Secretory gland	Role of hormone
Insulin	Pancreas	Regulates concentration of glucose in blood
Anti-diuretic hormone	Pituitary	Controls water balance of human body
Human growth hormone	Pituitary	Promotes growth of long bones

Table 4.2 Hormones composed of protein

Mutation

A **mutation** is a change in the structure or composition of an organism's genome. It varies in form from a tiny change in the DNA structure of a gene to a large-scale alteration in chromosome structure or number. When such a change in genotype produces a change in phenotype, the individual affected is called a **mutant**.

Frequency of mutation

In the absence of outside influences, gene mutations arise **spontaneously** and at **random** but only occur rarely. The mutation rate of a gene is expressed as the number of mutations that occur at that gene site per million gametes. Mutation rate varies from gene to gene and species to species.

Mutagenic agents

Mutation rate can be artificially increased by **mutagenic agents**. These include certain chemicals (e.g. mustard gas) and various types of radiation (e.g. gamma rays, X-rays and UV light). The resultant mutations are described as **induced**.

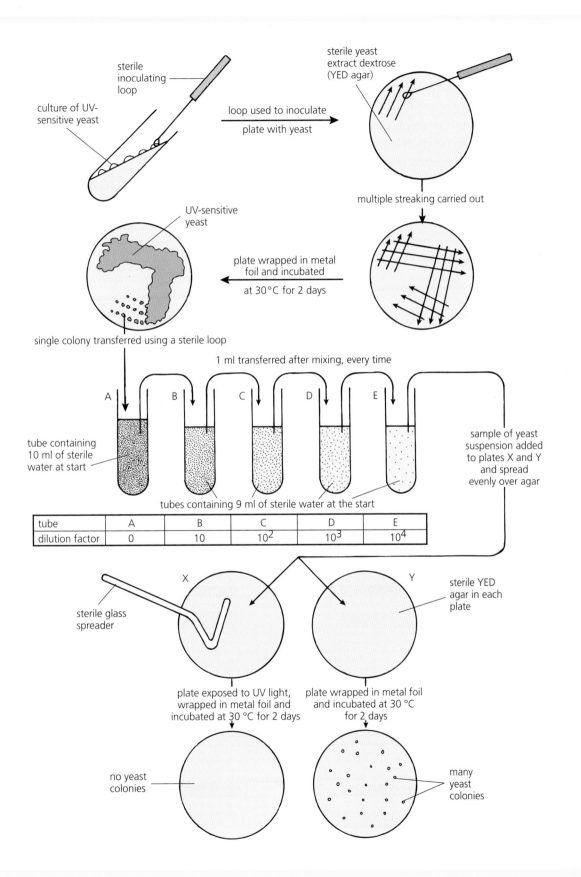

Figure 4.6 Effect of UV radiation on UV-sensitive yeast

Genetic disorder

A condition or disease that can be shown to be directly related to an individual's genotype is called a **genetic disorder**. For a protein to function properly it must possess the correct sequence of amino acids determined by the order of the nucleotide bases on a particular region of DNA in a gene. A change to the gene (or chromosome) caused by a mutation may result in the gene expressing a faulty version of the protein that does not function correctly. The mutated gene may even fail to express the protein at all.

Most proteins are indispensible to the organism. For example, an enzyme that controls a key step in a metabolic pathway is essential for the normal functioning of the body. Therefore the presence of an altered version of the protein (or its total absence) may result in disruption of the pathway and result in a genetic disorder. Many genetic disorders are **disabling** and some are **lethal**.

Single-gene mutation

This type of mutation involves an **alteration of a nucleotide sequence** in the gene's DNA.

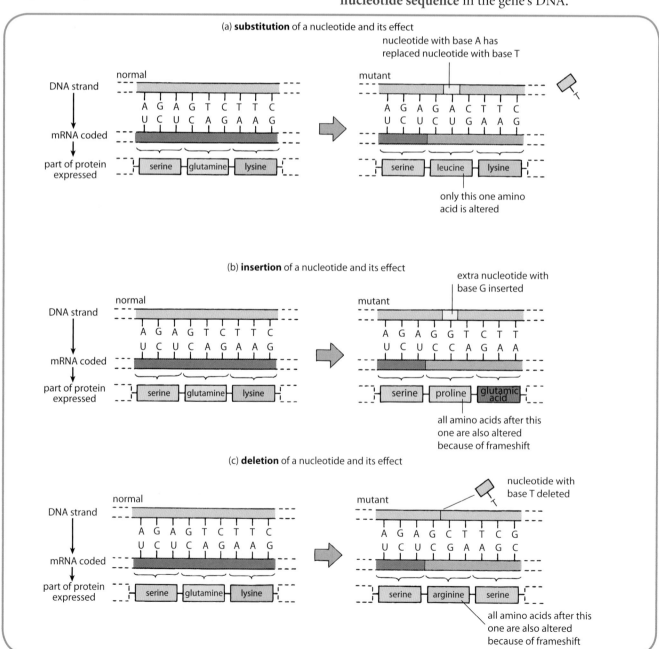

Figure 4.7 Types of point mutation

Point mutation

A point mutation involves a **change in one nucleotide** in the DNA sequence of a single gene. Three types of point mutation are shown in Figure 4.7. A single nucleotide is either **substituted, inserted** or **deleted**. In each case this results in one or more codons for one or more amino acids becoming altered.

Splice-site mutation

Before mRNA leaves the nucleus, introns (non-coding regions) are removed and exons (coding regions) are joined together. This process of post-transcriptional processing of mRNA is called **splicing** (see Figure 3.6 on page 37). Splicing is controlled by specific nucleotide sequences at **splice sites** on those parts of introns that flank exons. If a mutation occurs at one of these splice sites, the codon for an intron–exon splice may be affected and an intron may be retained in error by the modified mRNA (see Figure 4.8).

Nucleotide sequence repeat expansion

A gene mutation can also be the result of a **trinucleotide (triplet) repeat expansion**. Figure 4.9 shows a simplified version of this process. In reality, it normally involves the **insertion** of a large number of

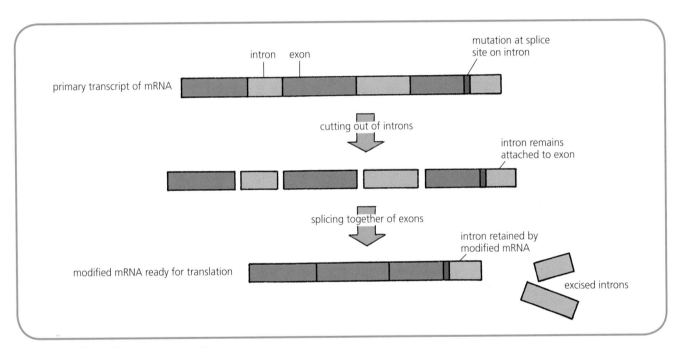

Figure 4.8 Effect of splice-site mutation

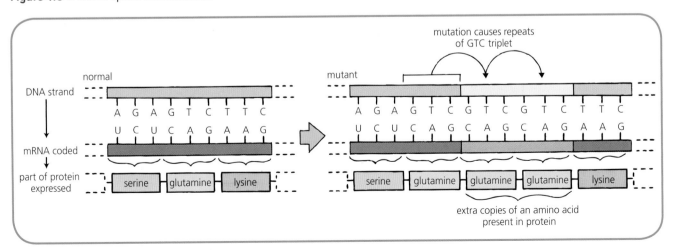

Figure 4.9 Trinucleotide repeat expansion mutation

copies of the nucleotide sequence (e.g. several hundred repeats) that makes the protein, if expressed, defective.

Impact on protein structure

Missense

Following a **substitution**, the altered codon codes for an amino acid that still makes sense but not the original sense (see Figure 4.10). This change in genome is called a **missense** mutation. (See the Case Studies on sickle-cell disease and phenylketonuria.)

Nonsense

As a result of a **substitution**, a codon that used to code for an amino acid is exchanged for one that acts as a premature **stop codon** (UAG, UAA or UGA). It causes

protein synthesis to be halted prematurely (see Figure 4.10) and results in the formation of a polypeptide chain that is shorter than the normal one and unable to function. This change in genome is called a **nonsense** mutation. (See the Case Study on Duchenne muscular dystrophy.)

Splice-site mutation

If one or more introns have been retained by modified mRNA, they may in turn be translated into an altered protein that does not function properly. (See the Case Study on beta (β) thalassemia.)

Frameshift

mRNA is read as a series of triplets (codons) during translation. Therefore, if one base pair is **inserted** or

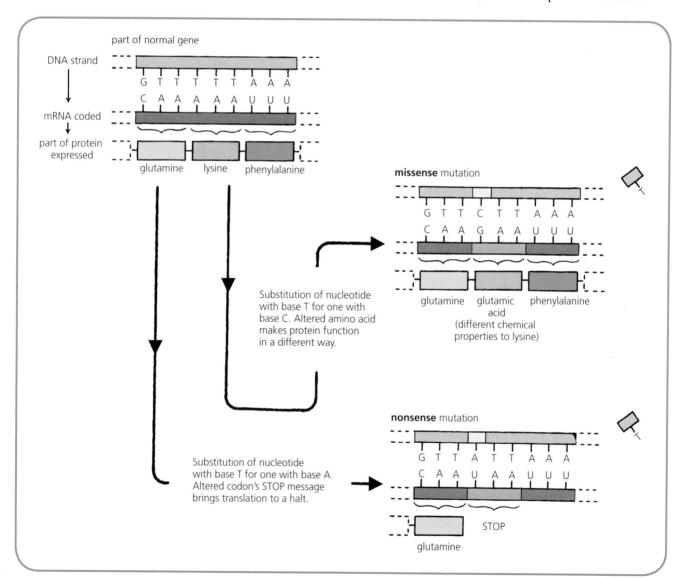

Figure 4.10 Possible effects of a base-pair substitution on sequence of amino acids

deleted (see Figure 4.7) this affects the reading frame (triplet grouping) of the genetic code. It becomes shifted in a way that alters every subsequent codon and amino acid coded all along the remaining length of the gene. The protein formed is almost certain to be non-functional. This change in genome is called a **frameshift** mutation. (See the Case Studies on Tay-Sachs syndrome and cystic fibrosis.)

Nucleotide sequence repeat expansion

This can result in the production of a defective protein possessing a string of **extra copies of one particular** amino acid. On the other hand, expansion of a nucleotide sequence repeat may occur to such an extent that the gene is **silenced** and fails to express any protein at all. (See the Case Studies on fragile X syndrome and Huntington's disease.)

Effect on sufferer

The effects of single gene mutations on the sufferers of the resultant genetic disorders are exemplified in the accompanying case studies. Almost without exception, a genetic disorder has an **adverse effect** on the individual affected.

Case Study   Sickle-cell disease

When one of the genes on chromosome 11 that codes for haemoglobin undergoes a **substitution** (see Figure 4.11), it becomes expressed as an unusual form of haemoglobin called **haemoglobin S**. This is an example of **missense**. Although haemoglobin S differs from normal haemoglobin by only one amino acid, that one tiny alteration leads to profound changes in the folding and ultimate shape of the haemoglobin S molecule, making it a very inefficient carrier of oxygen.

People who are homozygous for the mutant allele suffer drastic consequences. In addition to all of their haemoglobin being type S, which fails to perform the normal function properly, sufferers also possess distorted, sickle-shaped red blood cells (see Figure 4.12). These are less flexible than the normal type and tend to stick together and interfere with blood circulation. The result of these problems is severe shortage of oxygen followed by damage to vital organs and, in many cases, death. This potentially lethal genetic disorder is called **sickle-cell anaemia**.

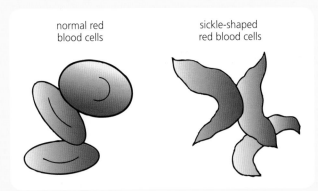

normal red blood cells

sickle-shaped red blood cells

Figure 4.12 Two types of red blood cell

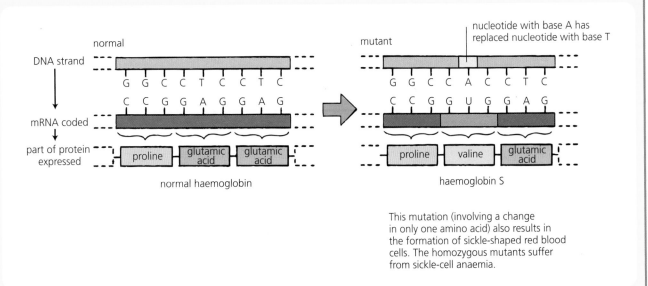

Figure 4.11 Mutation causing sickle-cell disease

normal

DNA strand

mutant

nucleotide with base A has replaced nucleotide with base T

G G C C T C C T C
C C G G A G G A G

G G C C A C C T C
C C G G U G G A G

mRNA coded

part of protein expressed

proline | glutamic acid | glutamic acid

proline | valine | glutamic acid

normal haemoglobin

haemoglobin S

This mutation (involving a change in only one amino acid) also results in the formation of sickle-shaped red blood cells. The homozygous mutants suffer from sickle-cell anaemia.

Sickle-cell trait

People who are heterozygous for the mutant allele do not suffer sickle-cell anaemia. Instead they are found to have a milder condition called **sickle-cell trait**. Their red blood cells contain both forms of haemoglobin but do not show 'sickling'. The slight anaemia that they tend to suffer does not prevent moderate activity.

Resistance to malaria

The sickle-cell mutant allele is rare in most populations. However, in some parts of Africa up to 40% of the population have the heterozygous genotype. This is because sickle-cell trait sufferers are **resistant to malaria**. The parasite cannot make use of the red blood cells containing haemoglobin S. This situation, where a genetic disorder confers an advantage on its sufferers, is very unusual.

Case Study   Phenylketonuria (PKU)

Phenylalanine and tyrosine are two amino acids that human beings obtain from protein in their diet. During normal metabolism, excess phenylalanine is acted on by an enzyme (enzyme 1 in the pathway shown in Figure 4.13).

Phenylketonuria is a genetic disorder caused by a mutation to a gene on chromosome 12 that normally codes for enzyme 1 in the pathway. Most commonly, the mutated gene has undergone a **substitution** of a nucleotide and **missense** occurs. The altered form of the protein expressed contains a copy of tryptophan in place of arginine and is non-functional. As a result

of this **inborn error of metabolism**, phenylalanine is no longer converted to tyrosine. Instead it accumulates and some of it is converted to **toxins**.

These poisonous metabolites inhibit one or more of the enzymes that control biochemical pathways in brain cells. The brain fails to develop properly, resulting in the person having severe learning difficulties. In Britain, newborn babies are screened for PKU and sufferers are put on a diet containing minimum phenylalanine. By this means, the worst effects of PKU are reduced to a minimum.

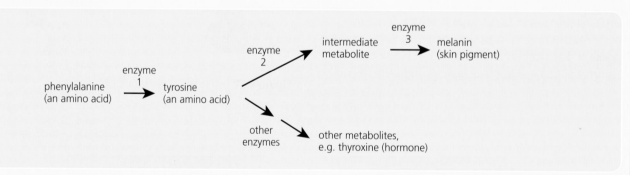

Figure 4.13 Normal fate of phenylalanine

Case Study — Duchenne muscular dystrophy (DMD)

Duchenne muscular dystrophy (DMD) is caused by any one of several types of mutation to a particular gene on chromosome X, such as a **deletion** or a **nonsense** mutation. The affected gene fails to code for a protein called dystrophin, which is essential for the normal functioning of muscles. In skeletal and cardiac muscle, for example, dystrophin is part of a group of proteins that strengthen muscle fibres and protect them from injury during contraction and relaxation.

Duchenne muscular dystrophy is the most common form of muscular dystrophy (muscle-wasting disease). In the absence of dystrophin, skeletal muscles become weak and lose their normal structure. This condition is accompanied by progressive loss of coordination. Sufferers are severely disabled from an early age and normally die young without passing the mutant allele on to the next generation. DMD is sex-linked and is almost entirely restricted to males, being passed on by carrier mothers to their sons.

Case Study — Beta (β) thalassemia

A molecule of haemoglobin is composed of two alpha-globin and two beta-globin polypeptide chains. These polypeptides are encoded by genes.

Beta (β) thalassemia is a genetic disorder caused by any one of several types of mutation that affect a gene on chromosome 11 that codes for beta-globin. One of the most common of these mutations is a **substitution** that occurs at a **splice site** on an intron and causes base G to be replaced by base A.

There are several forms of β-thalassemia, some more severe than others. One type, for example, is characterised by the complete lack of production of beta-globin; another by the production of an altered version of the protein. In either case, the sufferer has a relative excess of alpha-globin in their bloodstream, which tends to bind to, and damage, red blood cells. Patients with severe β-thalassemia require medical treatment such as blood transfusions.

Case Study — Tay-Sachs disease

Tay-Sachs disease is a genetic disorder resulting from a mutation to a gene on chromosome 15. Under normal circumstances the gene is responsible for encoding an enzyme that controls an essential biochemical reaction in nerve cells.

Changes to the gene take the form of point mutations such as **insertions** and **deletions**, which result

in the **frameshift** effect. The protein expressed is so different from the normal one that it is non-functional. As a result, the enzyme's unprocessed substrate accumulates in brain cells. This leads to neurological degeneration, generalised paralysis and death at about 4 years of age. (See page 255.)

Case Study | Cystic fibrosis

Cystic fibrosis is a genetic disorder caused by a three-base-pair **deletion** to a gene on chromosome 7. This type of mutation removes a codon for phenylalanine and causes the coded message to be seriously altered by the **frameshift** effect and produce a non-functional protein.

The normal allele for the gene codes for a **membrane protein** that assists in the transport of chloride ions into and out of cells. In the absence of this protein an abnormally high concentration of chloride gathers outside cells. Those regions of the body that coat their cells with mucus become affected because the high concentration of chloride causes mucus to become **thicker** and **stickier**. Organs such as the lungs, pancreas and alimentary canal become congested and blocked. Regular pounding on the chest to clear thick mucus (see Figure 4.14) and daily use of antibiotics can extend a sufferer's life into their thirties and beyond. Untreated, the sufferer normally dies at age 4–5 years.

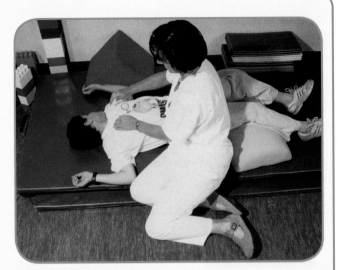

Figure 4.14 Easing symptoms of cystic fibrosis

Case Study | Fragile X syndrome

Fragile X syndrome is a leading cause of a wide spectrum of inherited mental disabilities. It is characterised by a variety of physical features and mental limitations, such as a very elongated face, low muscle tone, nervous speech, poor memory and a very short attention span.

This genetic disorder is caused by a **nucleotide sequence repeat expansion** of CGG coded from a region of the X chromosome. This mutation results in the failure of a gene to encode a protein that is normally active in brain synapses and involved in the control of synaptic plasticity (see page 265). Lack of this essential protein results in retarded neural development.

Whereas the gene of an unaffected individual contains 6–53 repeats of the CGG triplet, a sufferer of fragile X syndrome may have as many as 4000 repeats. Such expansion of the trinucleotide repeat brings about silencing of the affected gene. There is no treatment available for this condition.

Case Study   Huntington's disease

Huntington's disease is caused by a gene on chromosome 4 that has been affected by a **nucleotide sequence repeat expansion**. This type of mutation results in the codon CAG being repeated more than 35 times. The affected gene no longer encodes a certain protein essential for the normal functioning of the nervous system. Instead it codes for a defective form of the protein bearing a long chain composed of repeats of glutamine (the amino acid encoded by CAG). Lack of the correct protein leads to:

- premature death of neurons in regions of the brain
- decreased production of neurotransmitters
- progressive degeneration of the central nervous system.

Unlike all of the other genetic disorders described above, the mutant allele for Huntington's disease is **dominant** and therefore affects people with a heterozygous genotype. In addition, the symptoms of this genetic disorder often do not appear until the person reaches **early middle age**. Death usually follows 10–20 years later. Prior to the onset of the disease, each potential sufferer runs a 50% chance of passing the lethal allele on to each of their offspring before they themselves know if they are affected or not. The condition is incurable.

Chromosome structure mutations

This type of mutation involves the breakage of one or more **chromosomes**. A broken end of a chromosome is 'sticky' and it can join to another broken end. Three of the different ways in which this can occur are discussed below. Each brings about a change in the number or sequence of the genes in a chromosome.

Deletion

A **deletion** occurs when a chromosome breaks in two places and the segment in between becomes detached

(see Figures 4.15 and 4.16). The two ends then join up giving a shorter chromosome, which **lacks** certain genes. As a result, deletion normally has a drastic effect on the organism involved. (See the Case Study on Cri-du-chat syndrome.)

Duplication

A chromosome undergoes **duplication** when a segment of genes (e.g. deleted genes from its matching partner) becomes attached to one end of the first chromosome or becomes inserted somewhere along its length, as shown in Figure 4.17. This results in a set of genes

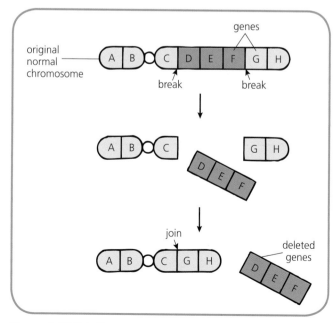

Figure 4.15 Deletion

Figure 4.16 'Look! Nessie's had a deletion!'

being **repeated**. Some duplications of genes may have a detrimental effect on the organism. For example the duplication of certain genes is a common cause of cancer.

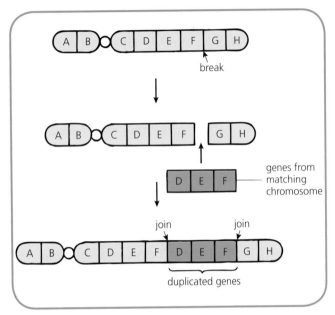

Figure 4.17 Duplication

Translocation

Translocation involves a section of one chromosome breaking off and becoming attached to another chromosome that is *not* its matching partner. Figure 4.18 shows two ways in which this can occur. Translocation is the most common type of mutation associated with cancer. (See the Case Study on chronic myeloid leukaemia.) A translocation can bring about a major change in an individual's phenotype. (See the Case Study on familial Down's syndrome.)

Lethal effect

A mutation to a chromosome often involves such a substantial change to the chromosome's structure (e.g. loss of several functional genes) that the mutation is **lethal**.

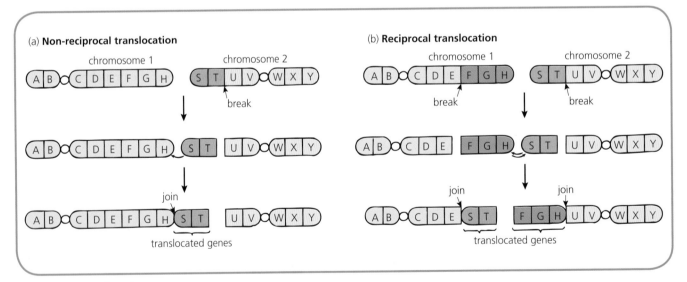

Figure 4.18 Translocation

Case Study	Cri-du-chat syndrome

Cri-du-chat syndrome is caused by a deletion of part of chromosome 5. Children born with this genetic disorder suffer severe learning difficulties. They develop a small head with unusual facial features and widely spaced eyes. The condition is so-called because the infant sufferer's crying resembles that of a distressed cat. Affected individuals usually die early in childhood.

Case Study	Chronic myeloid leukaemia (CML)

Chronic myeloid leukaemia is a form of cancer that affects some of the stem cells that give rise to white blood cells. These stem cells are affected by a **reciprocal translocation** involving genetic material on chromosomes 9 and 22, as shown in Figure 4.19. This translocation results in the formation of an **oncogene**. An oncogene encodes a protein that promotes uncontrolled cell growth (i.e. cancer). In CML the encoded protein is called tyrosine kinase.

CML is treated by using drugs that inhibit the effect of tyrosine kinase and reduce the number of white blood cells produced in the bone marrow. CML occurs most commonly in middle-aged and elderly people. Its incidence is increased by exposure to ionising radiation. The atomic bombing of Hiroshima and Nagasaki in Japan at the end of World War 2 resulted in greatly increased rates of CML among the population. The condition is lethal if left untreated.

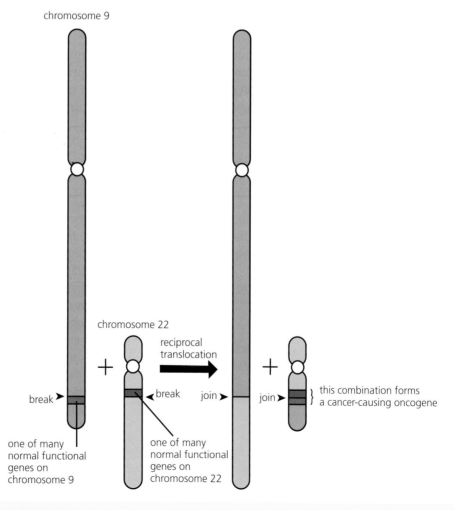

Figure 4.19 Mutation causing chronic myeloid leukaemia

Case Study   Familial Down's syndrome

The vast majority of cases of **Down's syndrome** are the result of a mutant gamete ($n = 24$), containing an extra copy of chromosome 21, fusing with a normal gamete ($n = 23$) at fertilisation to form an abnormal zygote ($2n = 47$). An individual affected in this way suffers Down's syndrome, which is characterised by severe learning difficulties and distinctive physical features.

About 5% of cases of Down's syndrome result from a type of chromosome mutation that does not alter the overall number of chromosomes present in the person's genotype. These individuals are said to suffer **familial Down's syndrome**. A **reciprocal translocation** between chromosomes 14 and 21, as shown in Figure 4.20, affects the genotype of one of the sufferer's parents. This individual is a carrier of the mutated chromosome but is phenotypically unaffected because he/she has two copies of all essential genetic material.

However, at gamete formation, some of the carrier parent's sex cells receive a copy of the mutated chromosome '14 + 21' **and** a copy of the normal chromosome 21. If one of these abnormal gametes meets a normal gamete containing a single copy

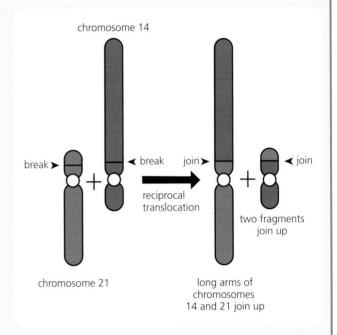

Figure 4.20 Mutation causing familial Down's syndrome

of every chromosome including chromosome 21, the zygote produced contains **three copies of the long arm** of chromosome 21. This abnormal zygote develops into a sufferer of familial Down's syndrome.

Testing Your Knowledge 2

1 Distinguish between the terms *mutation* and *mutant*. (2)
2 **a)** Name THREE types of point mutation that involve a change in one nucleotide in the DNA sequence of a gene. (3)
   b) Which of your answers to **a)** could result in:
   i) a frameshift mutation?
   ii) a missense mutation? (2)
3 **a)** Identify TWO possible effects that a nucleotide sequence repeat expansion mutation can have on the protein expressed. (2)
   b) Suggest why this type of mutation normally leads to a genetic disorder. (1)

4 Rewrite the following sentences using only the correct answer from each choice. (5)
   a) A mutation to a splice site on a gene may alter *pre-/post*-transcriptional processing of mRNA.
   b) An alteration in a chromosome's structure that involves a segment of genes being lost is called *deletion/translocation*.
   c) When a section of one chromosome breaks off and joins onto another non-matching chromosome, this type of mutation is called *duplication/translocation*.
   d) The type of chromosomal change involving a segment of genes from one chromosome becoming inserted somewhere along the length of its matching partner is called *deletion/duplication*.
   e) A substantial change to a chromosome's structure most often has an *adverse/beneficial* effect on the individual involved.

5 Human genomics

Sequencing DNA

Human **genomics** is the study of the human genome. It involves determining the sequence of the nucleotide base molecules all the way along the DNA (**genomic** sequencing) and then relating this genetic information about genes to their functions. Progress in this area has been accelerated by **bioinformatics** (see page 66), making genomics one of the major scientific advances of recent years.

Case Study — Human genome project (HGP)

A milestone in human history was reached in 2003 when the DNA sequence of the **human genome** was completed. It is based on the combined genome of a small number of donors and is regarded as the reference genome. This remarkable achievement was accomplished by adopting several procedures, including the following one.

Sequencing DNA

A portion of DNA with an unknown base sequence is chosen to be sequenced. Many copies of one of this DNA's strands (the template) are synthesised. Then, in order to make DNA strands that are complementary to these template strands, all the ingredients needed for synthesis are added to the preparation. These include DNA polymerase, primer and the four types of DNA nucleotide, as shown in Figure 5.1 on page 64. In addition the preparation receives a supply of **modified nucleotides** (ddA, ddT, ddG and ddC), each tagged with a different **fluorescent dye**.

Every so often during the synthesis process, a molecule of modified nucleotide just happens to be taken up instead of a normal one. However, when a modified nucleotide is incorporated into the new DNA strand, it brings the synthesis of that strand to a halt because a modified nucleotide does not allow any subsequent nucleotide to become bonded to it. Provided that the process is carried out on a large enough scale, the synthesis of a complementary strand will have been **stopped at every possible nucleotide position** along the DNA template.

The resultant mixture of DNA fragments of various lengths (each with its modified nucleotide and its unique fluorescent tag) are separated using **electrophoresis**. In this process the smallest (shortest) fragments travel the furthest distance. The identity and sequence of nucleotides (as indicated by their fluorescent dyes) is then read for the complementary DNA using this separation. From this information the sequence of the bases in the original DNA can be deduced.

This process has been automated and links the detection of the four fluorescent dyes to a computer. As these are monitored, the computer, working as an **automated sequence analyser**, processes the information and rapidly displays the sequence of bases in the DNA sample as a series of peaks (see Figure 5.2).

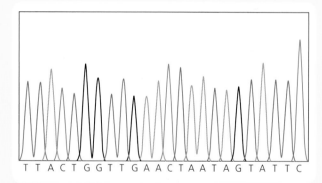

TTACTGGTTGAACTAATAGTATTC

Figure 5.2 Printout from a DNA sequence analyser

Results

After 13 years of work (principally by biologists in the USA and the UK) the sequence of the **three billion** nucleotide bases that make up the human genome was finally unravelled. However, this was by no means the end of the story. Having unravelled the molecular message, the challenge becomes understanding what the message means. One of the many goals of the HGP was to identify the molecular cause of diseases such as cancer in the hope that this knowledge would enable scientists to generate effective treatments. Some progress has already been made in this area.

→

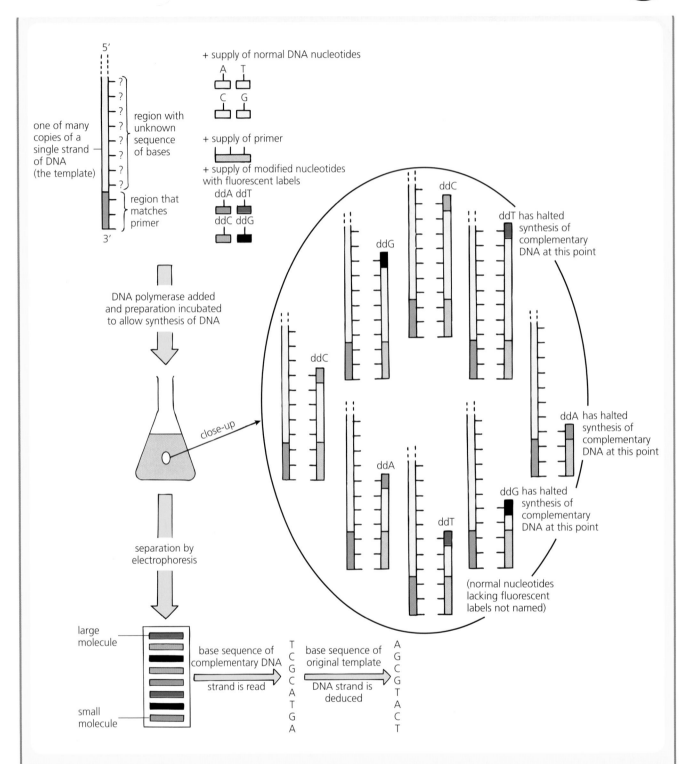

Figure 5.1 Sequencing DNA

It is now known that more than **300 disease-causing genes** exist and that over 4000 genes each express several different forms of the protein that they encode. However, much work remains to be done to accurately relate variation in **genomic structure** to variation in **phenotypic expression** and then to find cures for the genes that cause disorders.

Comparison of individual genomes

Differences in genome

A variation in DNA sequence that affects a single base pair in a DNA chain is called a **single nucleotide polymorphism** (**SNP**). SNPs are one of the ways in which genomes are found to differ from one individual to another. For example, the DNA of two people might differ by the SNP shown in Figure 5.3. This difference has arisen as a result of a point mutation where one base pair has been **substituted** for another. Two out of every three SNPs involve the replacement of cytosine (C) with thymine (T). SNPs can occur in coding and non-coding regions of the genome.

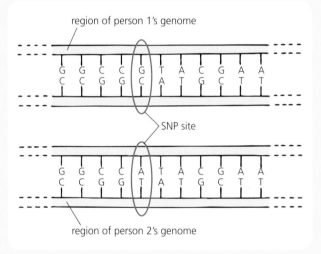

Figure 5.3 Single nucleotide polymorphism (SNP)

The use of **bioinformatics** (see page 66) has enabled scientists to catalogue more than a million SNPs, specify their exact locations in the human genome and use them as DNA markers. They believe that this **SNP map** will help them to identify and understand the workings of genes associated with diseases. Imagine, for example, that people affected by a certain disease always inherit a particular group of SNPs and that unaffected people do not. This would suggest that the gene responsible for the disease is located near the group of SNPs. If scientists know the exact location of these SNPs, then they can analyse nearby genes in the hope of finding the one responsible for the disease.

Single nucleotide polymorphisms are regarded as a valuable tool in biomedical research and may aid the development of future treatments for genetic disorders.

Likelihood of Alzheimer's disease

Some SNPs may indicate the likelihood of a person developing a particular illness. One of the genes associated with **Alzheimer's disease** is called **ApoE** (which codes for apolipoprotein E) and it illustrates how SNPs may be connected with the development of a disorder. The gene is affected by two SNPs and different combinations of these produce three alleles (ApoE2, ApoE3 and ApoE4). Each allele differs from the others by one or two bases and the protein expressed by each differs from the others by one or two amino acids, as shown in Table 5.1.

Allele of ApoE gene	Amino acid	
	Position of amino acid on expressed protein chain	
	112	158
ApoE2	cysteine	cysteine
ApoE3	cysteine	arginine
ApoE4	arginine	arginine

Table 5.1 Three alleles of ApoE gene

Research shows that an individual who inherits at least one ApoE4 allele has an increased chance of developing Alzheimer's disease. The change of one amino acid in the ApoE4 protein alters its structure and function sufficiently to make disease development more likely. On the other hand, inheriting the ApoE2 allele makes the person less likely to develop the condition.

Bioinformatics

The sequencing of the bases in DNA and the amino acids in proteins generates an enormous quantity of data. This information is analysed using computers and the results shared among the members of the molecular biology community over the internet.

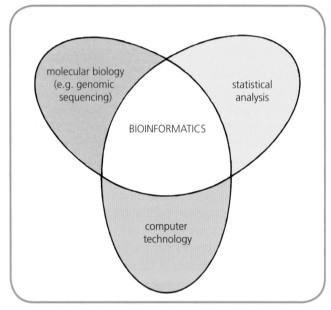

Figure 5.4 Bioinformatics

Bioinformatics is the name given to the fusion of molecular biology, statistical analysis and computer technology (see Figure 5.4). It is an ever-advancing area that enables scientists to carry out rapid mapping and analysis of DNA sequences on a huge scale and then compare them. Individual **gene sequences** (and their roles) can be identified by searching the complete DNA sequence of the target genome using a computer for:

- **protein-coding sequences** the same as, or very similar to, those present in known genes
- **start sequences** (because there is a good chance that each of these will be followed by a coding sequence)
- long sequences that lack **stop codons** (because a protein-coding sequence is normally a very long chain of base triplets containing no stop codons except the one at its end).

Similarly, a search for the identity (and role) of a **base sequence** can be mounted using a computer programme. This enables the scientist to find out if the base sequence matches a specific amino acid sequence already known to be typical of a certain protein.

Information about genetic sequences that used to take years to unravel is now obtained in days or even hours. Bioinformatics can be used to investigate evolutionary biology, inheritance and personalised medicine.

Case Study — **Bioinformatics over the internet**

Although many of the software packages on the internet are commercial, some offer scientists the use of open sources of data and analysis free of charge. The *European Molecular Biology Open Software Suite (EMBOSS)*, for example, is a free software package developed for the needs of molecular biologists. It contains hundreds of programs useful to scientists who want to make use of bioinformatics. These include:

- a comprehensive set of sequence analysis programs and tools
- an extensive library of programs for nucleotide analysis tasks
- a database that mounts searches based on sequence patterns
- programs that compare sequence alignments of chains of nucleotides (and amino acids).

Sequence alignments

Analysing differences at nucleotide (and amino acid) level between two sequences (e.g. from two different

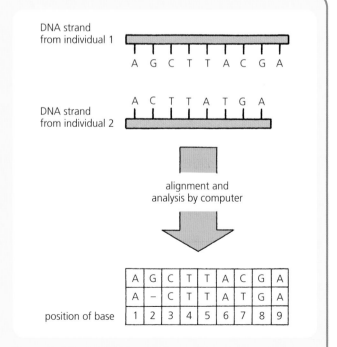

Figure 5.5 Nucleotide base alignment

individuals) begins with an **alignment** of the two sequences by a computer. Then an interpretation of the alignment is made. It would be concluded from the alignment shown in Figure 5.5, for example, that an insertion or deletion has occurred at the second position, a substitution has occurred at the seventh position and that the other nucleotide bases have not undergone any change.

BLAST (Basic Local Alignment Search Tool) is a further bioinformatics instrument that enables biologists to align new sequences with established ones. It allows a researcher to compare new sequences ('queries') with those held in a database and identify known sequences that match the ones being investigated.

Openness

EMBOSS and other suppliers of free software hope to encourage scientists to analyse their data and release the results online in a spirit of openness and universal availability. However, many scientists prefer to keep the results of their research projects under wraps in preparation for presentation at a conference and/or publication in a scientific journal.

Systematics

Systematics can be defined as the study of a group of living things with respect to their diversity, relatedness and classification. Data obtained by comparing human genome sequences are used in systematics to study the **origins** of modern humans and their **evolutionary relationships**. Unlike other primates such as orang-utans, whose DNA differs among the members of the species by around 5%, the mean difference in genomic sequence among the members of the human race is only about 0.3%. This high degree of similarity indicates that all humans are more closely related to one another than are other types of primate.

Careful examination of the genetic differences that do exist between different human populations shows that the **greatest variation** occurs among populations in Africa rather than those on other continents. Furthermore, genetic evidence indicates that all human populations outside Africa possess only a **small part** of the total genetic diversity found among African populations.

These findings support the **'out-of-Africa' theory**. This proposes that humans originated in Africa and underwent early evolutionary divergence in that continent over a very long period of time (e.g. millions of years) to form a variety of genetically different populations. Then small groups migrated out of Africa relatively recently (e.g. 100 000 years ago) and gave rise to all other human populations (see Figure 5.6).

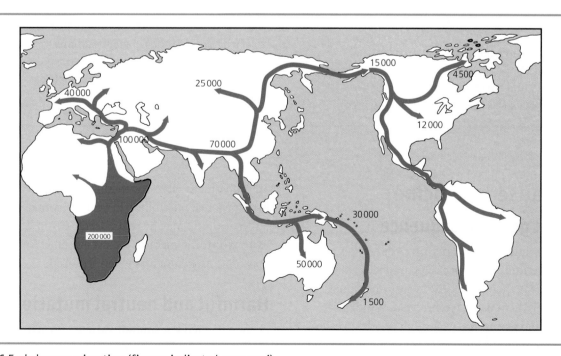

Figure 5.6 Early human migration (figures indicate 'years ago')

Related Topic

Mitochondrial DNA

Men and women both have DNA inside their **mitochondria** (see Figure 5.7) that is inherited from their mother and not mixed with other DNA. Men have DNA in their **Y chromosome** that does not mix with other DNA during gamete formation. Therefore both of these types of DNA retain an **accurate record of mutations** that have occurred over the generations in the individual's ancestral line. For example, those found in mitochondrial DNA can be traced back to a female common ancestor ('Mitochondrial Eve') who lived in Africa about 150 000 years ago.

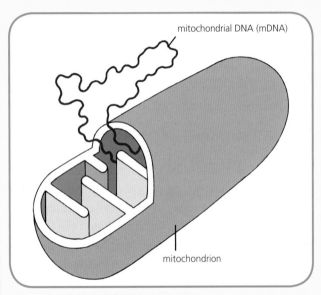

Figure 5.7 Mitochondrial DNA

Sequence data from both mitochondrial DNA and Y chromosomes from a variety of people native to different parts of the world are consistent with the 'out-of-Africa' theory. Although this theory is not (yet) accepted universally, human genome sequence data show that Africa has definitely played a major role in the origins of human beings.

Similarities in genome

In addition to displaying significant differences, a close comparison of genomes often reveals important similarities. For example they may show a high level of **conservation**. This means that the same or very similar DNA sequences are present in the genomes. Highly conserved DNA sequences can be used in comparisons of genomes of two different groups to find out how **close** or **distant** their relationship is. The greater the number of conserved DNA sequences that their genomes have in common, the more closely related the two groups that possess them.

Chimpanzees

The genomes of human beings and chimpanzees are very similar. Analysis of the base sequence of the two genomes reveals that the two groups have **98.5%** of their DNA in common. This makes the chimpanzee our closest living relative. Fossil evidence shows that humans and chimpanzees diverged from a **common ancestor** about six million years ago.

Scientists have used a combination of genome sequence data and fossil evidence to work out the **sequence** in which many key events in evolution have taken place. The evidence strongly supports the theory that living things have undergone a series of **modifications** from the first emergence of life on Earth through to the present day, gradually becoming more and more **complex** as evolution has progressed.

Personalised medicine

Personal genome sequence

A complete sequencing of a person's DNA bases is called a **personal genome sequence**. The branch of genomics involved in sequencing the genomes of individuals and analysing them using bioinformatics tools is called **personal genomics**. As a result of advances in computer technology, the process of sequencing DNA is rapidly becoming **faster** and **cheaper**. Sequencing an individual's DNA for medical reasons will soon become a real possibility. In years to come, a person's entire genome may be sequenced early in life and stored as an electronic medical record available for future consultation by doctors when required.

Harmful and neutral mutations

Having located the mutant variants present in the genome, it is important to distinguish between those altered genetic sequences that are genuinely **harmful** (e.g. fail to code for an essential protein) and those that are **neutral** (i.e. have no negative effect).

Genetic disorders

A **genetic disorder** or **disease** is the result of a variation in genomic DNA sequence. The challenge for scientists is to establish a **causal link** between a particular mutant variant in a genomic sequence and a specific genetic disease or disorder.

The causal genetic sequence has been identified, at least in part, for around 2200 genetic disorders and diseases in humans. However, this does not mean that it is a simple matter to produce treatments for these disorders. The nature of disease is highly complex. Most medical disorders depend on both **genetic** and **environmental** factors for their expression, though the specific effects of these are not fully understood.

Pharmacogenetics

Pharmacogenetics can be defined as the study of the effects (therapeutic, neutral or adverse) of pharmaceutical drugs on the genetically diverse members of the human population. Already it is known that one in ten drugs (e.g. the blood thinner warfarin) varies in effect depending on differences such as SNPs in the person's DNA profile.

In the future it may be possible to use genomic information and customise medical treatment to suit an individual's exact metabolic requirements. The most **suitable drug** and the **correct dosage** would be prescribed as indicated by personal genomic sequencing (and *not* as shown in Figure 5.8!). Ideally this advance would increase drug efficacy while reducing side effects and the 'one-size-fits-all' approach would be consigned to history.

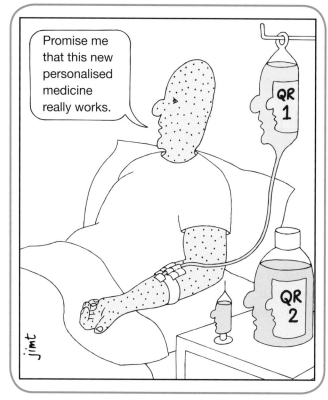

Figure 5.8 'Personalised' medicine

Related Topic

Rational drug design

Once information from DNA sequencing has been used to identify the genes involved in a disease, the next challenge is to establish the structure of the gene(s) and the protein(s) expressed. Pharmacogeneticists may then try to design a drug that will act as an effective treatment. The inventive process of creating a new medication based on knowledge of the structure of the target molecule (e.g. DNA or protein) is called **rational drug design**.

This process involves the synthesis of a specific chemical (normally an organic molecule) that is complementary in shape (and electrical charge) to its biomolecular target. If it is effective, it provides the patient with a therapeutic benefit by acting in one of the following ways:

- It binds to the particular region of DNA in the mutant gene that causes the genetic disorder and **prevents transcription** of abnormal mRNA.
- It binds to the abnormal mRNA that has been transcribed and **prevents it from being translated** into abnormal protein. (A type of RNA called interfering RNA induces post-transcriptional silencing of genes in this way and is therefore being used to design new drugs.)
- It binds to and **renders inactive the protein** whose presence would cause the genetic disorder.

An example of a medication produced as a result of rational drug design is a tyrosine kinase inhibitor (called imatinib). This is used to treat chronic myeloid leukaemia (see page 61) because it binds with tyrosine kinase and renders it inactive.

Rational drug design often benefits from **computer modelling** techniques. In addition, use of computers has accelerated discovery of new drugs by enabling scientists to make fairly accurate **predictions** about the affinity of a new compound for (and its likely effect on)

its target molecule before the new compound has even been synthesised. This is more efficient than traditional methods of drug discovery where the early stages of development are heavily dependent on results from trial-and-error testing on cultured cells and laboratory animals.

Future

Risk prediction

Already variations in DNA have been linked to conditions such as diabetes, heart disease, schizophrenia and cancer. In the future, when the locations in the human genome of many more markers for common diseases and disorders have been established, it should become possible to scan an individual's genome for **predisposition** to a disease and **predict risk** early enough to allow suitable action to be taken. Eventually reduction of risk may be achieved through appropriate drug treatment combined with a healthy lifestyle.

Ethical issues

If personal genomic sequencing becomes a routine predictive medical procedure then this raises many

ethical issues. For example, if a person's genome contains genetic markers indicating a high risk of a debilitating or fatal disease later in life, who should have access to this information?

- The person's employer? Perhaps the company will refuse to employ anyone who is at risk of the disorder.
- The person's offspring? Maybe this would tell them more than they want to know about their own genome.
- The person's life insurer? Perhaps the insurance company will insist on charging a much higher premium or refuse to provide cover at all.

Many people believe that laws should be introduced to prevent **genetic discrimination** based on information obtained from an individual's genome. These issues need to be addressed by society before genomic sequencing becomes inexpensive and widely available.

Testing Your Knowledge 1

1 a) i) What information is obtained from the process of genomic sequencing?
   ii) Give ONE example of a use to which this information can be put. (2)
   b) What is meant by the term *bioinformatics*? (1)
   c) Give TWO examples of the type of sequence that bioinformaticists would look for in a long chain of bases to identify gene sequences present. (2)
2 Briefly describe the 'out-of-Africa' theory as applied to human beings. (3)
3 a) What is meant by *personal genomics*? (2)
   b) Give TWO possible benefits of personalised medicine to patients of the future. (2)

4 Rewrite the following sentences, choosing only the correct answer from each underlined choice. (4)
   a) The role of a sequence of bases can be identified by finding the amino acid/nucleotide sequence of a known protein to which it corresponds.
   b) The study of a group of organisms with respect to their diversity, relatedness and classification is called pharmacogenetics/systematics.
   c) Comparison of human genome sequence data gives information about evolutionary relationships/ learned behaviour.
   d) An altered gene sequence that fails to code for a protein is more likely to be neutral/harmful than one that has no negative effect.

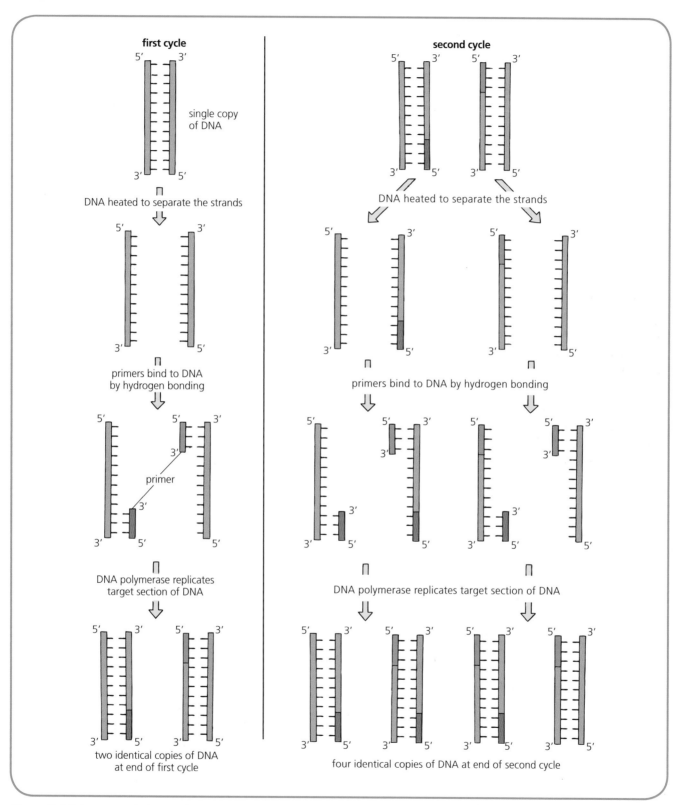

Figure 5.9 Polymerase chain reaction

Amplification and detection of DNA sequences

Polymerase chain reaction

The **polymerase chain reaction** (PCR) is a technique (see Figure 5.9) that can be used to create many copies of a piece of DNA in vitro (i.e. outside the body of an organism). This **amplification** of DNA involves the use of **primers**. In this case, each primer is a piece of single-stranded DNA complementary to a specific target sequence at the 3′ end of the DNA strand to be replicated.

The DNA is heated to break the hydrogen bonds between base pairs and separate the two strands. Cooling allows each primer to bind to its target sequence. During the next step, **heat-tolerant DNA polymerase** adds nucleotides to the primers at the 3′ end of the original DNA strands.

The first cycle of replication produces two identical molecules of DNA, the second cycle four identical molecules and so on, giving an exponentially growing population of DNA molecules. By this means a tiny quantity of DNA can be greatly amplified and provide sufficient material for forensic and medical purposes.

Case Study | Use of PCR

PCR is used to amplify DNA from sources such as embryonic cells for prenatal screening of genetic disorders and blood, semen and other tissues from a crime scene for DNA fingerprinting. PCR depends on a process called **thermal cycling**. A cycle consists of three steps each carried out at a different temperature. The earliest designs of this technique used three water baths and normal DNA polymerase. The latter was destroyed during the heating step in the cycle and had to be replaced for use in the next cycle.

The following two important innovations enabled PCR to become automated:

- the isolation of **heat-tolerant** DNA polymerase from a species of bacterium native to hot springs
- the invention of the **thermal cycler**, a computerised heating machine able to control the repetitive temperature changes needed for PCR.

Figure 5.10 shows a simplified version of the steps carried out during thermal cycling in order to amplify DNA.

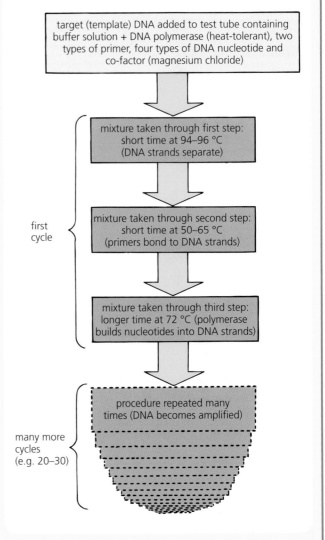

Figure 5.10 Amplification of DNA by thermal cycling

Related Topic

'Needles and haystacks'

Equally as impressive as the amplification of DNA by PCR is the **specificity** of the reaction. Each primer is a piece of single-stranded DNA synthesised as the exact complement of a short length of the DNA strand to which it is to become attached. This enables the primer to find 'the needle in the haystack'. In other words it is able to locate, among many different sites, the specific target DNA sequence that is to be amplified. The process then goes on to produce millions or even billions of copies of the DNA. Therefore this amplification of DNA by PCR is sometimes described as being like 'a haystack from the needle'.

DNA probes and arrays

One of the meanings of the word *array* is *an orderly arrangement of many items*. A **DNA microarray** is an orderly arrangement of thousands of different DNA probes as tiny spots attached to a glass slide. A **DNA probe** is a short, single-stranded fragment of DNA. It is used to detect the presence of a specific sequence of nucleotide bases in a sample of DNA. The DNA under investigation is called the **target DNA**.

A probe is able to carry out its function because its sequence of bases is complementary to the specific base sequence to be detected in the target DNA. **Fluorescent** labelling indicates those spots where a probe has successfully detected and combined with its complementary sequence on the target DNA.

Case Study **Medical uses of DNA probes**

Figure 5.11 shows a method used to carry out a gene expression microarray.

- **DNA probes** (each representing a known gene from the human genome) are synthesised and fixed into position on the glass slide.
- **Messenger RNA transcripts** made by active genes in the cells of the target tissue are isolated.
- These mRNA transcripts are used to make lengths of single-stranded, **complementary DNA (cDNA)** by a process called reverse transcription, using DNA nucleotides containing a fluorescent label.
- The cDNA mixture is added to the DNA microarray to allow **hybridisation** to take place. (Hybridisation is the pairing up and bonding that occurs if a length of cDNA comes into contact with a gene probe to which it is exactly complementary.)
- Following rinsing, the microarray is examined for **fluorescent spots**.

A spot only gives off fluorescence if hybridisation has taken place. Each fluorescent spot indicates a gene that is expressed by the cells in the target tissue.

Applications

Although each cell in the human body, almost without exception, contains a full set of identical genes, within each type of differentiated cell only a fraction of these genes are active and transcribing their message into mRNA. It is this subset of actively coding genes that gives the cell type its unique properties. A **gene expression microarray** containing DNA probes enables a comparison to be made between, for example, healthy and cancerous cells and shows which genes are active only in the diseased cells. It also allows the effect of a potential treatment on the expression of thousands of genes to be investigated simultaneously.

DNA probes can also be used in **genotype microarrays** to compare the **genomic content** of individuals (e.g. healthy individuals compared with those suffering a genetic disorder). In addition, the probes are employed in microarrays to detect the presence of certain **single nucleotide polymorphisms (SNPs)** in the genome of individuals where these may indicate predisposition to disease.

False positives and negatives

Some genetic tests are not 100% reliable. Most involve a very low incidence of error. If (on rare occasions) a test fails to detect a specific DNA sequence when in fact it is present, the result is called a **false negative**. If (on rare occasions) the test indicates the presence of a specific DNA sequence when, in reality, it is absent, the result is called a **false positive**. (See Appendix 4.)

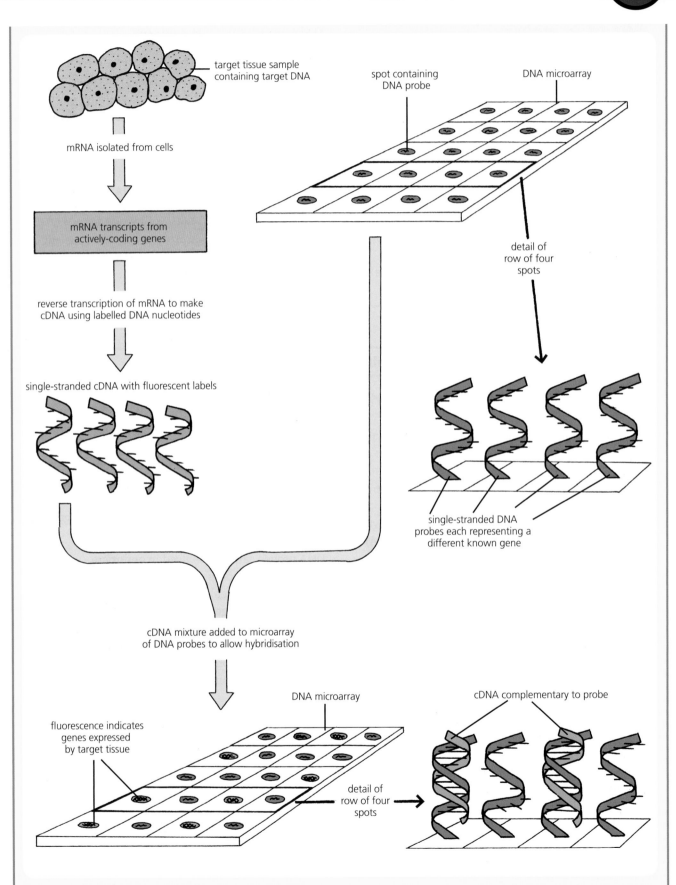

Figure 5.11 Use of DNA probes

Medical and forensic applications of amplified DNA

Medical

PCR can be used to amplify the genomic DNA from a cell sample taken from a patient. By this means sufficient DNA is generated to allow it to be **screened** for the presence or absence of a specific sequence known to be characteristic of a genetic disease or disorder. This enables medical experts:

- to estimate the risk of **disease onset** (for example, by identifying mutations in the person's genome that are known to increase the chance of developing a particular disease)
- to confirm a **diagnosis** of the genetic disorder if the condition is suspected (based on the patient's family history and/or the fact that the patient is showing early symptoms).

Genetic testing for cystic fibrosis

In the UK, it is estimated that 1 in 25 people carries a recessive mutant allele for **cystic fibrosis** (see page 58). Carriers do not suffer the disease but if two carriers produce a child, they risk a **1 in 4** chance of the child suffering cystic fibrosis. Therefore couples planning a family may decide to be tested for the presence of a mutant allele for cystic fibrosis in their genome. This can be done by **genetic testing** using blood cells. DNA is amplified and tested for the presence of a mutant allele using genetic probes.

Forensic

DNA profile

The human genome possesses many short, **non-coding regions of DNA** composed of a number of **repetitive sequences**. These regions are found to be randomly distributed throughout the genome and to differ in length and number of repeats of the DNA sequences from person to person. Each region of repetitive sequences is **unique** to the individual who possesses it. Therefore these regions of genetic material can be used to construct a **DNA profile** for that person.

Crime scene

Forensic scientists make use of the PCR reaction to amplify DNA samples from a **crime scene**. DNA

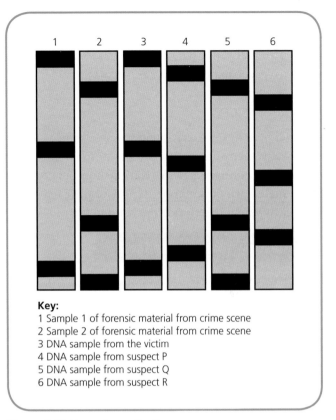

Key:
1 Sample 1 of forensic material from crime scene
2 Sample 2 of forensic material from crime scene
3 DNA sample from the victim
4 DNA sample from suspect P
5 DNA sample from suspect Q
6 DNA sample from suspect R

Figure 5.12 Forensic application

samples taken from the victim and the suspects are also amplified. Next the components of the samples are separated using gel electrophoresis and then compared.

Figure 5.13 Genetic 'fingerprints'

In the example shown in Figure 5.12, it is concluded that the DNA in sample 1 from the crime scene matches that of the victim and that the DNA in sample 2 from the crime scene matches that of suspect Q.

Paternity dispute

PCR followed by gel electrophoresis can also be employed to generate genetic profiles from DNA samples and **confirm genetic relationships** between individuals. Each person inherits 50% of their DNA from each parent therefore every band in their DNA profile ('genetic fingerprint') must match one in that of their father or their mother. The fact that each person has 50% of their bands in common with each of their parents (see Figure 5.13) allows **paternity disputes** to be settled.

Evidence based on DNA amplified by PCR has also been used to identify missing people from human remains left at the site of a disaster and to secure the release of innocent people who have been wrongly imprisoned.

Testing Your Knowledge 2

1 a) What can be produced *in vitro* by employing the polymerase chain reaction (PCR)? (1)
   b) In PCR, what is a *primer*? (2)
   c) Why is the DNA heated during the PCR process? (1)
   d) What is the purpose of cooling the DNA sample? (1)
   e) What characteristic of the DNA polymerase used in PCR prevents it from becoming denatured during the process? (1)

2 a) Briefly describe the structure of a genetic probe. (1)
   b) What is a genetic probe used for? (1)
3 a) What medical application does the amplification of DNA by PCR make possible? (1)
   b) What feature of the human genome makes each individual unique and allows genetic profiles to be constructed? (1)

What You Should Know Chapters 4–5

altered	evolutionary	personalised
amino	folded	probes
amplified	forensic	profiles
bioinformatics	frameshift	risk
chain	genomic	screened
chromosome	heat-tolerant	sequence
coding	insertion	three-dimensional
customised	lethal	
cycling	microarrays	translocated
deletion	missense	twenty
diagnosis	mutation	variable
disorder	origins	
environmental	peptide	

Table 5.2 Word bank for chapters 4–5

1 Proteins consist of subunits called _____ acids of which there are _____ different types.

2 Amino acid molecules are joined together by _____ bonds to form polypeptides. Polypeptides are coiled and _____ to form protein molecules whose _____ structure, which is maintained by cross-connections between amino acids, is directly related to their function.

3 A single-gene mutation involves the substitution, insertion or _____ of nucleotides in the DNA chain. Substitutions can result in _____ or nonsense mutations. Insertions or deletions lead to _____ mutations. An _____ can cause an expansion of a nucleotide sequence repeat.

4 A mutation can result in the production of an _____ protein that does not function properly, or in the failure of the gene to express the protein, thereby causing a genetic _____.

5 A _____ may undergo a structural mutation if one or more of its genes becomes deleted, duplicated or _____. A _____ to a chromosome involves a major change to the individual's genome and is often _____.

6 Determining the sequence of nucleotide bases for individual genes or for a person's entire genome is called _____ sequencing. Use is made of _____, involving computing and statistics, to compare sequence data.

7 Gene sequences can be identified by comparing them with those of known genes and looking for similar _____ sequences. Human genome sequence data are also compared to obtain information about human _____ and _____ relationships.

8 In the future, routine sequencing of an individual's genome may lead to _____ medicine. This could involve predicting _____ of disease through knowledge of a person's genome and administering _____ drugs in appropriate dosages. Diseases are complex and often affected by both genetic and _____ factors.

9 DNA can be _____ by the polymerase _____ reaction using primers, _____ DNA polymerase and repeated thermal _____.

10 DNA _____ are short, single-stranded fragments of DNA used in _____ to detect the presence of a specific base _____ in DNA samples.

11 In medicine, amplified DNA can be _____ for a specific genetic sequence associated with a disease to allow a _____ or an estimate of risk to be made.

12 The existence of _____ numbers of repetitive sequences of DNA makes each person's genome unique and allows DNA _____ of individuals to be constructed for _____ use.

6 Metabolic pathways

Cell metabolism

Cell metabolism is the collective term for the thousands of biochemical reactions that occur within a living cell. The vast majority of these are steps in a complex network of connected and integrated pathways that are catalysed by enzymes.

Metabolic pathways

The biochemical processes upon which life depends take the form of **metabolic pathways**, which fall into two categories:

- **Catabolic** pathways bring about the breakdown of complex molecules to simpler ones, usually releasing energy and often providing building blocks.
- **Anabolic** pathways bring about the biosynthesis of complex molecules from simpler building blocks and require energy to do so.

Such pathways are closely integrated and one often depends upon the other. For example, aerobic respiration in living cells is an example of **catabolism**, which releases the energy needed for the synthesis of protein from amino acids (an example of **anabolism**). This close relationship is shown in Figure 6.1. An important chemical called ATP (see chapter 7) plays a key role in the transfer of energy between catabolic and anabolic reactions.

Reversible and irreversible steps

Metabolic pathways are regulated by enzymes that catalyse specific reactions. A pathway often contains both **reversible** and **irreversible** steps, which allow the process to be kept under precise control. **Glycolysis** (see page 101) is the metabolic pathway that converts **glucose** to an intermediate metabolite called **pyruvate** at the start of respiration. Figure 6.2 shows the first three enzyme-controlled steps in a long pathway.

Glucose diffusing into a cell from a high concentration outside to a low concentration inside is irreversibly converted to intermediate 1 by enzyme A. This process is of advantage to the cell because it maintains a low concentration of glucose inside the cell and therefore promotes continuous diffusion of glucose into the cell from the high concentration outside.

The conversion of intermediate 1 to intermediate 2 by enzyme B is **reversible**. If more intermediate 2 is formed than the cell requires for the next step then some can be converted back to intermediate 1 and used in an alternative pathway (for example to build glycogen in animal cells or starch in plant cells). The conversion of intermediate 2 to intermediate 3 by enzyme C is **irreversible** and is a key **regulatory point** in the pathway. There is no going back for the substrate now. It is committed to following glycolysis through all the steps to pyruvate.

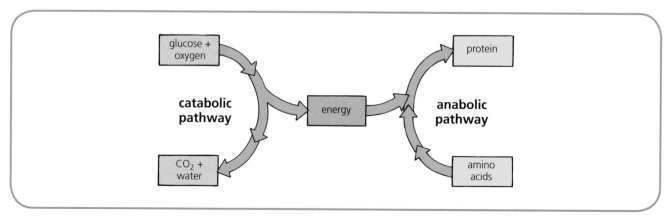

Figure 6.1 Two types of metabolic pathway

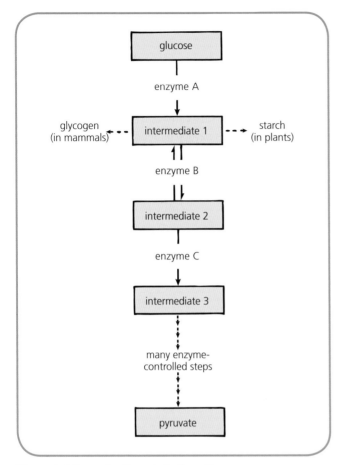

Figure 6.2 Example of a metabolic pathway

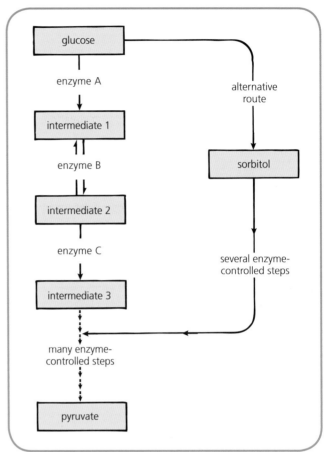

Figure 6.3 Alternative route

Alternative routes

Metabolic pathways can also contain **alternative routes** that allow steps in the pathway to be bypassed. Figure 6.3 shows a pathway from glucose via an intermediate (called sorbitol) that bypasses the steps controlled by enzymes A, B and C but returns to glycolysis later in the pathway. This bypass is used when the cell has a plentiful supply of sugar.

Activation energy and enzyme action

The rate of a chemical reaction is indicated by the amount of chemical change that occurs per unit time. Such a change may involve the joining together of simple molecules into more complex ones or the splitting of complex molecules into simpler ones. In either case the energy needed to break chemical bonds in the reactant chemicals is called the **activation energy**.

The bonds break when the molecules of reactant have absorbed enough energy to make them unstable. They are now in the **transition state** and the reaction can occur. This energy input often takes the form of heat energy and the reaction only proceeds at a high rate if the chemicals are raised to a high temperature (see Figure 6.4).

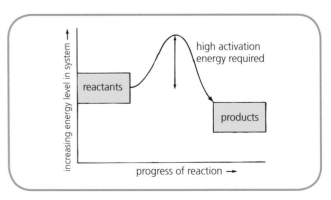

Figure 6.4 Uncatalysed reaction

Related Activity

Investigating the effect of heat on the breakdown of hydrogen peroxide

Hydrogen peroxide is a chemical that breaks down into water and oxygen as shown in the following equation:

hydrogen peroxide   →   water   +   oxygen

$$2H_2O_2 \quad \rightarrow \quad 2H_2O \; + \; O_2$$

In the experiment shown in Figure 6.5, test tubes containing hydrogen peroxide and drops of detergent are placed in five water baths at different temperatures. The detergent is used to sustain any oxygen bubbles that are released as a froth.

After 30 minutes the tubes are inspected for the presence of a froth of oxygen bubbles which indicates the breakdown of hydrogen peroxide. The diagram shows a typical set of results where the volume of froth is found to increase with increase in temperature.

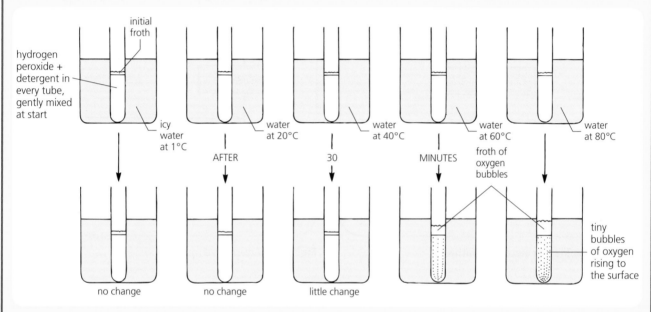

Figure 6.5 Investigating the effect of heat on the breakdown of hydrogen peroxide

Related Activity

Investigating the effect of manganese dioxide on the breakdown of hydrogen peroxide

In the experiment shown in Figure 6.6, the bubbles forming the froth in tube A are found to relight a glowing splint. This shows that oxygen is being released during the breakdown of hydrogen peroxide. In tube B, the control, the breakdown process is so slow that no oxygen can be detected.

It is concluded therefore that manganese dioxide (which remains chemically unaltered at the end of the reaction) has increased the rate of this chemical reaction which would otherwise have only proceeded very slowly. A substance that has this effect on a chemical reaction is called a **catalyst**.

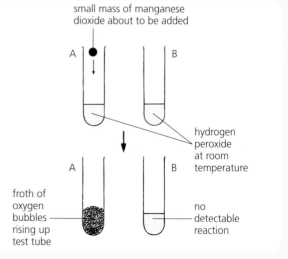

Figure 6.6 Effect of a catalyst

Properties and functions of a catalyst

A **catalyst** is a substance that:

- lowers the activation energy required for a chemical reaction to proceed (see Figure 6.7)
- speeds up the rate of a chemical reaction
- takes part in the reaction but remains unchanged at the end of it.

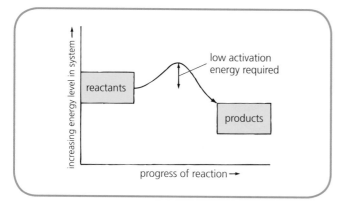

Figure 6.7 Catalysed reaction

Importance of enzymes

Living cells cannot tolerate the high temperatures needed to make chemical reactions proceed at a rapid rate. Therefore they make use of **biological catalysts** called **enzymes**.

Enzymes speed up the rate of the reactions in a metabolic pathway by **lowering the activation energy** needed by the reactant(s) to form the transition state. It is from this unstable state that the end products of the reaction are produced.

By this means biochemical reactions are able to proceed rapidly at the relatively low temperatures (e.g. 5–40°C) needed by living cells to function properly. In the absence of enzymes, biochemical pathways such as respiration and photosynthesis would proceed so slowly that life as we know it would cease to exist.

Enzyme action

Enzyme molecules are made of **protein**. Somewhere on an enzyme's surface there is a groove or hollow where its **active site** is located. This site has a particular shape

---

Related Activity

Investigating the effect of catalase on the breakdown of hydrogen peroxide

Catalase is an enzyme made by living cells. It is especially abundant in fresh liver cells. In the experiment shown in Figure 6.8, the bubbles produced in tube C are found to relight a glowing splint. This shows that oxygen is being released during the breakdown of hydrogen peroxide as follows:

$$\text{hydrogen peroxide} \quad \xrightarrow{\text{catalase}} \quad \text{water} \quad + \quad \text{oxygen}$$
$$\text{(substrate)} \quad \text{(enzyme)} \quad \text{(end products)}$$

In tube D, the control, the breakdown process is so slow that no oxygen can be detected. It is concluded that the enzyme catalase has increased the rate of this chemical reaction, which would otherwise have proceeded only very slowly.

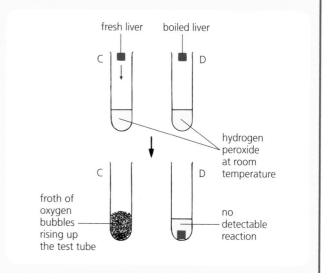

Figure 6.8 Effect of catalase

that is determined by the chemical structure of, and bonding between, the amino acids in the polypeptide chains that make up the enzyme molecule.

Specificity

An enzyme acts on one type of substance (its **substrate**) whose molecules exactly fit the enzyme's active site. The enzyme is **specific** to its substrate and the molecules of

substrate are complementary to the enzyme's active site for which they show an **affinity** (chemical attraction).

Induced fit

The active site is not a rigid structure. It is **flexible** and **dynamic**. When a molecule of substrate enters the active site, the shape of the enzyme molecule and

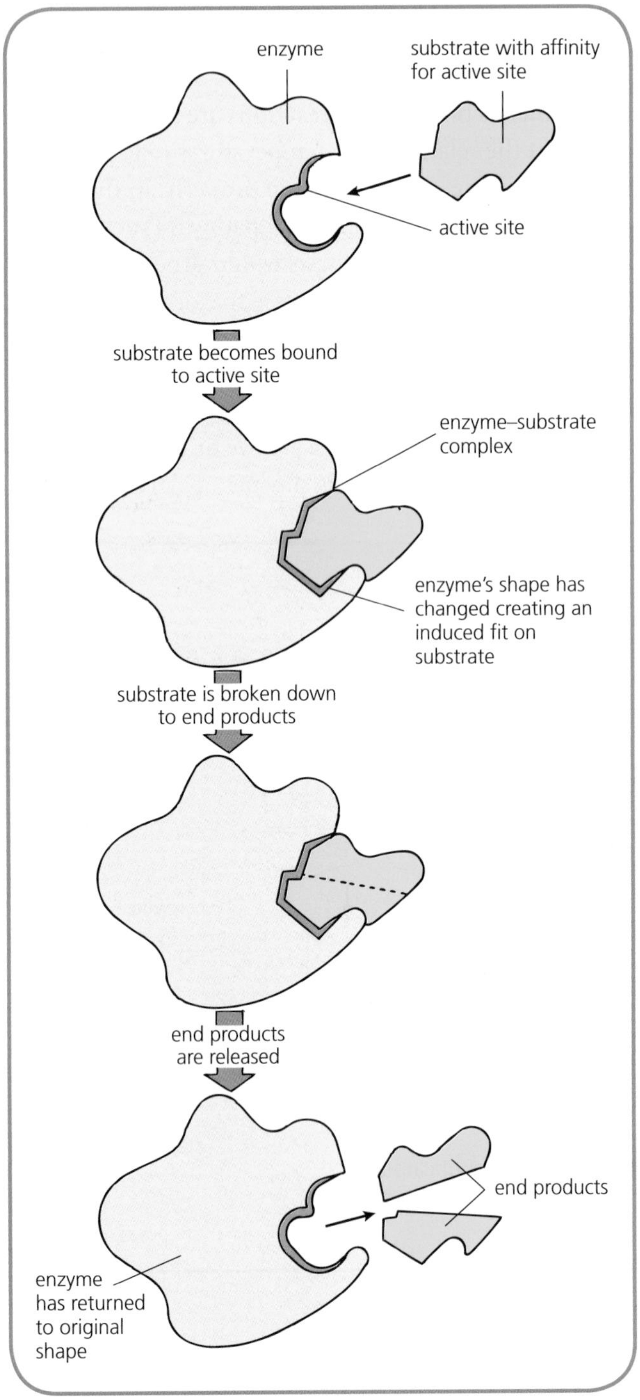

enzyme

substrate with affinity for active site

active site

substrate becomes bound to active site

enzyme–substrate complex

enzyme's shape has changed creating an induced fit on substrate

substrate is broken down to end products

end products are released

end products

enzyme has returned to original shape

Figure 6.9 Induced fit during an enzyme-catalysed reaction

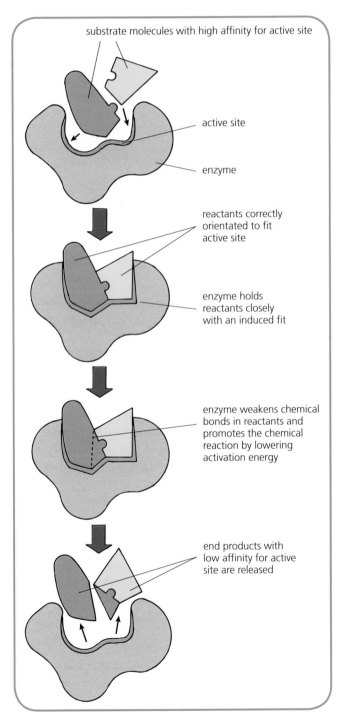

substrate molecules with high affinity for active site

active site

enzyme

reactants correctly orientated to fit active site

enzyme holds reactants closely with an induced fit

enzyme weakens chemical bonds in reactants and promotes the chemical reaction by lowering activation energy

end products with low affinity for active site are released

Figure 6.10 Orientation of reactants during an enzyme-catalysed reaction

the active site change slightly, making the active site fit very closely round the substrate molecule. This is called **induced fit** (see Figure 6.9). The process is like a rubber glove, slightly too small, exerting a very tight fit round a hand. Induced fit ensures that the active site comes into very close contact with the molecules of substrate and increases the chance of the reaction taking place.

Orientation of reactants

When the reaction involves two (or more) substrates (see Figure 6.10), the shape of the active site determines the **orientation** of the reactants. This ensures that they are held together in such a way that the reaction between them can take place.

First the active site holds the two reactants closely together in an induced fit. Then it acts on them to weaken chemical bonds that must be broken during the reaction. This process **reduces the activation energy** needed by the reactants to reach the **transition state** that allows the reaction to take place.

Once the reaction has occurred, the products have a **low affinity** for the active site and are released. This leaves the enzyme free to repeat the process with new molecules of substrate.

Factors affecting enzyme action

To function efficiently, an enzyme requires a suitable temperature, an appropriate pH and an adequate supply of substrate. Inhibitors (see page 91) may slow down the rate of an enzyme-controlled reaction or bring it to a halt.

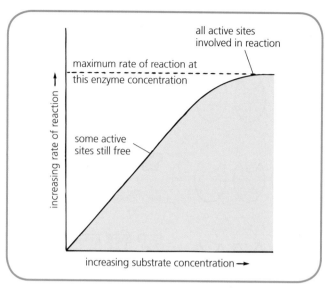

Figure 6.11 Effect of increasing substrate concentration

Effect of substrate concentration on enzyme activity

The graph in Figure 6.11 shows the effect of increasing substrate concentration on the rate of an enzyme-controlled reaction for a limited concentration of enzyme. At low concentrations of substrate, the reaction rate is low since there are too few substrate molecules present to make maximum use of all the active sites on the enzyme molecules. An increase in substrate concentration results in an increase in reaction rate since more and more active sites become involved.

This upward trend in the graph continues as a straight line until a point is reached where further increase in substrate concentration fails to make the reaction go any faster. At this point all the active sites are occupied (the enzyme concentration has become the limiting factor). The graph levels off since there are now more substrate molecules present than there are free active sites with which to combine. The effect of increasing substrate concentration is summarised at molecular level in a simplified way in Figure 6.12.

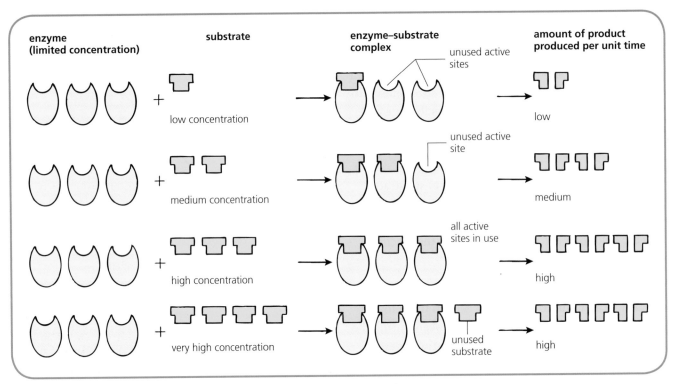

Figure 6.12 Effect of increasing substrate concentration at a molecular level

Investigating the effect of increasing substrate concentration

Liver cells contain the enzyme catalase which catalyses the breakdown of hydrogen peroxide to water and oxygen. In the experiment shown in Figure 6.13, the one variable factor is the concentration of the substrate (hydrogen peroxide). When an equal mass of fresh liver is added to each cylinder, the results shown in the diagram are produced. The height of the froth of oxygen bubbles indicates the activity of the enzyme at each concentration of substrate.

From the experiment it is concluded that increase in substrate concentration results in increased enzyme activity until a point is reached (in cylinder G) where some factor other than substrate concentration has become the limiting factor.

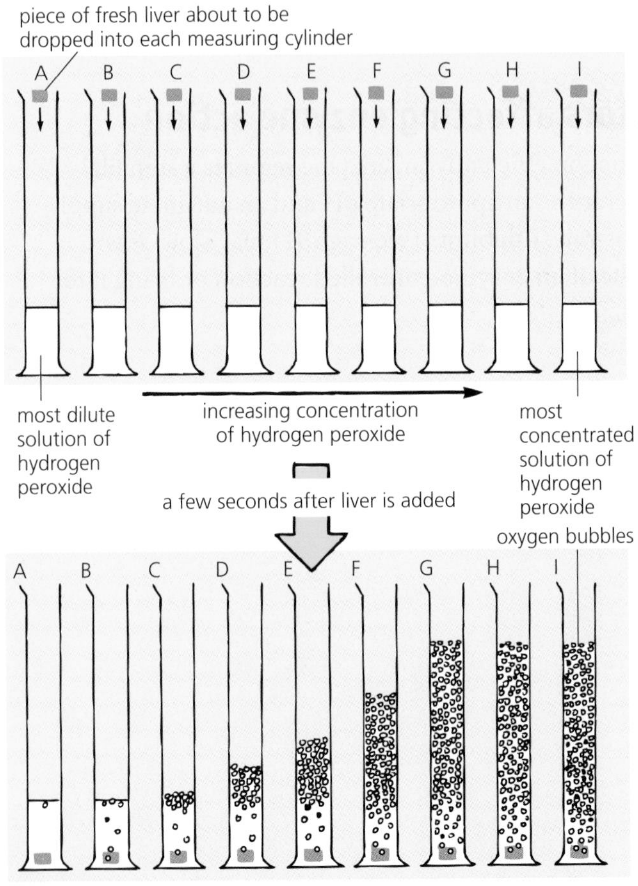

piece of fresh liver about to be dropped into each measuring cylinder

A B C D E F G H I

most dilute solution of hydrogen peroxide

increasing concentration of hydrogen peroxide

most concentrated solution of hydrogen peroxide

a few seconds after liver is added

oxygen bubbles

A B C D E F G H I

Figure 6.13 Effect of substrate concentration on enzyme activity

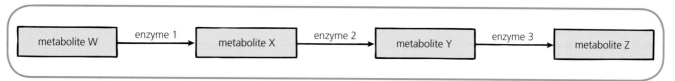

Figure 6.14 Action of a group of enzymes

Direction of enzyme action

A metabolic pathway usually involves a **group** of enzymes as shown in Figure 6.14. As substrate W becomes available, enzyme 1 becomes active and converts W to X. In the presence of metabolite X, enzyme 2 becomes active and converts X to Y and so on. A continuous supply of W entering the system drives the sequence of reactions in the direction W to Z with the product of one reaction acting as the substrate of the next.

Reversibility

Most metabolic reactions are **reversible**. Often an enzyme can catalyse a reaction in both a forward and a reverse direction. The actual direction taken depends on the relative concentrations of the reactant(s) and product(s).

A metabolic pathway rarely occurs in isolation. If, as a result of related biochemical pathways, the concentration of metabolite Y in Figure 6.14 were to increase to an unusually high level and that of X were to decrease, then enzyme 2 could go into reverse and convert some of Y back to X until a balanced state (equilibrium) was restored once more.

Multi-enzyme complexes

Enzymes are often found to act in groups. A group may take the form of a **multi-enzyme complex**. Enzymes such as DNA polymerase (see chapter 2) and RNA polymerase (see chapter 3) form parts of multi-enzyme complexes.

Testing Your Knowledge 1

1. a) Define the term *metabolism*. (2)
   b) Describe TWO ways in which the two types of metabolic pathway differ from one another. (2)
2. Give THREE reasons why enzymes are referred to as *biological catalysts*. (3)
3. a) What determines the structure of an enzyme's active site? (1)
   b) What is meant by the *affinity* of substrate molecules for an enzyme's active site? (1)
   c) What term means 'the change in shape of an active site to enable it to bind more snugly to the substrate'? (1)
   d) Rewrite the following sentences, choosing the correct answer from each underlined choice.

   The shape of the active site ensures that the reactants are correctly <u>orientated/denatured</u> so that the reaction can take place. This is made possible by the fact that the enzyme <u>increases/decreases</u> the activation energy needed by the reactants to reach the <u>transitory/transition</u> state. (3)
4. a) What is meant by the term *rate of reaction*? (See page 79 for help.) (1)
   b) i) What effect does an increase in concentration of substrate have on reaction rate when a limited amount of enzyme is present?
      ii) Explain why. (4)

Control of metabolic pathways

A **metabolic pathway** normally consists of several stages, each of which involves the conversion of one metabolite to another. Each step in a metabolic pathway is driven by a specific enzyme (see Figure 6.15). Each enzyme is coded for by a gene (though complex enzymes composed of several polypeptides require several genes to be involved in their production). If the appropriate enzymes are present, the pathway proceeds. If one enzyme is absent, the pathway comes to a halt. Enzyme action can be regulated at the level of **gene expression** (see below) and at the level of **enzyme action** (see page 91).

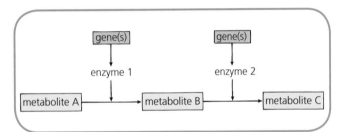

Figure 6.15 Control of two steps in a metabolic pathway

Control by switching genes on and off

Some metabolic pathways are only required to operate under certain circumstances. To prevent resources being wasted, the genes that code for the enzymes controlling each of their stages are 'switched on' or 'switched off' as required.

Lactose metabolism in *Escherichia coli*

Lactose is a sugar found in milk. Each molecule of lactose is composed of a molecule of glucose and a molecule of galactose (see Figure 6.16).

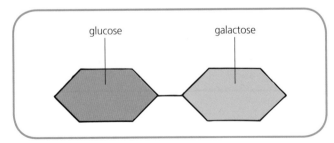

Figure 6.16 Molecule of lactose

Gene action in *Escherichia coli*

Background information

- Glucose is a simple sugar used in respiration by the bacterium *E. coli* for energy release.
- *E. coli* can only make use of the glucose in lactose if it is released from the galactose.
- Lactose is digested to glucose and galactose by the enzyme β-galactosidase.
- *E. coli*'s chromosome has a gene that codes for β-galactosidase.
- *E. coli* is found to produce β-galactosidase only when lactose is present in its nutrient medium and fails to do so when lactose is absent.
- Somehow the gene that codes for β-galactosidase is switched on in the presence of lactose and switched off in the absence of lactose.
- The process of switching on a gene only when the enzyme that it codes for is needed is called **enzyme induction**.

Lac operon of *E. coli*

An **operon** (see Figure 6.17) consists of one or more **structural genes** (containing the DNA code for the enzyme in question) and a neighbouring **operator** gene, which controls the structural gene(s). The operator gene is, in turn, affected by a **repressor** molecule coded for by a **regulator** gene situated further along the DNA chain.

Absence of lactose

Environments inhabited by *E. coli* normally contain glucose but not lactose. The bacterium would waste some of its resources if it made the lactose-digesting enzyme when no lactose was present to digest. Figure 6.17 shows how the bacterium is prevented from doing so by the repressor molecule combining with the operator gene. The structural gene remains switched off and its DNA is not transcribed.

Presence of lactose

When the bacterium finds itself in an environment containing lactose, the events shown in Figure 6.18 take place and lead to the **induction** of β-galactosidase. Lactose (the **inducer**) prevents the repressor molecule from binding to the operator gene. The system is no longer blocked, the structural gene becomes switched on and production of β-galactosidase proceeds.

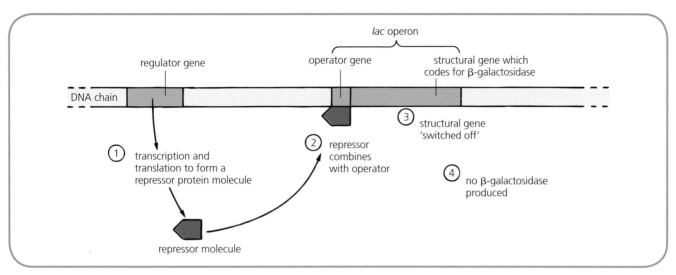

Figure 6.17 Effect of repressor in absence of lactose

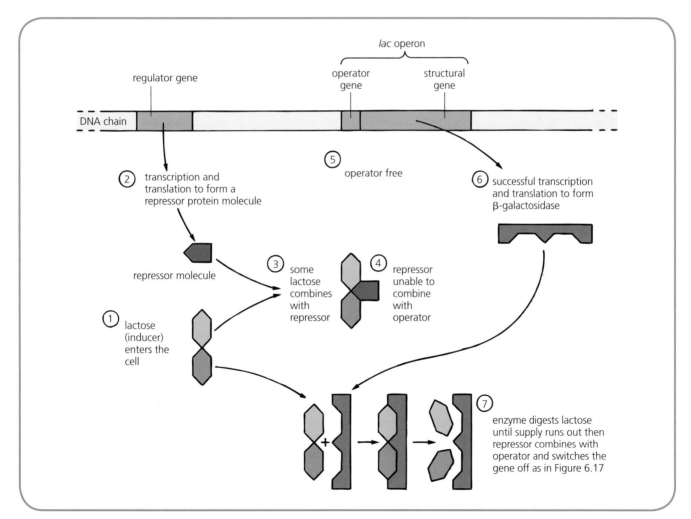

Figure 6.18 Effect of inducer (lactose)

When all the lactose has been digested, the repressor molecule becomes free again and combines with the operator as before. The structural gene becomes switched off and the waste of valuable resources (such as amino acids and energy) is prevented.

Hypothesis

This hypothesis of gene action was first put forward by two scientists called Jacob and Monod. It is now supported by extensive experimental evidence from work done using bacteria.

Related Activity

Investigating the *lac* operon of *E. coli*

ONPG is a colourless synthetic chemical that can be broken down by the enzyme β-galactosidase as follows:

β-galactosidase

ONPG ⟶ galactose + yellow compound

The presence of the yellow colour indicates activity by β-galactosidase. The experiment is set up as shown in Figure 6.19.

From the results it is concluded that:

- in tube 1, lactose has acted as an inducer and switched on the gene in *E. coli* that codes for β-galactosidase; this enzyme has acted on the ONPG, forming the yellow colour
- in tubes 2 and 4, no yellow colour was produced because ONPG was absent
- in tube 3, β-galactosidase has acted on ONPG, forming the yellow compound.

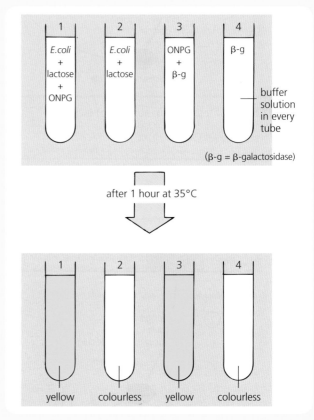

Figure 6.19 Investigating the *lac* operon

Related Topic

The *ara* operon of *E. coli*

Arabinose is a sugar that can be used by *E. coli* as its sole source of carbon and energy in the absence of glucose and lactose. To be of use to the bacterium, arabinose must be broken down by enzymes. The genes for these enzymes (the *ara* operon) are **switched off** in the absence of arabinose (see Figure 6.20) and **switched on** in its presence as a result of arabinose acting as an inducer (see Figure 6.21).

Transformation of the *ara* operon

E. coli cells can be genetically transformed using a modified plasmid called pGLO. This procedure brings about the replacement of the *ara* operon's structural genes with a gene (originally from a fluorescent jellyfish) that codes for **green fluorescent protein** (GFP) and a second gene that makes the transformed cells resistant to an antibiotic (called **ampicillin**). In the presence of arabinose, the inducer, these transformed bacteria produce GFP instead of arabinose-digesting enzymes, as shown in Figure 6.22.

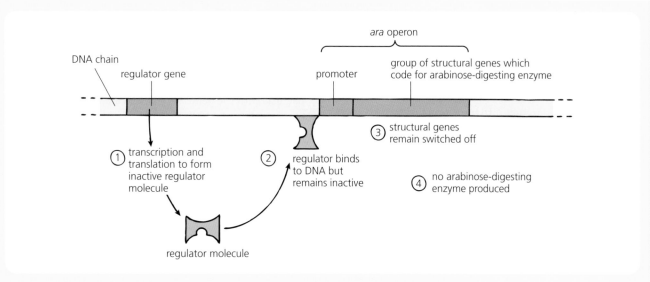

Figure 6.20 Situation in a normal bacterium in the absence of arabinose

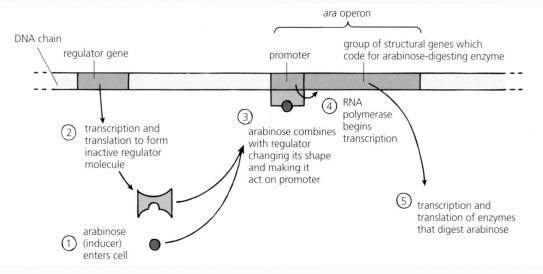

Figure 6.21 Situation in a normal bacterium in the presence of arabinose

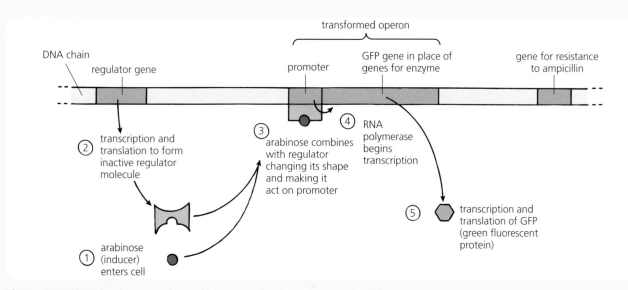

Figure 6.22 Situation in a transformed bacterium in the presence of arabinose

Related Activity

Investigating the transformed *ara* operon

This experiment is set up as shown in Figure 6.23. pGLO plasmids are added to tube 1 to transform the bacteria into cells containing the genes for GFP and ampicillin resistance. Tube 2 is the control.

After appropriate treatment to bring about the transformation, samples of bacteria are subcultured onto four nutrient agar plates, as shown. Only transformed bacteria with the genes for GFP and ampicillin resistance grow on plates A and B. No bacteria grow on plate C because they are not resistant to ampicillin antibiotic.

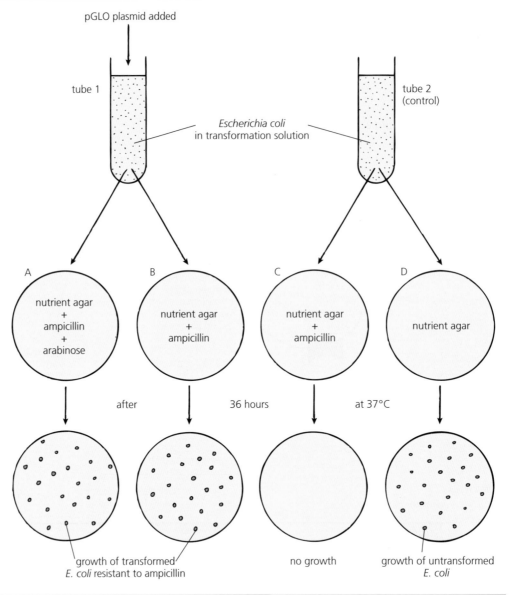

result in UV light	fluorescence	no fluorescence	no fluorescence	no fluorescence
reason	presence of arabinose (inducer) has switched GFP gene on	cells lack arabinose (inducer) needed to switch GFP gene on	cells have been killed by ampicillin	cells lack GFP gene and arabinose

Figure 6.23 Investigating the transformed *ara* operon

Untransformed bacteria grow on plate D, which lacks ampicillin.

When the plates are exposed to ultraviolet light, fluorescence occurs in plate A only (see Figure 6.24). It is concluded that the presence of arabinose, the inducer, has switched on the GFP gene in these cells, resulting in the production of green fluorescent protein.

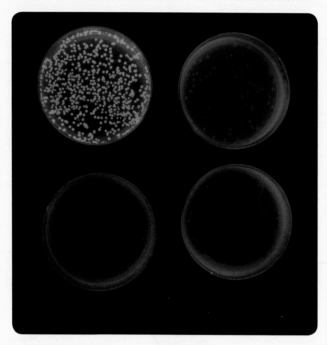

Figure 6.24 Expression of the GFP gene

Regulation of gene expression by signal molecules

In the *lac* operon system of control in *E. coli* (and the *ara* operon – see Related Activity), a gene that had been switched off becomes switched on in response to a **signal molecule** from the cell's environment. In the *lac* operon system, the signal molecule is **lactose**. It combines with the product of the regulator gene (the repressor molecule) enabling the structural gene to be expressed and lead to the production of the required enzyme.

Control by regulation of enzyme action

Some metabolic pathways (e.g. glycolysis – see page 101) are required to operate continuously. The genes that code for their enzymes are always switched on and the enzymes that they code for are always present in the cell. Control of these metabolic pathways is brought about by **regulating the action of their enzymes** as follows.

Effect of signal molecules

The activity of some enzymes is controlled by **signal molecules**. For example, the hormone **epinephrine** (adrenaline), released into the bloodstream by the adrenal glands, binds to receptors in the membrane of liver cells where it acts as a signal molecule. It triggers a series of events in the liver cells that results in the activation of an enzyme that converts glycogen to glucose when the body needs energy urgently.

Signal molecules that affect a cell's metabolism and originate within the cell itself are called **intracellular** signal molecules. On the other hand, signal molecules such as epinephrine that come from the cell's environment (e.g. from other cells) are called **extracellular** signal molecules.

Effect of inhibitors

An **inhibitor** is a substance that decreases the rate of an enzyme-controlled reaction.

Competitive inhibitors

Molecules of a **competitive inhibitor** compete with molecules of the substrate for the active sites on the enzyme. The inhibitor is able to do this because its

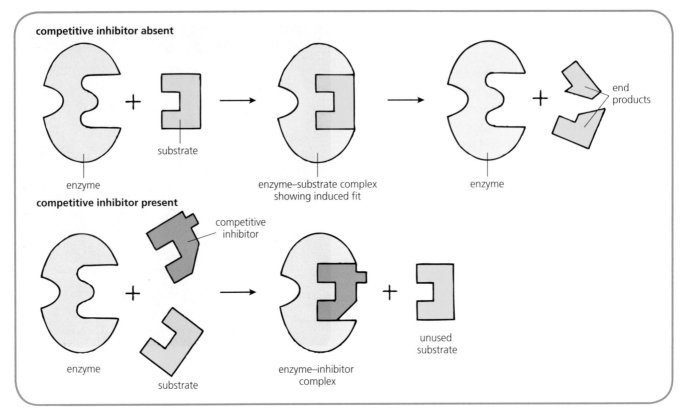

competitive inhibitor absent

substrate

enzyme

enzyme–substrate complex
showing induced fit

enzyme

end
products

competitive inhibitor present

competitive
inhibitor

enzyme

substrate

enzyme–inhibitor
complex

unused
substrate

Figure 6.25 Effect of a competitive inhibitor

molecular structure is **similar** to that of the substrate and it can attach itself to the enzyme's active site as shown in Figure 6.25. Since active sites **blocked** by competitive inhibitor molecules cannot become occupied by substrate molecules, the rate of the reaction is reduced.

As substrate molecules increase in concentration and outnumber those of the competitive inhibitor, more and more active sites become occupied by true substrate rather than inhibitor molecules. The reaction rate continues to increase until all the active sites are occupied (almost all of them by substrate).

Effect of increasing substrate concentration

The graph in Figure 6.26 shows the effect of increasing substrate concentration on rate of reaction for a limited amount of enzyme affected by a limited amount of inhibitor. In graph line 1 (the control), increase in substrate concentration brings about an increase in reaction rate until a point is reached where all active sites on the enzyme molecules are occupied and then the graph levels off.

In graph line 2, increase in substrate concentration brings about a gradual increase in reaction rate. Although the competitive inhibitor is competing for and occupying some of the enzyme's active sites, the true substrate is also occupying some of the sites.

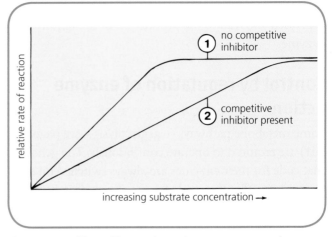

relative rate of reaction

increasing substrate concentration →

1 no competitive
inhibitor

2 competitive
inhibitor present

Figure 6.26 Effect of increasing substrate concentration on competitive inhibition

Investigation

Inhibition of β-galactosidase by galactose

Normally the enzyme β-galactosidase catalyses the reaction:

$$\text{lactose} \xrightarrow{\text{β-galactosidase}} \text{glucose + galactose}$$

However, it is also able to break down a colourless, synthetic compound called ONPG, as follows:

$$\text{ONPG} \xrightarrow{\text{β-galactosidase}} \text{galactose + yellow compound}$$

The experiment shown in Figure 6.27 is set up to investigate the inhibitory effect of galactose on the action of β-galactosidase as the concentration of the substrate, ONPG, is increased. The **independent variable** in this experiment is substrate concentration.

At the end of the experiment, an increasing intensity of yellow colour (indicating products of enzyme activity) is found to be present in the tubes, with tube 1 the least yellow and tube 4 the most yellow. The intensity of colour can be measured quantitatively using a **colorimeter**. This allows the results to be displayed as a graph.

A possible explanation for these results is that galactose acts as a **competitive inhibitor**, having most effect at low concentrations of substrate. As the concentration of substrate increases, more and more active sites on the enzyme become occupied by substrate, not inhibitor, and reaction rate increases.

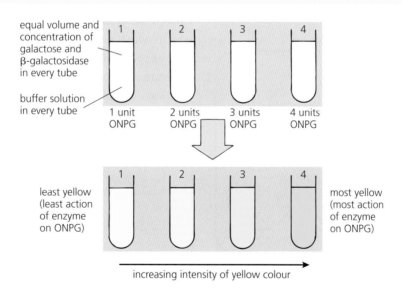

Figure 6.27 Investigating the inhibitory effect of galactose

Regulation by changing the shape of the active site

Non-competitive inhibitors

A **non-competitive inhibitor** does not combine directly with an enzyme's active site. Instead it becomes attached to a non-active (**allosteric**) site and changes the shape of the enzyme molecule. This results in the active site becoming **altered indirectly** and being unable to combine with the substrate as shown in Figure 6.28. The larger the number of enzyme molecules affected in this way, the slower the enzyme-controlled reaction. Therefore, the non-competitive inhibitor acts as a type of regulator.

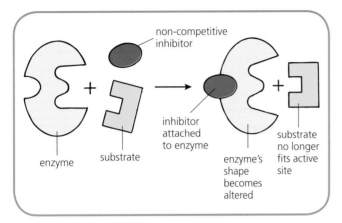

Figure 6.28 Effect of a non-competitive inhibitor

Some enzyme molecules are composed of several polypeptide subunits and each subunit has its own active site. The enzyme molecule also has several non-active (allosteric) sites. Depending on circumstances, the enzyme molecule may exist as an **active** form or an **inactive** form. These have different shapes, as shown in Figure 6.29.

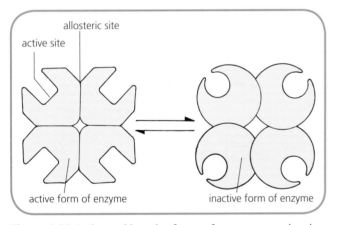

Figure 6.29 Active and inactive forms of an enzyme molecule

The enzyme molecule changes shape if a **regulatory molecule** becomes bound to one of its allosteric sites (see Figure 6.30). If the regulatory molecule is an **activator**, the enzyme adopts its active form and enzyme activity is stimulated. If the regulatory molecule is a **non-competitive inhibitor**, the enzyme changes to its inactive state and enzyme action is inhibited. The more enzyme molecules affected by activators, the faster the reaction rate; the more enzyme molecules affected by inhibitors, the slower the reaction rate.

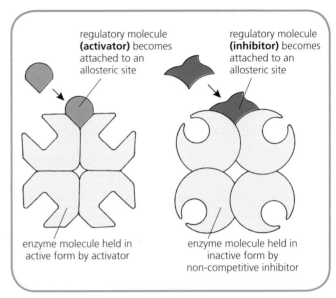

Figure 6.30 Enzyme regulation by an activator and an inhibitor

Feedback inhibition by an end product

End-product inhibition (see Figure 6.31) is a further way in which a metabolic pathway can be regulated. As the concentration of end product (metabolite Z) builds up, some of it binds to molecules of enzyme 1 in the pathway. This slows down the conversion of metabolite W to X and in turn regulates the whole pathway.

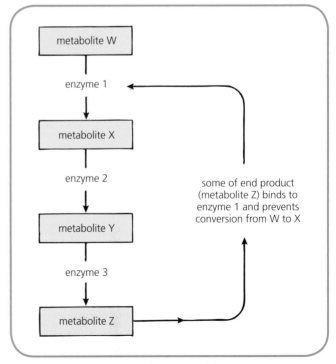

Figure 6.31 Regulation by feedback inhibition

As the concentration of Z drops, fewer molecules of enzyme 1 are affected and more of W is converted to X and so on. The pathway is kept under **finely tuned** **control** by this means (called **negative feedback control**) and wasteful conversion and accumulation of intermediates and final products are avoided.

Related Activity

Investigating the effect of phosphate on phosphatase

Phosphatase is an enzyme that releases the phosphate group from its substrate for use in cell metabolism. Phosphatase is present in the extract obtained from ground-up mung bean sprouts. **Phenolphthalein phosphate** is a chemical that can be broken down by phosphatase as follows:

phenolphthalein phosphate $\xrightarrow{\text{phosphatase}}$ phenolphthalein + phosphate (pink in alkaline conditions)

The experiment is set up as shown in Figure 6.32. At the end of the experiment a decreasing intensity of pink colour is found to be present in the tubes. Tube 1 is the most pink and tube 5 is the least pink. From these results it is concluded that tube 1 contains most free phenolphthalein as a result of most enzyme activity and that tube 5 contains least free phenolphthalein as a result of least enzyme activity. In other words as phosphate concentration increases, the activity of the enzyme phosphatase decreases. A possible explanation for this effect is that phosphate acts as an **end-product inhibitor** of the enzyme phosphatase.

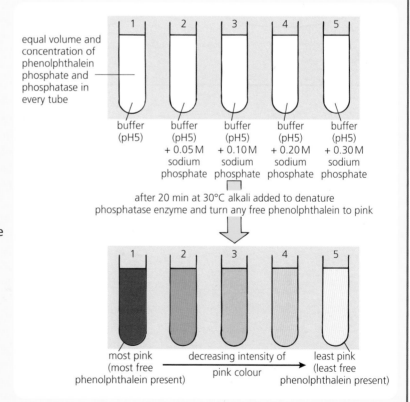

Figure 6.32 Investigating the effect of phosphate on phosphatase

Testing Your Knowledge 2

1. a) Briefly explain why *E. coli* is only able to produce the enzyme β-galactosidase when lactose is present in its food. (2)
   b) What is the benefit of this on/off mechanism to the bacterium? (1)
   c) In this example, does lactose act as an intracellular or an extracellular signal molecule? (1)

2. a) What property of a competitive inhibitor enables it to compete with the substrate? (1)
   b) i) What effect does an increase in concentration of substrate have on rate of reaction when a limited amount of competitive inhibitor and enzyme are present?
      ii) Explain why. (3)

3. An enzyme molecule may possess several active sites and several allosteric sites and exist in an active or an inactive form. Explain how molecules of such an enzyme could be controlled so that they could bring about:
   a) an increase in reaction rate (1)
   b) a decrease in reaction rate. (1)

4. Figure 6.33 shows a metabolic pathway where metabolites P, Q and R are present in equal quantities at the start.
   a) Name enzyme X's **i)** substrate, **ii)** product. (2)
   b) Name enzyme Y's **i)** substrate, **ii)** product. (2)
   c) In which direction will the pathway proceed if more of metabolite P is added to the system? (1)
   d) i) Metabolite R can act as an end-product inhibitor. Describe how this would work.
      ii) What is the benefit of end-product inhibition? (3)

Figure 6.33

7 Cellular respiration

Cellular respiration is a series of metabolic pathways that brings about the release of energy from a foodstuff and the regeneration of the high-energy compound **adenosine triphosphate (ATP)**.

Adenosine triphosphate

A molecule of **ATP** is composed of adenosine and three inorganic phosphate (P_i) groups, as shown in Figure 7.1. Energy held in an ATP molecule is released when the bond attaching the terminal phosphate is broken by enzyme action. This results in the formation of **adenosine diphosphate (ADP)** and inorganic phosphate (P_i). On the other hand, energy is required to regenerate ATP from ADP and inorganic phosphate. This relationship is summarised in Figure 7.2.

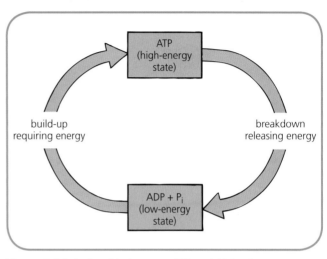

Figure 7.2 Relationship between ATP and ADP + P_i

Transfer of energy via ATP

ATP is important because it acts as the **link** between catabolic energy-releasing reactions (e.g. respiration) and anabolic energy-consuming reactions (e.g. synthesis of proteins). It provides the means by which chemical energy is transferred from one type of reaction to the other in a living cell (see Figures 7.3 and 7.4).

Turnover of ATP molecules

It has been estimated that an active cell (e.g. a bacterium undergoing cell division) requires approximately two million molecules of ATP per second to satisfy

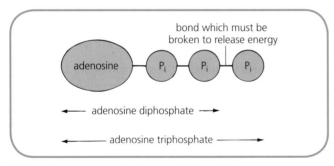

Figure 7.1 Structure of ATP

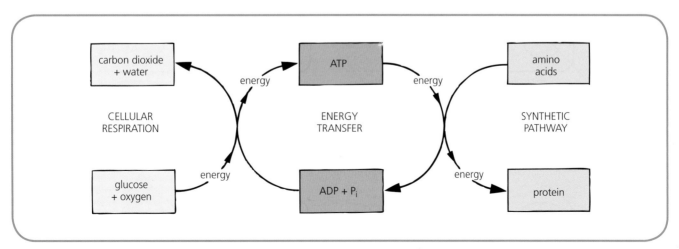

Figure 7.3 Transfer of chemical energy by ATP

Figure 7.4 'What do you mean, I need **ATP**? I thought you said **a teepee**!'

its energy requirements. This is made possible by the fact that a **rapid turnover** of ATP molecules occurs continuously in a cell. At any given moment some ATP molecules are undergoing breakdown and releasing the energy needed for cellular processes while others are being regenerated from ADP and P_i using energy released during cell respiration.

Fixed quantity of ATP

Since ATP is manufactured at the same time as it is used up, there is no need for a living organism to possess a vast store of ATP. The quantity of ATP present in the human body, for example, is found to remain fairly **constant** at around 50 g despite the fact that the body may be using up *and* regenerating ATP at a rate of about 400 g/h.

Related Activity

Measuring ATP using luciferase

Background

- **Luciferase** is an enzyme present in the cells of fireflies. It is involved in the process of bioluminescence (the production of light by a living organism).
- Luciferase catalyses the following reaction:

$$\text{luciferin} + \text{ATP} \xrightarrow{\text{luciferase}} \text{end products} + \text{light energy}$$

- The presence of **ATP** is essential for the production of light energy and the reaction does not proceed in its absence.
- When luciferin and luciferase are plentiful and ATP is the limiting factor, the intensity of light emitted is proportional to the concentration of ATP present.

The experiment is carried out as shown in Figure 7.5 and the results used to draw a graph of known values of ATP concentration (see Figure 7.6). When the experiment is

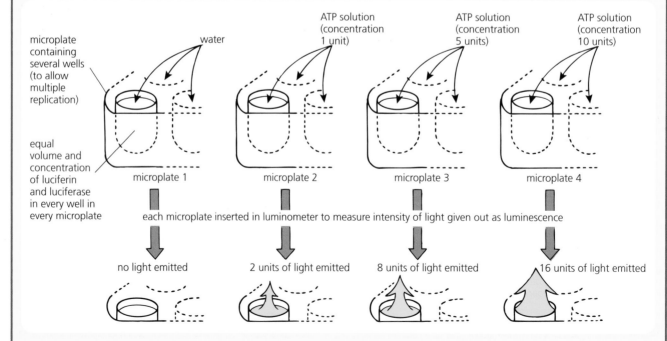

Figure 7.5 Measuring light emitted from known concentrations of ATP

repeated using material of unknown ATP content, the ATP concentration can be determined from the graph. For example, the sample shown in Figure 7.7 would contain 7.5 units of ATP.

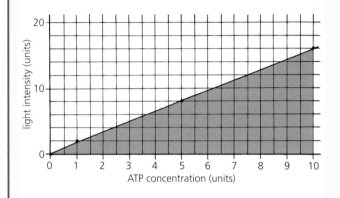

Figure 7.6 Graph of luciferase results

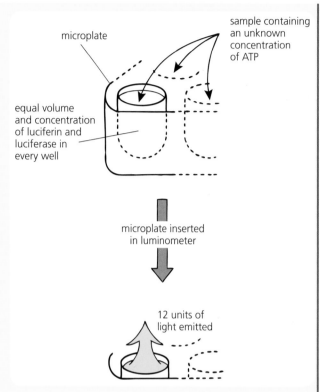

Figure 7.7 Measuring light emitted from unknown concentrations of ATP

Phosphorylation

Phosphorylation is an enzyme-controlled process by which a phosphate group is added to a molecule. Phosphorylation occurs, for example, when the reaction shown in Figure 7.2 goes from bottom to top and inorganic phosphate (P_i) combines with low-energy ADP to form high-energy ATP.

Phosphorylation of a reactant in a pathway

Phosphorylation also occurs when phosphate (P_i) and energy are transferred from ATP to the molecules of a reactant in a metabolic pathway, making them **more reactive**. Often a step in a pathway can proceed only if a reactant becomes **phosphorylated** and energised. In the early stages of cellular respiration, for example, some reactants must undergo phosphorylation during what is called the energy investment phase. One of these steps is shown in Figure 7.8.

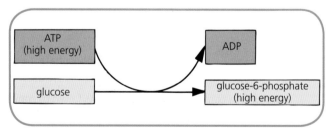

Figure 7.8 Phosphorylation of glucose

Effect of phosphorylase on a phosphorylated substrate

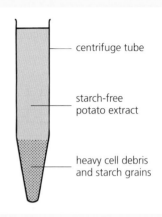

Figure 7.9 Preparation of potato extract

Background

- **Glucose-1-phosphate** is a phosphorylated form of glucose.
- A molecule of starch is composed of many glucose molecules linked together in a long chain.
- Potato tuber cells contain **phosphorylase**, an enzyme that promotes the synthesis of starch.
- Potato extract containing phosphorylase is prepared by liquidising a mixture of potato tuber and water and then centrifuging the mixture until the potato extract (see Figure 7.9) is **starch-free**.

The experiment is set up at room temperature on a spotting tile, as shown in Figure 7.10.

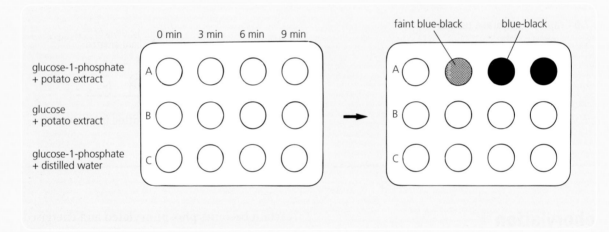

Figure 7.10 Investigating a phosphorylated substrate

One dimple from each row is tested at 3-minute intervals with iodine solution. Starch is found to be formed in row A only. It is concluded that in row A phosphorylase has promoted the conversion of the **phosphorylated** (and more reactive) form of the substrate, glucose-1-phosphate, to starch, as in the following equation:

$$\underset{\text{(phosphorylated substrate)}}{\text{glucose-1-phosphate}} \xrightarrow[\text{(enzyme)}]{\text{phosphorylase}} \underset{\text{(end product)}}{\text{starch}}$$

In row B (a control), phosphorylase has failed to convert the more stable (and less reactive) form of the substrate, glucose, to starch.

In row C (a control), the molecules of glucose-1-phosphate have failed to become bonded together into starch without the aid of phosphorylase.

Positive and negative controls

A **positive control** is set up to assess the validity of a testing procedure or design and ensure that the equipment and materials being used are in working order and appropriate for use in the experiment being carried out.

For example, a positive control for the above experiment could be set up as row D, which would contain starch in every dimple. If the addition of iodine solution at each 3-minute interval gave a blue-black colour, this would confirm that:

● the iodine solution being used was working properly as a testing reagent for starch
● the experiment was not adversely affected in some way, for example by the contamination of the spotting tile or by changes in room temperature.

If a positive control does not produce the expected result, then this indicates that there is something wrong with the design of the testing procedure or with the materials being used.

A **negative control** is one that should not work. It is a copy of the experiment in which all factors are kept exactly the same except the one being investigated. When the results are compared, any difference found between the experiment and a negative control must be due to the factor being investigated.

In the above investigation, starch is not synthesised in row B, showing that the glucose must be in a phosphorylated state to become converted to starch. If row B had not been set up, it would be valid to suggest that starch would have been formed whether or not the glucose was phosphorylated. Similarly, starch was not formed in row C showing that phosphorylase (in potato extract) must be present for phosphorylated glucose to be converted to starch. If row C had not been included, it would be valid to suggest that phosphorylated glucose would have become starch whether or not phosphorylase was present. Therefore rows B and C in this investigation are negative controls.

Synthesis of ATP

When an energy-rich substance such as glucose is broken down during cellular respiration in a living cell, it releases energy that is used to synthesise ATP from ADP and P_i. Cellular respiration consists of many enzyme-controlled steps, which ensure that energy release occurs in an orderly fashion.

A flow of high-energy electrons from the respiratory pathway is used to pump hydrogen ions (H^+) across the inner mitochondrial membrane and maintain a region of higher concentration of hydrogen ions on one side of the membrane (see Figure 7.11).

Embedded in the membrane are molecules of the protein **ATP synthase**, which is an enzyme. The return flow of hydrogen ions from the region of higher concentration to the region of lower concentration takes place via ATP synthase molecules. This makes part of the ATP synthase rotate and catalyse the synthesis of **ATP** from ADP and P_i. ATP synthase molecules operate in a similar way in the membranes of chloroplasts.

Metabolic pathways of cellular respiration

Glycolysis

The process of cellular respiration begins in the cytoplasm of a living cell with a molecule of **glucose** being broken down to form **pyruvate**. This process of 'glucose-splitting' is called **glycolysis**. It consists of a series of enzyme-controlled steps. Those in the first half of the chain make up the **energy investment phase** (where 2ATP are used up per molecule of glucose); those in the second half of the chain make up an **energy payoff phase** (where 4ATP are produced per molecule of glucose), as shown in Figure 7.12.

Phosphorylation of intermediates occurs twice during the first phase:

● in step 1, where an intermediate is formed that can connect with other metabolic pathways
● in step 3, which is an irreversible reaction leading only to the rest of the glycolytic pathway and catalysed by the enzyme **phosphofructokinase**.

The generation of 4ATP that occurs during the second half of the pathway gives a **net gain of 2ATP** per molecule of glucose during glycolysis. In addition, during the energy payoff phase, H^+ ions are released from the substrate by a dehydrogenase enzyme. These H^+ ions are passed to a coenzyme molecule called

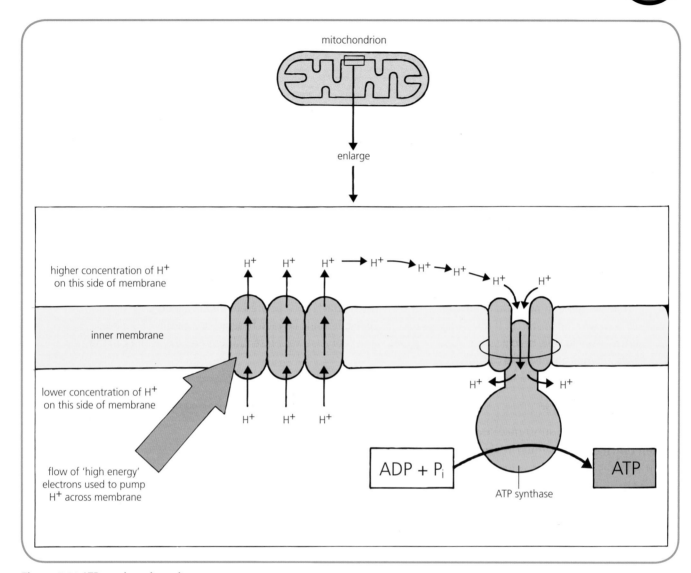

Figure 7.11 ATP synthase in action

NAD (full name, nicotinamide adenine dinucleotide), forming **NADH**.

The process of glycolysis does not require oxygen. However, NADH only leads to the production of further molecules of ATP at a later stage in the respiratory process if oxygen is present. In the absence of oxygen, fermentation occurs (see page 113).

Citric acid cycle

If oxygen is present, aerobic respiration proceeds and pyruvate is broken down into carbon dioxide and an **acetyl group**. Each acetyl group combines with **coenzyme A** to form **acetyl coenzyme A**. During this process, further H$^+$ ions are released and become bound to NAD, forming NADH. A simplified version of the metabolic pathway is shown in Figure 7.13.

The acetyl group of acetyl coenzyme A combines with **oxaloacetate** to form **citrate** and enter the **citric acid cycle**. This cycle consists of several enzyme-mediated stages, which occur in the central matrix of mitochondria and result finally in the regeneration of oxaloacetate.

In three steps in the cycle, dehydrogenase enzymes remove **H$^+$ ions** from the respiratory substrate along with associated **high-energy electrons**. These H$^+$ ions and high-energy electrons are passed to the coenzyme NAD to form NADH. In one other step a similar reaction occurs but the coenzyme is **FAD** (full name, flavine adenine dinucleotide). On accepting H$^+$ ions and electrons it becomes **FADH$_2$**. In addition, ATP is produced in one of the steps and carbon dioxide is released in two of the steps.

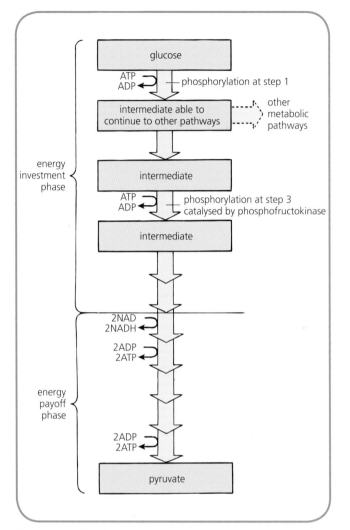

Figure 7.12 Glycolysis

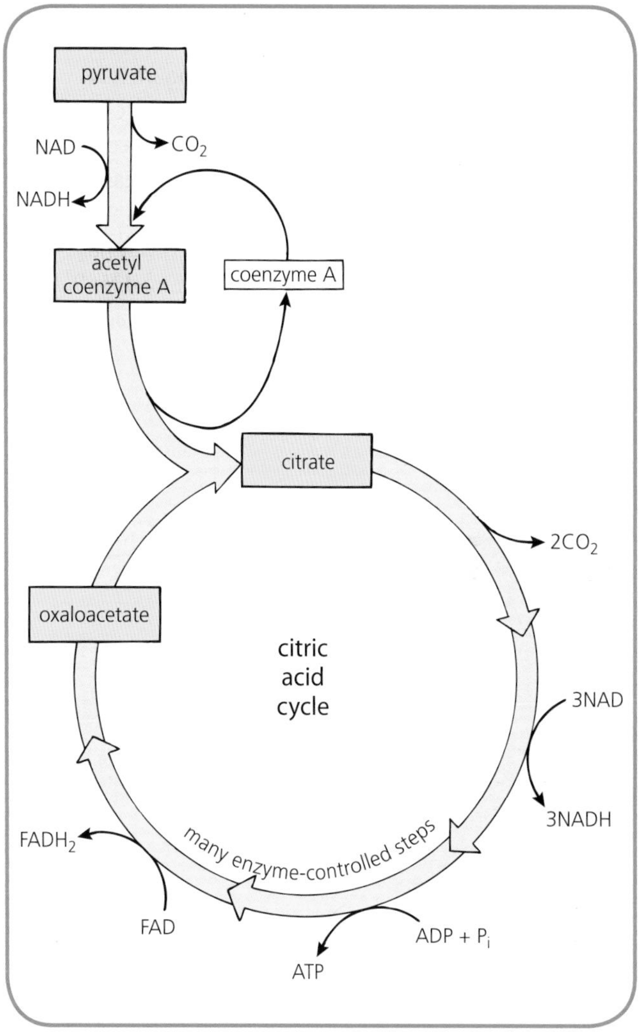

Figure 7.13 Citric acid cycle

Related Activity

Demonstrating the effect of malonic acid

Background

- Succinate and fumarate are two of the intermediates in the citric acid cycle.
- When succinate is converted to fumarate during respiration in a living cell, hydrogen is released and passed to the coenzyme FAD. The reaction is catalysed by the enzyme **succinic dehydrogenase** as follows:

$$\text{succinate} + \text{FAD} \xrightarrow{\text{succinic dehydrogenase}} \text{fumarate} + \text{FADH}_2$$

- **Malonic acid** is a chemical that inhibits the action of succinic dehydrogenase.
- In this investigation a chemical called DCPIP is used as the hydrogen acceptor. DCPIP changes colour upon gaining hydrogen as follows:

dark blue (lacks hydrogen) ⟶ colourless (has gained hydrogen)

➜

The experiment is carried out as shown in Figure 7.14. From the results it is concluded that in tube A succinic dehydrogenase in respiring mung bean cells has converted succinate to fumarate and that the hydrogen released has been accepted by DCPIP, turning it colourless. It is concluded that in tube B, the respiratory pathway has been blocked and no hydrogen has been released for DCPIP to accept because malonic acid has inhibited the action of succinic dehydrogenase.

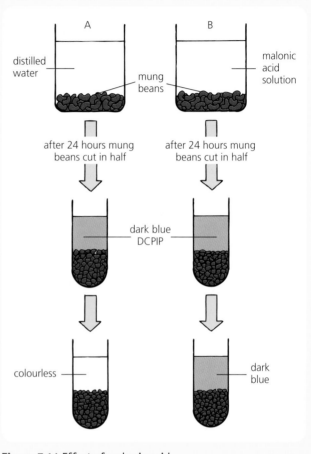

Figure 7.14 Effect of malonic acid

Electron transport chain

An **electron transport chain** consists of a group of protein molecules. There are many of these chains in a cell. They are found attached to the **inner membrane** of mitochondria. NADH and $FADH_2$ from the glycolytic and citric acid pathways release **high-energy electrons** and pass them to the electron transport chains (see Figure 7.15).

The electrons begin in a high-energy state. As they flow along a chain of electron acceptors, they release energy. This is used to pump **hydrogen ions** across the membrane from the inner cavity (matrix) side to the intermembrane space, where a higher concentration of hydrogen ions is maintained. The return flow of hydrogen ions to the matrix (the region of lower H^+ concentration) via molecules of **ATP synthase** drives this enzyme to synthesise ATP from ADP and P_i. Most of the ATP generated by cellular respiration is produced in mitochondria in this way.

When the electrons come to the end of the electron transport chain, they combine with **oxygen**, the final electron acceptor. At the same time, the oxygen combines with a pair of hydrogen ions to form **water**. In the absence of oxygen, the electron transport chains do not operate and this major source of ATP becomes unavailable to the cell.

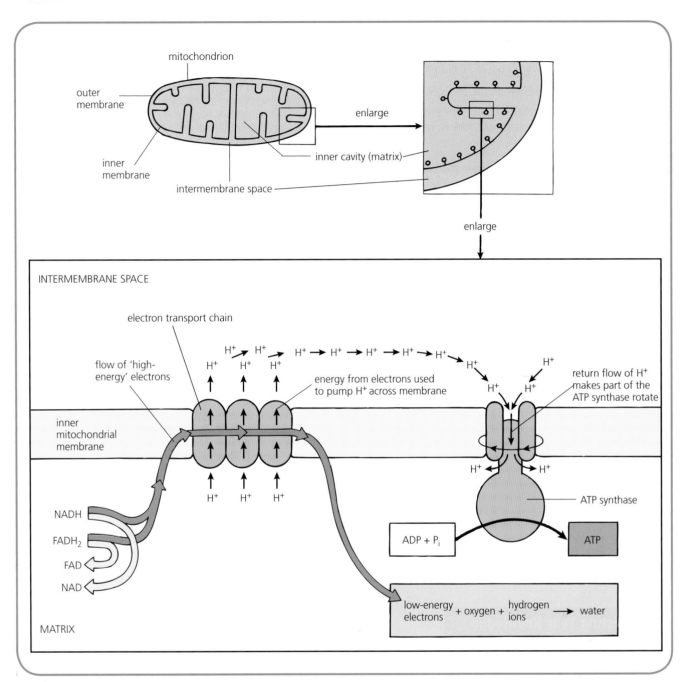

Figure 7.15 Electron transport chain

Related Activity

Investigating the activity of dehydrogenase enzyme in yeast

Background

● During respiration, glucose is gradually broken down and hydrogen released at various stages along the pathway. Each of these stages is controlled by an enzyme called a **dehydrogenase**.

● Yeast cells contain small quantities of stored food that can be used as a respiratory substrate.

● Resazurin dye is a chemical that changes colour upon gaining hydrogen, as follows:

blue ⟶ pink ⟶ colourless
(lacks   (some   (much
hydrogen) hydrogen hydrogen
          gained)   gained)

➜

- Before setting up the experiment shown in Figure 7.16, dried yeast is added to water and aerated for an hour at 35°C to ensure that the yeast is in an active state.

Once the experiment has been set up, the contents of tube A are found to change from blue via pink to colourless much faster than those of tube B. Tube C, the control, remains unchanged.

It is concluded that in tube A, hydrogen has been rapidly released and has acted on and changed the colour of the resazurin dye. For this to be possible, dehydrogenase enzymes present in the yeast cells must have acted on glucose, the respiratory substrate.

In tube B, the reaction was slower because no glucose was added and the dehydrogenase enzymes could only act on any small amount of respiratory substrate already present in the yeast cells.

In tube C, boiling has killed the cells and denatured the dehydrogenase enzymes.

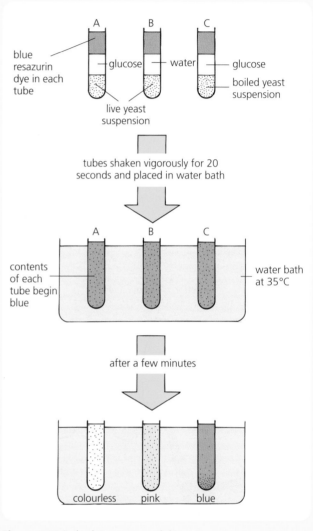

Figure 7.16 Dehydrogenase activity

Testing Your Knowledge

1   **a)** What compound is represented by the letters ATP? (1)
    b) What is the structural difference between ATP and ADP? (1)
    c) Give a simple equation to indicate how ATP is regenerated in a cell. (2)

2   Explain each of the following:
    a) During the glycolysis of one molecule of glucose, the net gain is two and not four molecules of ATP. (1)
    b) Living organisms have only small quantities of oxaloacetate in their cells. (1)
    c) A human body can produce ATP at a rate of around 400 g/h, yet at any given moment there are only about 50 g present in the body. (2)

3   Using the letters G, C and E, indicate whether each of the following statements refers to glycolysis (G), citric acid cycle (C) or electron transport chain (E). (Some statements may need more than one letter.) (8)
    a) It brings about the breakdown of glucose to pyruvate.
    b) It ends with the production of water.
    c) It begins with acetyl from acetyl coenzyme A combining with oxaloacetate.
    d) It involves a cascade of electrons, which are finally accepted by oxygen.
    e) It has an energy investment and energy payoff phase.
    f) It results in the production of NADH.
    g) It involves the release of carbon dioxide.
    h) It results in the production of ATP.

Substrates for respiration

Carbohydrates

Starch (a complex carbohydrate stored by plants) and **glycogen** (a complex carbohydrate stored by animals) are composed of chains of **glucose** molecules. They act as respiratory substrates since they can be broken down to release glucose as required (see Figure 7.17). Other sugar molecules such as maltose and sucrose can also be converted to glucose or intermediates in the glycolytic pathway and used as respiratory substrates.

Fats

When required for use as a respiratory substrate, a molecule of **fat** is broken down into **glycerol** and **fatty acids**. These products then become available for use in cellular respiration. Glycerol is converted to a glycolytic intermediate (see Figure 7.18) and fatty acids are metabolised into molecular fragments that enter the pathway as acetyl coenzyme A for use in the citric acid cycle.

Proteins

Proteins in the diet are broken down to their component **amino acids** by the action of digestive enzymes. Amino acids in excess of the body's

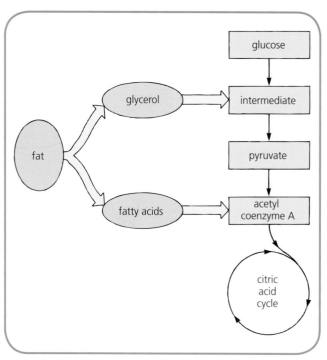

Figure 7.18 Fat as a respiratory substrate

requirements for protein synthesis undergo deamination, forming urea and respiratory pathway intermediates, as shown in Figure 7.19. These intermediates then enter the metabolic pathway and act as respiratory substrates, regenerating ATP as before.

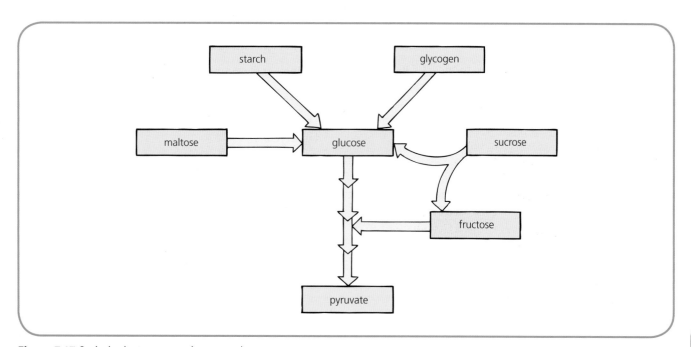

Figure 7.17 Carbohydrates as respiratory substrates

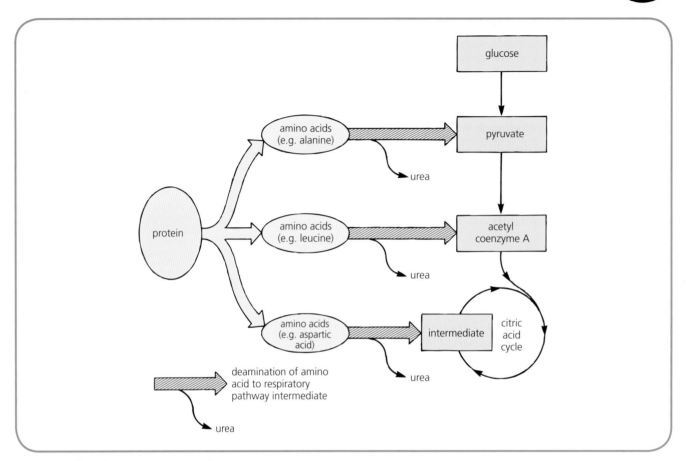

Figure 7.19 Protein as a respiratory substrate

Practical Activity and Report

Investigating the use of three different sugars as respiratory substrates for yeast

Information

- Figure 7.20 shows, in a simple way, the molecular structure of three types of sugar and the digestive enzymes needed to break down maltose and sucrose.

- Strictly speaking, this activity is really three investigations being carried out simultaneously.

- In each case the independent variable is time.

- The dependent variable that you are going to measure is the volume of carbon dioxide released as a result of yeast using a particular type of sugar as its respiratory substrate.

You need

- 3 graduated tubes

- 3 large beakers (e.g. 500 ml) of coloured tap water

- 3 clamp stands

- 1 container of glucose solution (10 g in 90 ml water)

- 1 container of maltose solution (10 g in 90 ml water)

- 1 container of sucrose solution (10 g in 90 ml water)

- 3 conical flasks (250 ml) each with rubber stopper and delivery tube

- 3 labels

- 3 portions of dried yeast, each 1 g

- clock

What to do

1 Read all of the instructions in this section and prepare your results table before carrying out the experiment.

2 Fill each graduated tube with coloured tap water and clamp it in an inverted position in a beaker of coloured water, as shown in Figure 7.21.

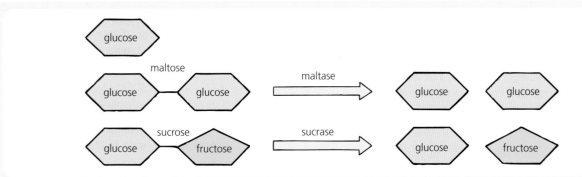

Figure 7.20 Relationship between three sugars

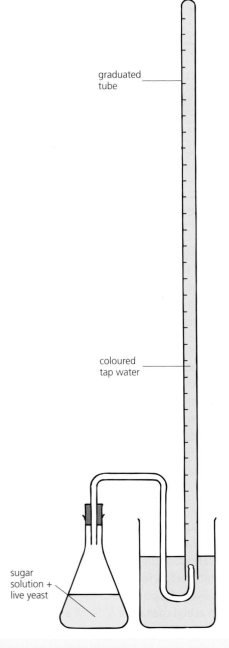

graduated tube

coloured tap water

sugar solution + live yeast

Figure 7.21 Yeast investigation set-up

3 Label the conical flasks 'glucose', 'maltose' and 'sucrose' respectively and add your initials.

4 Pour the appropriate sugar solution into each conical flask and add a portion of dried yeast.

5 Assemble the stoppers and delivery tubes as shown in Figure 7.21.

6 Start the clock and record, at 5-minute intervals, the total volume of carbon dioxide that has been released for each flask over a period of 2 hours.

7 If other students have carried out the same experiment, pool the results.

Reporting

Write up your report by doing the following:

1 Rewrite the title given at the start of this activity.

2 Write the subheading '**Aim**' and state the aim of your experiment.

3 a) Write the subheading '**Method**'.

b) Draw a diagram of your apparatus set-up at the start of the experiment after the yeast has been added and bubbles of carbon dioxide are being released.

c) Briefly describe the experimental procedure that you followed using the impersonal passive tense. *Note: The impersonal passive tense avoids the use of 'I' and 'we'. Instead it makes the apparatus the subject of the sentence. In this experiment, for example, you could begin your report by saying 'Three graduated tubes were filled with coloured water... (not 'I filled three graduated tubes with coloured water...).*

d) Continuing in the impersonal passive tense, state how your results were obtained.

4 Write the subheading '**Results**' and draw a final version of your table of results.

5 Write the subheading '**Analysis and Presentation of Results**'. Present your results as three line graphs with shared axes on the same sheet of graph paper.

6 Write the subheading '**Conclusion**' and a short paragraph to state what you have found out from a study of your results. This should include answers to the following questions:

a) Which respiratory substrate(s) was yeast able to use effectively?

b) Which enzyme (see Figure 7.20) is probably produced by yeast cells in adequate quantities to digest its substrate before its use in cellular respiration?

c) Which respiratory substrate(s) was yeast not able to use effectively?

d) Which enzyme is probably not produced by yeast in adequate quantities within the 2-hour timescale to digest its substrate and make it suitable for use in cellular respiration?

7 Write the final subheading '**Evaluation of Experimental Procedure**'. Give an evaluation of your experiment (keeping in mind that you may comment on any stage of the experiment that you wish). Try to incorporate answers to the following questions in your evaluation, making sure that at least one of your answers includes a supporting statement.

a) Why is the same mass of yeast and the same mass of sugar used in every flask?

b) Why is the same genetic strain of yeast used in each flask?

c) Why must the rubber stoppers be tightly fitting?

d) Why should a control flask containing distilled water and yeast have been included in this investigation?

e) What is the purpose of pooling results with other groups?

Research Topic | **Use of respiratory substrates during exercise and starvation**

Carbohydrate, fat and protein can all be used as **respiratory substrates**. Their individual contributions to the body's overall energy supply depends upon the body's circumstances.

Exercise

For several minutes, from the start of **aerobic exercise**, the body burns primarily carbohydrates. After 20–30 minutes of continuous exercise, respiratory substrate usage shifts to a balance of around 50% carbohydrate and 50% fat. During the first hour of exercise, protein makes up less than 2% of the respiratory substrate but its utilisation increases during prolonged exercise. It can reach 5–15% of fuel usage during the latter stages of prolonged exercise lasting 5 hours or more.

Marathon running

The respiratory substrates used in this lengthy athletic event (42.2 km) are glucose, glycogen and fat, as shown in Figure 7.22. During the first few minutes of the race, readily available glucose from **muscle glycogen** is the main fuel used to generate energy. However, as the race continues and the rate of blood flow increases, blood-borne fuels carried to the exercising muscles become the dominant sources of energy. **Blood glucose** (largely from liver glycogen) and slower-acting **fatty acids** provide most of the energy over the next 30 minutes or so. In the later stages of the race, fatty acids become increasingly important as supplies of glucose decrease.

A marathon runner therefore depends on a combination of carbohydrate and fat. The relative contribution made by each fuel depends on availability. The athlete might decide to 'load up' with carbohydrate during pre-race meals. He or she might consume an **approved refreshment** of glucose solution after 11 km and thereafter at intervals of 5 km. Under these circumstances the degree of dependency on fat reserves is greatly reduced.

Starvation

Starvation results when the body continuously expends more energy than it takes in as food. During the early stages of starvation, the body uses up its store of **glycogen** and then mobilises its **fat** reserves. As starvation becomes prolonged, liver cells continue to use fatty acids from stored fats to form **acetyl coenzyme A**. Some acetyl coenzyme A enters the citric acid cycle (see Figure 7.18) and is used in the normal way; some becomes converted to water-soluble

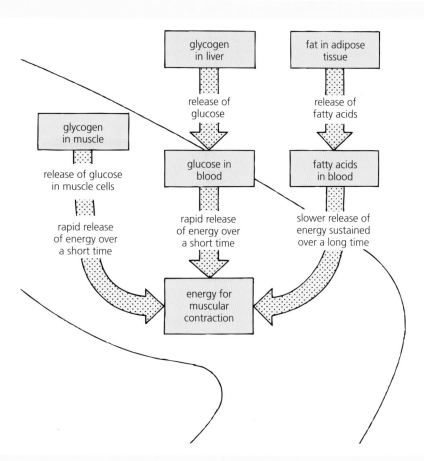

Figure 7.22 Fuelling a marathon

ketones, which are transported in the bloodstream to the brain and provide it with a vital source of energy.

Tissue protein is used as a source of energy only during prolonged starvation when the reserves of glycogen and fat have become exhausted. Then skeletal muscle and other tissues rich in protein are used up to provide energy during the crisis. Eventually the person becomes emaciated and death soon follows.

Regulation of the respiratory pathway

Refer back to Figure 7.12, which shows glycolysis. The third step in this process is catalysed by the enzyme **phosphofructokinase**. It is an **irreversible** step and the intermediate formed as a result of enzyme action is committed to continuing glycolysis. This third step is regarded as a **key regulatory point** in the pathway because the activity of the enzyme can be regulated as follows.

When the cell contains more **ATP** than it needs to meet its current demands, the high concentration of ATP **inhibits** phosphofructokinase (see Figure 7.23) and slows down glycolysis. When the concentration of ATP decreases again, the enzyme is no longer inhibited and glycolysis speeds up. Phosphofructokinase is similarly inhibited by high concentrations of citrate. However, if the citrate concentration drops (because other pathways are using citric-acid-cycle intermediates) then the enzyme is no longer inhibited and the rate of glycolysis increases.

This process of **feedback inhibition** regulates and synchronises the rates of the glycolytic and citric acid cycle pathways. It is important because:

- the needless build-up of an intermediate is prevented

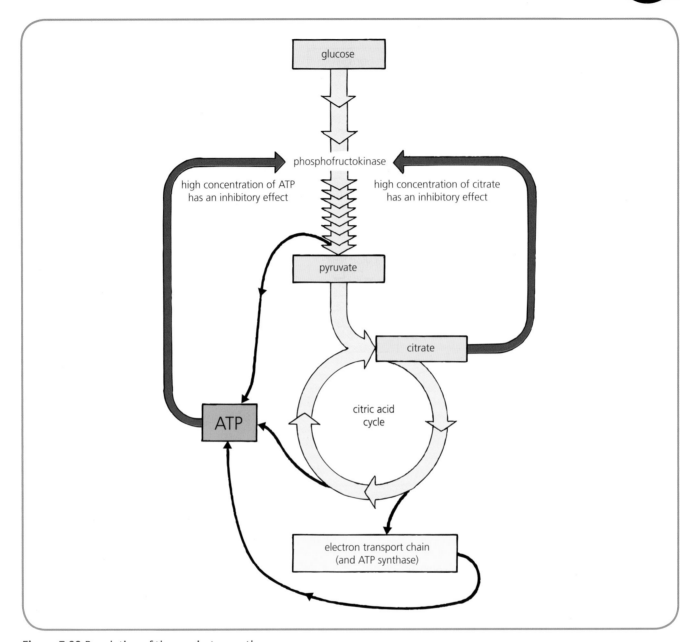

Figure 7.23 Regulation of the respiratory pathway

- ATP is only produced from respiratory pathways as required
- resources are conserved.

8 Energy systems in muscle cells

Creatine phosphate system

ATP is the source of immediately-available chemical energy needed for muscular contraction. During intense muscular activity, muscle cells break down ATP to ADP and phosphate. This process is accompanied by the release of energy to do work. However, each muscle cell only stores sufficient ATP for a few contractions. Much of the energy needed for repetitive muscular contraction comes from a chemical called **creatine phosphate**, which plays a key role in the coupled reactions shown in Figure 8.1.

During strenuous muscular activity, creatine phosphate in muscle cells breaks down, releasing energy and

phosphate, which are used to convert ADP to ATP by **phosphorylation** at a fast rate. The ATP formed is used to sustain maximal muscular contraction for **around ten seconds** during intense muscular effort. Therefore sufficient energy is generated to fuel a short burst of intense activity such as a 100-metre sprint, a golf swing or the lifting and lowering of a heavy weight.

During a rest period when demand for energy by muscle cells is low, little of the ATP produced by cellular respiration is required for muscular contraction. Under these circumstances ATP is used to generate creatine phosphate. Enzymes bring about the breakdown of ATP, releasing phosphate and energy, which are used to produce creatine phosphate by phosphorylation. This creatine phosphate acts as a **high-energy reserve** available to muscle cells during the next period of strenuous activity.

Lactic acid metabolism

During strenuous exercise, the tiny quantity of ATP present in muscle cells lasts for a few seconds. Resynthesis of ATP from creatine phosphate continues to provide energy for a few more seconds until the creatine phosphate store is depleted. As the intensive exercise continues, the cells respire by **fermentation** because they do not receive an adequate supply of oxygen from the bloodstream to support an increased level of **aerobic** respiration.

Neither the citric acid cycle nor the electron transport chain can generate the additional ATP required; only glycolysis is able to provide more ATP. It generates **2NADH** and **2ATP** from each molecule of glucose as it is broken down to pyruvic acid (see Figure 8.2). This process is followed by the conversion of pyruvic acid (pyruvate) to **lactic acid (lactate)** accompanied by the transfer of hydrogen from the NADH and the regeneration of **NAD**. NAD must be present to enable glycolysis to continue and produce more ATP.

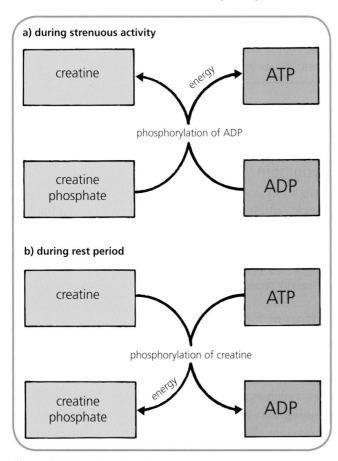

Figure 8.1 Creatine phosphate system

Case Study | Creatine supplements

Creatine present **naturally** in muscle cells has been manufactured by liver cells or has been derived from items in the diet such as meat. A 70 kg adult has about 120 g of creatine in their muscles.

Supplements and sporting performance

Increased intake of creatine as supplements has been shown to increase the store in the body by up to an upper limit of around 20%. Excess creatine is excreted in urine. Various studies have been carried out to investigate if a relationship exists between increased creatine levels in muscle cells (following supplements taken over a 5–6 day loading period) and **improved athletic performance**.

Creatine supplements were found to improve the performance of many 'power' athletes engaged in exercise of a brief, high-intensity nature such as sprinting and weight-lifting. However, those studies that examined the effect of creatine supplements on performance of events involving endurance exercise such as long-distance running showed no significant improvement.

It is now known that different people respond to creatine supplements in different ways. For example, those who have a naturally high store of creatine in their muscles do not receive an energy-boosting effect from a supplement. Many medical experts consider that claims for creatine as a muscle performance enhancer have been greatly exaggerated.

Side effects

No adverse effects have been reported for **short-term** studies. However, many athletes are taking creatine supplements over **extended periods** and no long-term studies have been carried out. It is known that the kidneys of athletes taking the supplements often have to produce urine containing creatine at concentrations of up to **90** times the normal level. This may be doing their kidneys long-lasting damage. At present no one can state confidently that taking creatine supplements over an extended period is safe. People are advised to avoid doing so until more research has been carried out.

However, as lactic acid gathers in muscle cells, it causes **fatigue** and an **oxygen debt** builds up. This is repaid when exercise comes to a halt (see Figure 8.3). Energy generated by aerobic respiration is now used to convert lactic acid (transported in the bloodstream to the liver) back to pyruvic acid and glucose.

ATP totals

Lactic acid metabolism (fermentation) is a less efficient process since it produces only **2 ATP** per molecule of glucose compared with around **38 ATP** formed by cellular respiration in the presence of oxygen (aerobic respiration).

Types of skeletal muscle fibre

All physical activities require parts of the body to move. These movements are brought about by the action of skeletal muscle fibres, which fall into two categories based on the duration of their twitches:

- type 1 – **slow**-twitch muscle fibres
- type 2 – **fast**-twitch muscle fibres

The two types differ in many ways, as shown in Table 8.1.

Myoglobin

Myoglobin is an oxygen-storing protein present in muscle cells. It has a stronger affinity for oxygen than haemoglobin. (Affinity means tendency to combine with a substance.) Therefore myoglobin is able to **extract oxygen** from blood for use by muscle cells, especially those in slow-twitch muscle fibres.

Different fibres for different events

Slow-twitch muscle fibres depend on aerobic respiration to generate most of their ATP and are especially effective when put to use during **endurance** events such as rowing, cycling and long-distance running. Fast-twitch muscle fibres depend on glycolysis to generate ATP and are especially effective when put to use during **power** events such as weight-lifting and sprinting.

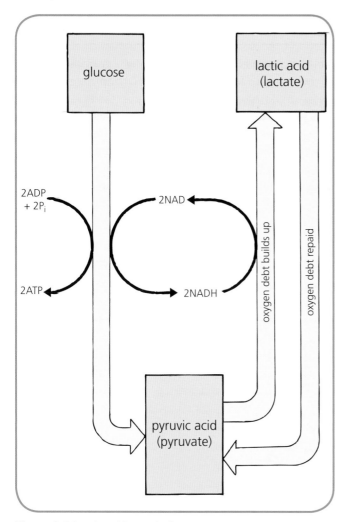

Figure 8.2 Lactic acid metabolism

Figure 8.3 Repayment of oxygen debt

Skeletal muscles contain a **genetically determined** mixture of slow-twitch and fast-twitch fibres. In most muscles, a fairly even balance exists between the two but in some muscles, one type predominates over the other. For example, the muscles in the back responsible for maintaining posture contain mostly slow-twitch fibres, whereas the muscles that move the eyeballs are made up mainly of fast-twitch fibres.

Feature	Type of skeletal muscle fibre	
	1 (slow-twitch)	2 (fast-twitch)
Speed of contraction	Slow	Fast
Length of time for which contraction is sustained	Long	Short
Speed at which fibre becomes fatigued	Slow	Fast
Respiratory pathway(s) normally used to generate ATP	Glycolysis and aerobic pathways	Glycolysis only
Number of mitochondria present	Large	Small
Density of blood capillaries associated with fibre	High	Low
Concentration of myoglobin in cells	High	Low
Major storage fuels used	Fats	Glycogen and creatine phosphate

Table 8.1 Comparison of slow- and fast-twitch muscle fibres

Case Study — Muscle fibre types in elite athletes

Skeletal muscles normally contain a balanced mixture of slow- and fast-twitch fibres. The actual composition of an individual's muscles is genetically determined. Some people inherit a higher than average percentage of type 1; others a higher than average percentage of type 2. It is thought therefore that the former are more suited to endurance sports whereas the latter have an edge in power sports.

It is certainly the case that a correlation exists between the type of sporting event performed by athletes who eventually excel in their field at a

world-class level and the genetically determined ratio of their slow to fast muscle fibres. Olympic sprinters, for example, have been shown to possess up to **80% fast-twitch** fibres in their leg muscles whereas elite athletes who excel in marathons often have around **80% slow-twitch** fibres in the same muscles. It must also be kept in mind that athletic success is also closely related to the quantity and quality of intensive practice undertaken by the athlete during training.

Testing Your Knowledge Chapters 7 and 8

1 Name TWO complex carbohydrates composed of chains of glucose molecules. (2)
2 Figure 8.4 shows the relationship between carbohydrate and two other classes of food that can act as alternative sources of energy. Identify blanks 1–5. (5)
3 The equation in Figure 8.5 represents the creatine phosphate system.
  a) Identify which chemical undergoes phosphorylation during i) reaction X, ii) reaction Y. (2)
  b) Which of reactions X and Y is the period of rest and which is the period of strenuous activity? (1)
4 Compare fast- and slow-twitch muscle fibres with reference to major storage fuels used, relative number of mitochondria and relative concentration of myoglobin. (3)

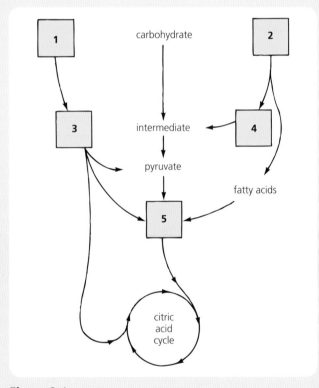

Figure 8.4

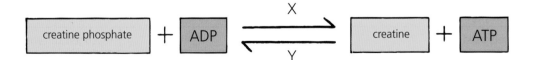

Figure 8.5

What You Should Know Chapters 6–8

acetyl	competitive	FAD
activation	complex	fatigue
ADP	concentration	fibres
affinity	creatine	genetic
anabolism	debt	glycolysis
ATP	electron	hydrogen
ATP synthase	endurance	induced
break	energy	inhibit
catabolism	environment	inhibition
citrate	enzymes	investment

irreversible	phosphate	substrate
lactic	phosphofructo-kinase	synchronise
lowering		transferred
metabolism	phosphorylation	transition
NADH	product	transport
negative	proteins	twitch
off	pyruvate	wasted
on	regulated	water
orientation	shape	
oxygen	structure	

Table 8.2 Word bank for chapters 6–8

1   Cell _____ encompasses all the enzyme-catalysed reactions that occur in a cell.

2   _____ consists of biosynthetic metabolic pathways that build up _____ molecules from simpler constituents and need a supply of energy; _____ consists of metabolic pathways that _____ down larger molecules into smaller ones and usually release _____.

3   For a metabolic reaction to occur, _____ energy is needed to form a _____ state from which end products are produced. _____ catalyse biochemical reactions by _____ the activation energy needed by the reactants to form their transition state.

4   Substrate molecules have an _____ for the active site on an enzyme. The active site's shape determines the _____ of the reactants on it and it binds to them closely with an _____ fit.

5   The enzymes controlling a metabolic pathway usually work as a group. Although some steps are _____, most metabolic reactions are reversible. The direction in which the reaction occurs depends on factors such as concentration of the _____ and removal of a _____ as it becomes converted to another metabolite.

6   Each step in a metabolic pathway is _____ by an enzyme that catalyses a specific reaction. Each enzyme is under _____ control.

7   Some metabolic pathways are required continuously and the genes that code for their enzymes are always switched _____. Other pathways are only needed on certain occasions. To prevent resources being _____,

the genes that code for their enzymes are switched on or _____ as required in response to signals from within the cell and from its _____.

8   Molecules of a _____ inhibitor resemble the substrate in _____. They become attached to the active site and slow down the reaction. Their effect is reversed by increasing the _____ of substrate.

9   Some regulatory molecules stimulate enzyme activity or _____ it non-competitively by changing the _____ of the enzyme molecule and its active site(s).

10 Some metabolic pathways are controlled by end product _____, a form of _____ feedback control.

11 _____ is a high-energy compound able to release and transfer energy when it is required for cellular processes. ATP is regenerated from _____ and P_i by phosphorylation using energy released during cellular respiration. _____ also occurs when P_i and energy are _____ from ATP to a reactant in a pathway.

12 Cellular respiration begins with _____, the breakdown of glucose to _____. This consists of an energy _____ phase and an energy payoff phase with a net gain of two molecules of ATP.

13 In the presence of oxygen, pyruvate is broken down into carbon dioxide and an _____ group. With the help of coenzyme A, the acetyl group enters the citric acid cycle by combining with oxaloacetate to become _____.

14 As one respiratory substrate is converted to another in the citric acid cycle, carbon dioxide is released, ATP is

formed and pairs of _____ ions along with high-energy electrons are removed and passed to coenzymes NAD and _____, forming NADH and FADH$_2$.

15 _____ and FADH$_2$ pass their high-energy electrons to _____ transport chains where the energy released is used to pump hydrogen ions across inner mitochondrial membranes. The return flow of these hydrogen ions makes part of each _____ molecule rotate and catalyse the synthesis of ATP. _____, the final electron acceptor, combines with hydrogen ions and electrons to form _____.

16 Complex carbohydrates, _____ and fats can all be used as respiratory substrates if they are first converted to suitable intermediates able to enter the pathway.

17 _____ is an enzyme that catalyses an irreversible step in glycolysis. Its activity is inhibited by high levels of ATP

and citrate. These inhibitory feedback mechanisms help to _____ and regulate the respiratory pathway.

18 During strenuous activity, _____ phosphate in muscle cells breaks down, releasing energy and _____ used to convert ADP to ATP. The ATP formed fuels muscle contraction for a few seconds.

19 During a period of vigorous exercise, muscle cells do not receive the adequate supply of oxygen needed for the electron _____ chain to operate. Instead of entering the citric acid cycle, pyruvate is converted to _____ acid, which causes _____ of muscles. Any oxygen _____ that develops is repaid when the vigorous exercise comes to a halt.

20 Skeletal muscle contains fast-twitch _____, good for short-lived power events, and slow-_____ fibres, good for _____ activities.

Applying Your Knowledge and Skills

Chapters 1–8

1 Figure 8.6 shows a possible use of stem cells in the future.

a) Match blank boxes P, Q, R, S, T and U with the following possible answers. (5)

i) stem cells induced to differentiate

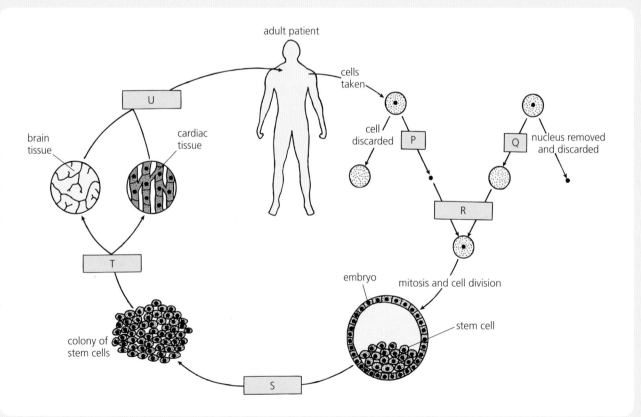

adult patient

cells taken

cell discarded

U

brain tissue

cardiac tissue

P

Q nucleus removed and discarded

R

T

embryo

mitosis and cell division

stem cell

colony of stem cells

S

Figure 8.6

ii) nucleus removed and retained

iii) stem cells removed and cultured in laboratory

iv) egg lacking nucleus retained

v) matching tissue transplanted to patient without fear of rejection

vi) nucleus inserted into egg

b) Which of the following is **not** represented in Figure 8.6? (1)

  A cytoplasmic hybrid cell

  B amniotic stem cell line

  C undifferentiated stem cells

  D nuclear transfer technique

c) **i)** Name a source of the donor egg cells used at present in this line of research.

  ii) Why does this prevent the series of events shown in the diagram being put into practice? (2)

2 Refer back to Figure 2.8 on page 19, which shows Griffith's transformation experiment. Avery continued this line of investigation and isolated a chemically purified form of the 'principle' that transformed live R to live S cells. He then carried out the experiment shown in Figure 8.7.

a) Find out and state the effect that:

  i) the enzyme DNAase has on DNA, its substrate

  ii) the enzyme protease has on protein, its substrate. (2)

b) Why did the transforming principle treated with DNAase have no effect on the mouse? (2)

c) Why did the transforming principle treated with protease still work and affect the mouse? (2)

d) Identify TWO factors from Figure 8.7 that must be kept constant throughout the experiment. (2)

e) State the means by which the reliability of the experiment could be checked. (1)

3 Refer back to Figure 2.19 on page 27 and then draw a labelled diagram to show both the DNA strands and the test tube contents that would result after three DNA replications involving semi-conservative replication only. (4)

4 Figure 8.8 on page 120 shows mRNA's codons and the amino acids that they code for.

a) Which letters should have been inserted at positions **X** and **Y** in the wheel? (2)

b) How many codons are able to trigger the process of translation? (1)

c) Identify the codons that can bring translation to a halt. (3)

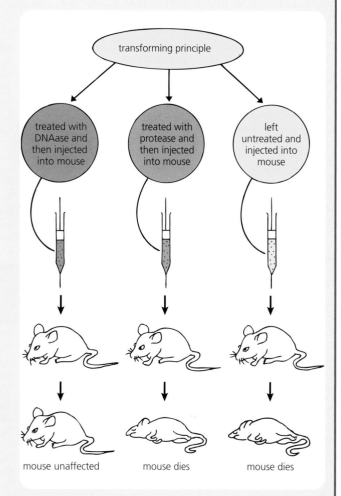

Figure 8.7

d) Which amino acid is coded for by codon:

  i) UUU; **ii)** ACC; **iii)** GGU? (3)

e) Name all the codons that code for leucine. (2)

f) Refer back to Figure 3.9 on page 38 and then state which amino acid would become attached to this tRNA's site of attachment. (Remember that tRNA has an anticodon and that the wheel shows mRNA codons.) (1)

5 Figure 8.9 on page 120 shows the base sequence on a region of DNA undergoing a type of point mutation.

a) **i)** Identify the type of point mutation that occurred.

  ii) Describe the way in which the DNA has been altered. (2)

b) Refer to Table 3.2 on page 39 and Appendix 1 and work out the amino acid sequence for:

  i) the original DNA

  ii) the mutant DNA. (2)

c) **i)** State whether this mutation would be missensical or nonsensical.

  ii) Explain your choice. (2)

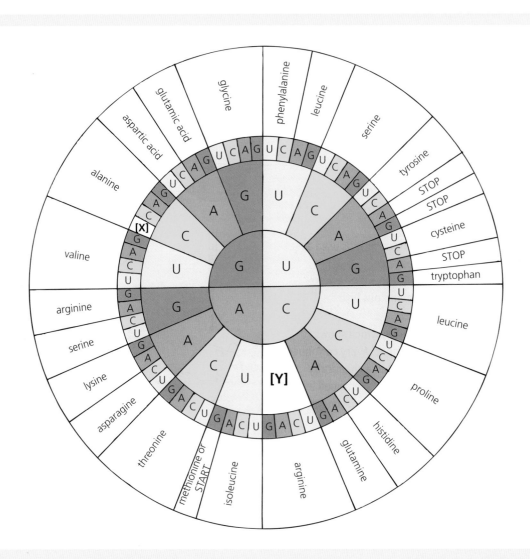

Figure 8.8

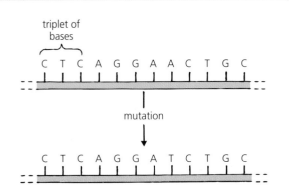

triplet of bases

Figure 8.9

6 Copy and complete Table 8.3 on page 121, which refers to three genetic disorders described in case studies chapter 4 (pages 56–59). (12)

7 The DNA fragments shown in Figure 8.10 on page 121 were formed during the type of sequencing technique illustrated in Figure 5.1 on page 64. Each fluorescent tag indicates the point on the strand where replication of complementary DNA was brought to a halt by a modified nucleotide.

   a) Work out the sequence of the bases in the complementary DNA strand. (1)

   b) Deduce the sequence of bases in the original DNA strand. (1)

8 The graph in Figure 8.11 on page 121 shows the expected number of copies of DNA that would be generated by the polymerase chain reaction (PCR) under ideal conditions.

   a) i) What name is given to the type of graph paper used here to present the data?

      ii) Why has this type of graph paper been used? (2)

Genetic disorder	Type of single-gene mutation responsible	Effect of mutation on structure of protein expressed	Effect of mutation on functioning of protein	Effect of mutation on phenotype of untreated, affected individual
Phenylketonuria (PKU)				
Cystic fibrosis				
Huntington's disease				

Table 8.3

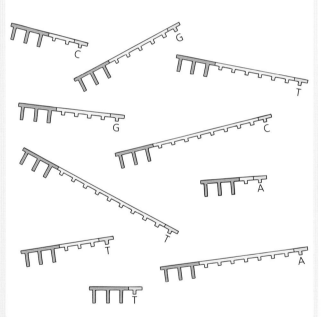

Figure 8.10

b) How many cycles of PCR are required to produce 10 000 000 copies of the DNA? (1)

c) How many cycles are required to increase the number of copies of DNA already present at ten cycles by a factor of 10^3? (1)

d) i) How many copies of the DNA were present after 30 cycles?
   ii) Now state your answer in words. (2)

e) Refer back to Figure 5.9 on page 71 and, with the aid of coloured pencils, draw a diagram of the DNA

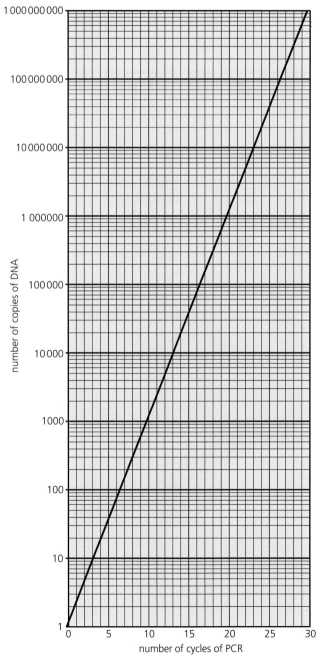

Figure 8.11

that would be present at the end of the third cycle of PCR. (4)

9 ONPG is a chemical that can be broken down as follows:

β-galactosidase

ONPG $\longrightarrow$ galactose + yellow compound

Figure 8.12 shows an experiment set up to investigate the *lac* operon in *E. coli* (where β-g = β-galactosidase).

a) Explain the yellow colour in **i)** tube C, **ii)** tube E. (2)

b) Explain the faint yellow colour in tube A. (1)

c) It could be argued that ONPG acts as an inducer.

Draw a diagram of the control that should be set up to check out this possibility. (1)

d) How could the reliability of the results be improved? (1)

e) Return to page 90 and consider the experiment shown in Figure 6.23. Imagine that bacteria from plates B and D have been grown on plates E and F respectively, which both contain nutrient agar and arabinose.

i) Predict the effect of exposing plates E and F to ultraviolet light.

ii) Explain your answer. (3)

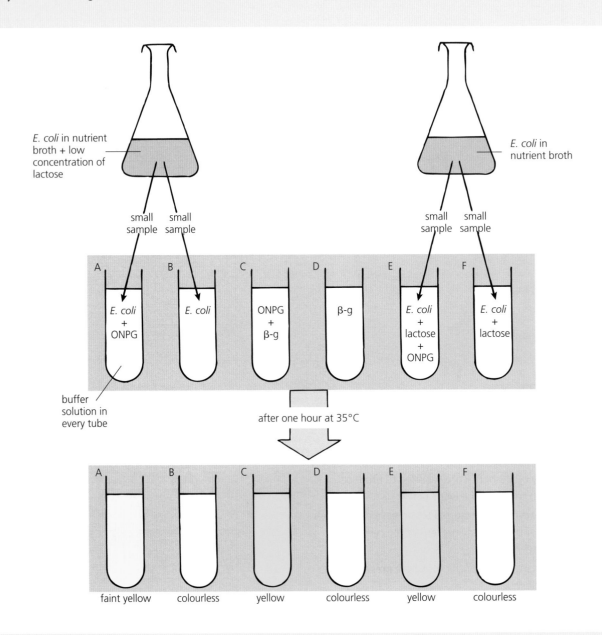

Figure 8.12

10 Give an account of enzyme activity under the headings:

   a) induced fit (3)
   b) activation energy (3)
   c) effect of substrate concentration. (3)

11 Refer back to the investigation shown in Figure 7.14 on page 104. Based only on these results, it could be argued that malonic acid has simply killed the cells.

   a) How could the experiment be adapted to investigate if malonic acid really does act as a competitive inhibitor? (1) (Hint: see chapter 6, pages 91–92.)

   b) Explain how you would know from the results of your redesigned experiment whether or not malonic acid had acted as a competitive inhibitor. (2)

Note: Since this group of questions does not include examples of every type of question found in SQA exams, it is recommended that students also make use of past exam papers to aid learning and revision.

Physiology and Health

9 The structure and function of reproductive organs and gametes and their role in fertilisation

Testes

The reproductive system of the human male is shown in Figure 9.1. The testes are the site of **sperm** production and the manufacture of the male sex hormone **testosterone**. Sperm (full name spermatozoa) are male gametes formed from germline cells in tiny tubes called **seminiferous tubules**. These tubules unite to form coiled tubes that connect to the **sperm duct**. It

is by the sperm duct that free-swimming sperm leave the testis. Testosterone is produced by **interstitial cells** located in the tissue between the seminiferous tubules. Testosterone passes directly into the bloodstream.

Sperm are **motile**. On being released during sexual intercourse inside the female body, they move through the uterus and along the oviducts where they may meet an egg. The process of fertilisation takes place in the

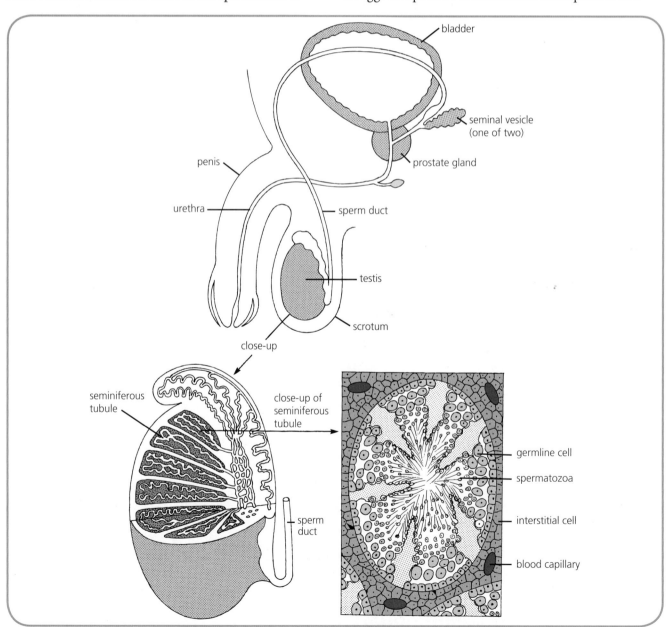

Figure 9.1 Male reproductive system

oviduct and is dependent upon the motility of sperm, which brings the two gametes together to form a zygote. Such motility requires a **fluid** medium and a source of **energy**.

Accessory glands

The **seminal vesicles** (see Figure 9.1) secrete a liquid rich in fructose. This sugar provides sperm with the energy needed for motility following their release by the male at ejaculation. The liquid secreted by the seminal vesicles also contains hormone-like compounds that stimulate contractions of the female reproductive tract. These movements help the sperm to reach the oviduct at a much faster rate than could be achieved by swimming alone.

The **prostate gland** (see Figure 9.1) secretes a thin, lubricating liquid containing **enzymes** whose action maintains the fluid medium at the optimum viscosity for sperm motility. **Semen** is the collective name given to the milky liquid released by the male. It contains sperm from the testes and the fluid secretions from the seminal vesicles and prostate gland that maintain the mobility and viability of sperm.

Ovaries

The reproductive system of the human female is shown in Figure 9.2. **Eggs (ova)** are formed from germline cells in the **ovaries**. The ovaries contain immature eggs

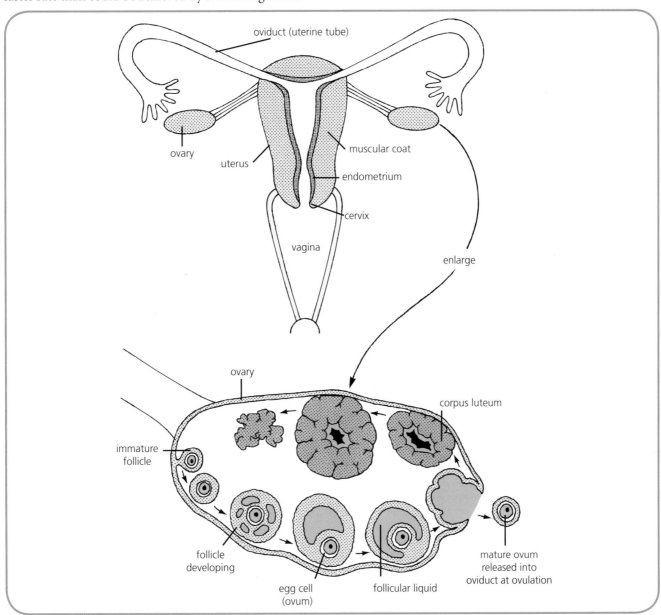

Figure 9.2 Female reproductive system

127

at various stages of development. Each egg (ovum) is surrounded by a **follicle**, which secretes the hormone **oestrogen** and protects the developing egg. Following ovulation (release of an egg), the follicle develops into a **corpus luteum** (see Figure 9.2), which secretes the hormone **progesterone** (see chapter 10).

10 Hormonal control of reproduction

Hormonal control

Hormones are chemical messengers produced by an animal's **endocrine** (ductless) glands and secreted directly into the bloodstream. When a hormone reaches a certain **target tissue**, it brings about a specific effect. Hormones control the onset of puberty, sperm production and the menstrual cycle.

Hormonal onset of puberty

At puberty, the **hypothalamus** (see Figure 10.1) secretes a releaser hormone whose target is the **pituitary gland**. On being stimulated the pituitary responds by producing two hormones. The first of these is called **FSH** (follicle-stimulating hormone); the second is known in men as **ICSH** (interstitial cell-stimulating hormone) and in women as **LH** (luteinising hormone). The release of these hormones at puberty triggers the onset of sperm production in men and the menstrual cycle in women.

Hormonal control of sperm production

Influence of pituitary hormones on testes

The two functions of the testes are regulated by the pituitary hormones. When FSH arrives in the bloodstream, it promotes **sperm production** in the seminiferous tubules. When ICSH arrives, it stimulates interstitial cells to produce the male sex hormone **testosterone**.

Influence of testosterone

Testosterone stimulates **sperm production** in the seminiferous tubules. It also activates the prostate gland and seminal vesicles to produce their secretions.

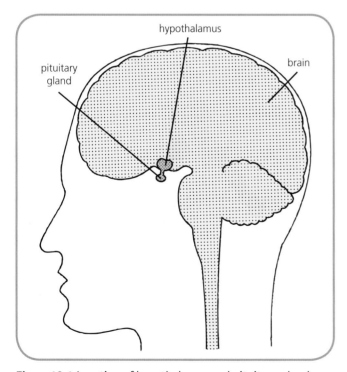

Figure10.1 Location of hypothalamus and pituitary gland

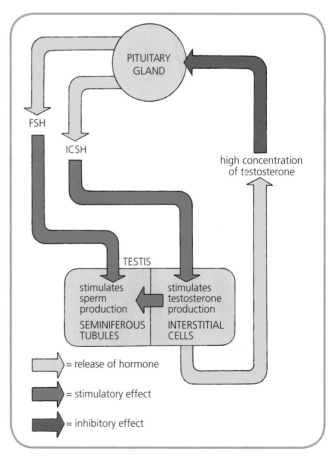

Figure 10.2 Self-regulation of testosterone production

Self-regulation of testosterone production

As the concentration of testosterone builds up in the bloodstream, it reaches a level where it **inhibits** the secretion of FSH and ICSH by the pituitary gland (see Figure 10.2). Since this leads in turn to a decrease in testosterone concentration, it is soon followed by a resumption of activity by the pituitary gland, which makes FSH and ICSH again, and so on. This type of self-regulating mechanism is called **negative feedback control**.

Hormonal control of the menstrual cycle

Influence of pituitary hormones on ovaries

FSH and LH from the pituitary gland affect the ovaries in several ways. FSH stimulates the development and maturation of each **follicle** (see Figure 10.3). It also

stimulates ovary tissue to secrete the sex hormone **oestrogen**.

LH triggers **ovulation**. It also brings about the development of the **corpus luteum** from the follicle and then stimulates the corpus luteum to secrete the sex hormone **progesterone**. Oestrogen and progesterone are known as the **ovarian** hormones.

Influence of ovarian hormones on uterus and pituitary gland

Oestrogen

Oestrogen stimulates **proliferation** (cell division) of the inner layer of the uterus, called the **endometrium**, thereby effecting its repair following menstruation and preparing it for implantation. This ovarian hormone also stimulates the secretion of LH by the pituitary gland (see Figure 10.4).

Progesterone

Progesterone promotes the **further development** and **vascularisation** of the endometrium into a spongy layer rich in blood vessels, making it ready to receive a blastocyst (embryo) if fertilisation occurs. Progesterone

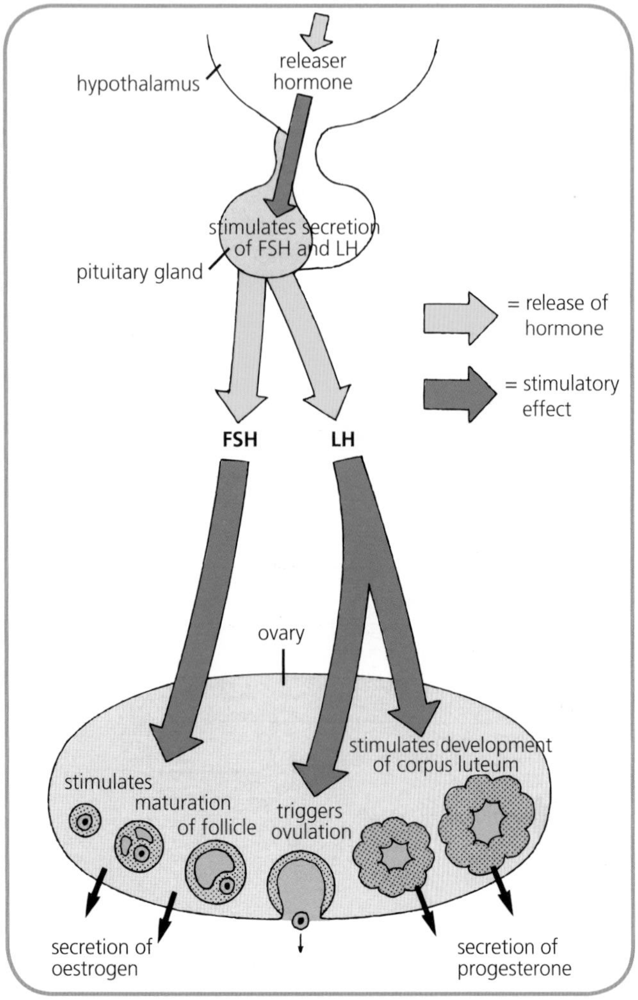

Figure 10.3 Effect of pituitary hormones on ovary

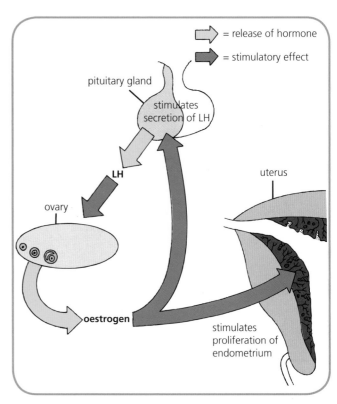

Figure 10.4 Effect of oestrogen on uterus and pituitary

also **inhibits** the secretion of FSH and LH by the pituitary gland (see Figure 10.5).

The menstrual cycle

The events described above that are under hormonal control in the human female fit together as interacting parts of a synchronised system – the **menstrual cycle**.

A cycle takes about 28 days though this can vary from woman to woman. Each cycle is continuous with the one that went before and the one about to follow. For convenience the first day of menstruation (as indicated by menstrual blood flow) is regarded as 'day one' of the cycle. The menstrual cycle is summarised in Figure 10.6. It is made up of two phases.

Follicular phase

During this first half of the cycle, FSH from the pituitary gland stimulates:

- the development and maturation of a follicle
- the production of oestrogen by the ovarian tissues.

As the concentration of the oestrogen builds up, it brings about the repair and proliferation of the endometrium. Eventually the high concentration of oestrogen triggers a surge in the production of LH (and FSH) by the pituitary gland at about day 14. This **surge of LH** is the direct cause of **ovulation** since it makes the blister-like wall of the follicle rupture and release the egg. The egg is then moved slowly along the oviduct. During a short period of about 3–4 days, fertilisation may occur if the egg meets a sperm.

Luteal phase

During this second half of the cycle (following ovulation), LH stimulates the follicle to become the corpus luteum. This gland-like structure secretes progesterone and oestrogen. The subsequent rise in progesterone concentration stimulates further

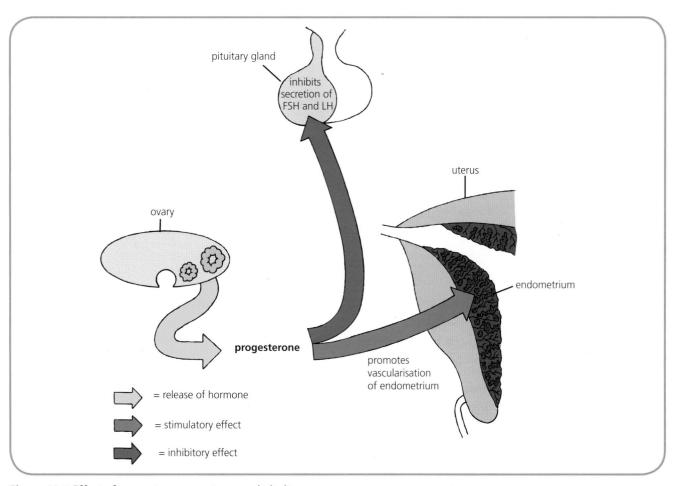

Figure 10.5 Effect of progesterone on uterus and pituitary

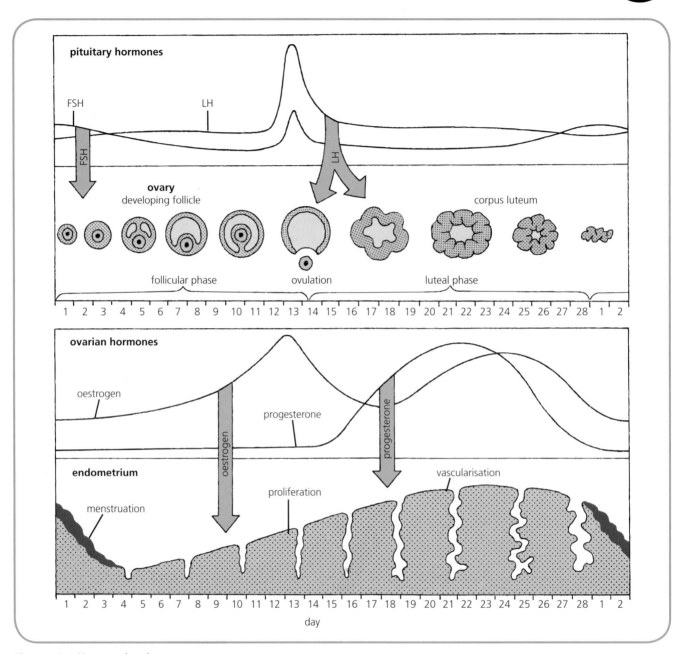

Figure 10.6 Menstrual cycle

development of the endometrium. It becomes thick (with an increase in vascular blood vessels) and spongy, ready to accept and nourish a blastocyst if fertilisation takes place and the blastocyst becomes implanted.

The combined high levels of oestrogen and progesterone during this luteal phase also trigger an **inhibitory effect** on the pituitary gland. Concentrations of FSH and LH drop as a result and no new follicles develop at this time. This is a further example of negative feedback control.

If fertilisation does not occur, **lack of LH** leads, in turn, to the **degeneration** of the corpus luteum by about day 22 in the cycle. This is followed by a **rapid drop** in progesterone (and oestrogen). By day 28 in the cycle, these ovarian hormones are at such a low level that the endometrium can no longer be maintained and **menstruation** begins. This involves the loss of the inner layer of the endometrium accompanied by a small volume of blood. This stage continues for a few days.

Related Topic

Fertilisation

If fertilisation does occur, the embryo secretes a hormone called human chorionic gonadotrophin (**HCG**) which has the same effect as LH. HCG maintains the corpus luteum which continues to secrete progesterone and prevent menstruation from taking place. After about 6 weeks the placenta takes on the job of secreting progesterone. The presence of HCG in urine is the basis of pregnancy testing.

Role of the cervix in fertility

The cells lining the cervix at the neck of the uterus (see Figure 9.2) secrete **mucus**, which lubricates the vagina. These cells are stimulated by high levels of oestrogen to secrete watery mucus easily penetrated by sperm. Since the highest concentration of oestrogen immediately precedes ovulation, the secretion of thin mucus at this time increases the chance of fertilisation by facilitating the entry of sperm into the female reproductive tract.

High levels of progesterone cause the cervical mucus to become viscous and in the event of pregnancy change into a semi-solid 'plug'. This protects the fertilised egg from possible infection.

Changes in body temperature

Body temperature rises by about 0.5°C at ovulation and remains at this higher level during the luteal phase of the cycle.

Testing Your Knowledge Chapters 9 and 10

1 a) i) Where in a testis are sperm produced?
       ii) Which hormone is produced by interstitial cells? (2)
    b) i) Which accessory glands secrete a liquid rich in hormone-like chemicals?
       ii) Describe the contribution to fertilisation made by these chemicals. (2)

2 a) Name the structure that surrounds an egg in an ovary. (1)
    b) What does this structure develop into, following ovulation? (1)

3 Copy and complete Table 10.1, which refers to four hormones in the female body. (8)

4 Decide whether each of the following statements is true or false and then use T or F to indicate your choice. Where a statement is false, give the word that should have been used in place of the word in bold print. (6)
    a) The menstrual cycle consists of the **endometrial** phase and the luteal phase.
    b) FSH stimulates ovary tissue to secrete **oestrogen**.
    c) Oestrogen stimulates the **proliferation** of the endometrium.
    d) LH triggers the process of **menstruation**.
    e) Progesterone inhibits secretion of FSH and LH by the **ovaries**.
    f) Lack of LH leads to **degeneration** of the corpus luteum.

Hormone	Site of production	One function of the hormone
		Stimulates the development and maturation of each follicle
		Brings about development of the corpus luteum
		Stimulates secretion of LH by the pituitary gland
		Promotes vascularisation of the endometrium

Table 10.1

11 Biology of controlling fertility

Knowledge of the biology of fertilisation is put to effective use when designing treatments for infertility and devising methods of contraception.

Fertile periods

Continuous versus cyclical fertility

The negative feedback effect of testosterone (see Figure 10.2, page 129) maintains a relatively constant level of the pituitary hormones (FSH and ICSH) in the bloodstream of men. This results in a fairly steady quantity of testosterone being secreted and sperm being produced. Therefore men are **continuously** fertile.

This contrasts markedly with the **cyclical** fertility of women. The delicate interplay of pituitary and ovarian hormones that occurs in their body normally results in the **period of fertility** being restricted to the 1–2 days immediately following ovulation.

Calculation of the fertile period

Temperature

Within the menstrual cycle, the alternating processes of menstruation and ovulation are separated by intervals of about 2 weeks. Approximately 1 day after the LH surge that triggers ovulation, the woman's **body temperature** rises by about 0.2–0.5°C under the action of progesterone. It remains at this elevated level for the duration of the luteal phase of the cycle (see Figure 11.1).

The period of fertility lasts for about 1–2 days. The infertile phase is resumed, on average, after the third daily recording of the higher temperature by which time the unfertilised egg has disintegrated.

Mucus

The **cervical mucus** secreted into the vagina during the fertile period is thin and watery to allow sperm easy access to the female reproductive system. However, after ovulation, the mucus gradually increases in viscosity under the action of progesterone, showing that the system has returned to the infertile phase.

Use of indicators

The above indicators can be used by a woman to calculate her fertile period. This knowledge is useful to a couple who wish to have a child and want to know when sexual intercourse is most likely to achieve fertilisation.

Treatments for infertility

Stimulating ovulation

A woman may fail to ovulate because of an underlying factor, such as failure of the pituitary gland to secrete adequate FSH or LH. In such cases, ovulation can be successfully stimulated by:

- drugs that mimic the normal action of FSH and LH
- drugs that prevent the negative feedback effect of

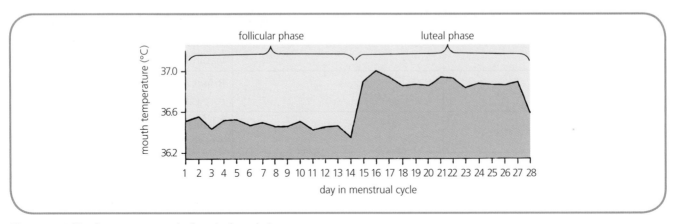

Figure 11.1 Rise in temperature during the luteal phase

oestrogen on FSH secretion during the luteal phase of the menstrual cycle.

On some occasions these drugs are so effective that they bring about 'super-ovulation' and this can lead to a multiple birth such as quintuplets (see Figures 11.2 and 11.3). Drugs that cause super-ovulation are also used to promote the release of eggs to be used for IVF (see below).

Artificial insemination

Insemination is the introduction of semen into the female reproductive tract. It occurs naturally as a result of sexual intercourse.

Figure 11.3 Effect of 'super-ovulation'

Figure 11.2 Quintuplets

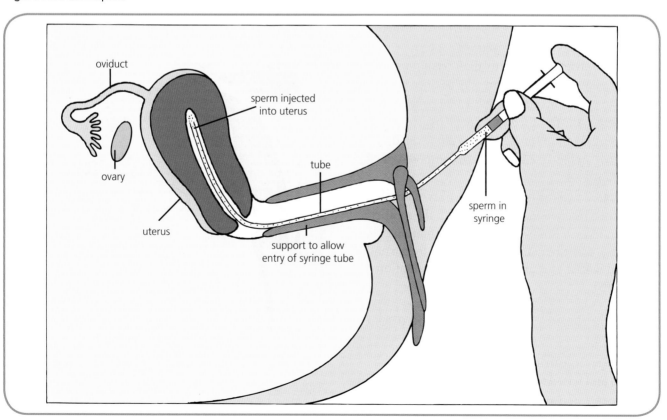

Figure 11.4 Artificial insemination

Artificial insemination is the insertion of semen into the female tract by some means other than sexual intercourse. Artificial insemination may be employed as a method of treating infertility. If a man has a **low sperm count**, several samples of his semen can be collected over a period of time and each preserved by freezing until required. They are then defrosted and released together into his partner's cervical region at the time when she is most likely to be fertile (see Figure 11.4).

Artificial insemination can also be used to insert semen of a **donor** who has a normal sperm count into the cervical region of a woman whose partner is sterile.

In vitro fertilisation (IVF)

This method of treatment attempts to solve the problem of infertility caused by a blockage of the oviducts (uterine tubes). It enables fertilisation to occur outside the bodies of the would-be parents in a culture dish. Figure 11.5 shows some of the stages involved in the procedure.

- At stage 1, the woman is given hormonal treatment to stimulate multiple ovulation.
- At stage 2, a surgical procedure is employed to remove several of these eggs from around her ovary using a piece of equipment similar to a syringe.
- At stage 3, the eggs are mixed with sperm in a culture dish of nutrient medium to allow fertilisation to occur. Alternatively a sperm may be injected directly into an egg at this stage (see ICSI on page 137).
- At stage 4, the fertilised eggs are incubated in the nutrient medium for 2–3 days to allow cell division to occur and form embryos each composed of eight (or more) cells.
- At stage 5, two (or three) of the embryos are chosen and then inserted via the vagina into the mother's uterus (which is now ready for implantation).
- At stage 6, the remaining embryos are frozen and stored in case a second attempt at implantation is required.

Pre-implantation genetic screening and diagnosis

Before stage 5 is carried out, one or two cells may be removed and tested for genetic abnormalities. The test may take one of two forms:

- **pre-implantation genetic screening (PGS)** – a non-specific approach that checks the embryo for

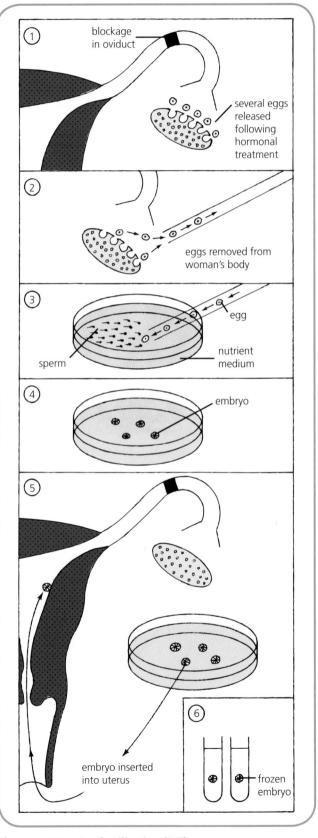

Figure 11.5 *In vitro* fertilisation (IVF)

single gene disorders and common chromosomal abnormalities in general

- **pre-implantation genetic diagnosis (PGD)** – a specific approach that is used to check for a *known* chromosomal or gene defect.

These tests enable experts to select which embryos should and which should not be allowed to become implanted in the mother's endometrium.

Related Topic

Ethics of PGS and PGD

Some people strongly support the practices of PGS and PGD because they offer **reassurance** to couples who would otherwise be at high risk of producing children with serious genetic disorders. In the absence of these techniques, many of these couples would probably choose to remain childless. The supporters also claim that in addition to helping the individual families affected, a reduced frequency of genetic diseases and disorders is of **great benefit** to society as a whole.

Other people are opposed to PGS and PGD. They insist that it is morally wrong to interfere with the process of conception by making it **selective**. They argue that these procedures are the start of **eugenics**, whereby the human race would be subjected to selective breeding in order to 'improve its quality'. They speculate that this route could lead to a world of 'designer' children in which genetic engineering of offspring would become routine practice.

Intracytoplasmic sperm injection

During IVF, eggs and sperm are mixed together in a culture dish. There is only a good chance of fertilisation occurring if a large number of active sperm are present. In those cases where the man's sperm count is low or many of his mature sperm are defective in some way, **intracytoplasmic sperm injection (ICSI)** can be employed.

This procedure involves drawing a healthy sperm into a syringe needle and then injecting it directly into an egg to bring about fertilisation (see Figure 11.6). During the procedure the egg is held in place by a holding tool. ICSI is commonly used as part of IVF treatment.

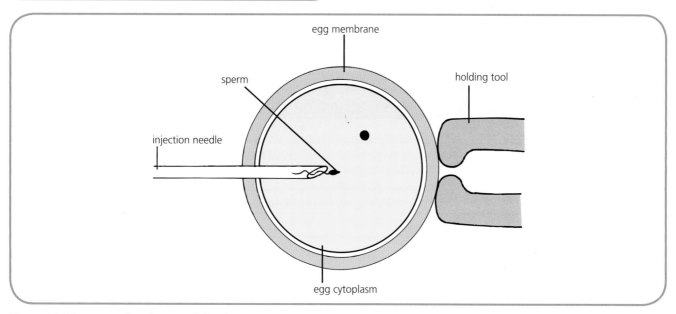

Figure 11.6 Intracytoplasmic sperm injection

Related Activity

Examining data on success rate for IVF and its effect on long-term health

Success rate

Tables 11.1, 11.2 and 11.3 show data that refer to IVF success rates for the UK.

From Table 11.1 it can be concluded that the number of patients receiving IVF treatment and the number of babies born as a result of it increased significantly over the period of time considered. These changes were accompanied by a slight increase in IVF success rate when measured as percentage live birth rate per cycle.

	Year		
	2006	**2007**	**2008**
Number of patients	34 855	36 861	39 879
Number of cycles	44 275	46 829	50 687
Number of births as a result of IVF	10 242	11 091	12 211
Number of babies born	12 596	13 672	15 082
IVF success rate as live birth rate per cycle (%)	23.1	23.7	24.1

Table 11.1 Success rate for IVF over 3 years

Age of patient (years)	IVF success rate as live birth rate per cycle (%)	
	Year 2007	**Year 2008**
Under 35	32.3	33.1
35–37	27.7	27.2
38–39	19.2	19.3
40–42	11.9	12.5
43–44	3.4	4.9
Over 44	3.1	2.5

Table 11.2 Effect of age on IVF success rate

		State of embryo used in IVF	
		Fresh	**Frozen**
Number of patients		33 520	7 792
Number of cycles		39 334	8 959
Number of births as a result of IVF		10 010	1 618
Number of babies born		12 480	1 855
IVF success rates as percentage of live birth rates by age	Under 35 years	33.1	22.2
	35–37	27.2	17.8
	38–39	19.3	15.8
	Over 39 years	10.7	11.9

Table 11.3 Effect of freezing on success rate for IVF

(It should be noted that the number of successful births is smaller than the number of babies born because some births are multiple.)

From Table 11.2 it can be concluded that the chance of successfully giving birth following IVF treatment is highest for women under the age of 35 and decreases steadily with increasing age of patient.

From Table 11.3 it can be concluded that during the period studied more fresh than frozen embryos were used. In addition the success rate for IVF using fresh embryos was much higher for women of 39 and under but slightly lower for women over the age of 39.

Long-term health issues of IVF

Mother

Many thousands of children are born annually using IVF treatment involving drugs that stimulate the mother's ovaries to release a large number of eggs. About 6% of patients undergoing this form of IVF are found to suffer hyperstimulation of their ovaries. In addition, medical experts are concerned that stimulated-cycle IVF may be exposing many women to an increased risk of uterine cancer in later life and urge that research be carried out in this area. They also recommend that patients consider **natural-cycle** IVF in place of stimulated-cycle IVF despite the fact that the former's success rate is only 7–10% compared with 25–30% for the latter.

Child

Most children conceived through IVF tend to have a body mass at birth that is **significantly lower** than the average for full-term babies conceived naturally, and often closer to that of premature babies.

In general, children born with a low body mass are more likely to suffer long-term **health problems** such as obesity, diabetes, heart conditions and hypertension in later life at around 50 or more years of age. Experts are concerned that children conceived through IVF will also be more prone to these conditions. However, the first successful IVF treatment was not carried out until 1978. Therefore no IVF-born people are old enough yet to allow a survey to be conducted and tentative conclusions to be drawn.

Contraception

Contraception is the intentional prevention of conception or pregnancy by natural or artificial means.

Physical methods of contraception

Barrier methods

A **barrier method** makes use of a device that physically blocks the ability of sperm to reach an ovum. These devices include:

- the **condom** – a rubber sheath that fits over the man's penis
- the **diaphragm** – a dome-shaped rubber cap that is inserted into the woman's vagina to block the cervix before each act of sexual intercourse
- the **cervical cap** – a rubber structure that fits tightly round the cervix and can be left in position for a few days.

These methods are very effective but they are not as successful as chemical methods (see page 140).

Intra-uterine devices

An **intra-uterine device** (**IUD**) is a T-shaped structure (see Case Study – Example of a physical contraceptive) that is fitted into the uterus for many months or even years to prevent the implantation of an embryo into the endometrium.

Sterilisation procedures

In men, **vasectomy** involves the cutting and tying of the two sperm ducts, thereby preventing sperm being released during sexual activity. (Sperm produced after this sterilisation procedure normally undergo phagocytosis and are destroyed.)

In women, **tubal ligation** involves the cutting and tying of the two oviducts (see Figure 11.7) to prevent eggs meeting sperm and reaching the uterus. Sterilisation is a highly effective means of contraception but it is normally irreversible.

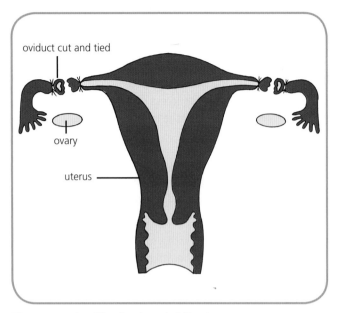

Figure 11.7 Sterilisation by tubal ligation

Chemical methods of contraception

Pills containing a combination of hormones

These **oral contraceptive pills** normally contain **synthetic progesterone** combined with **synthetic oestrogen**. One common method requires the woman to take a pill every day without fail for 3 weeks from the final day of the previous menstrual period. This procedure makes the concentration of progesterone and oestrogen in her bloodstream increase and exert negative feedback control.

Therefore, secretion of FSH and LH by the pituitary gland is inhibited. Since little or no FSH is present, follicle maturation remains inhibited and ovulation fails to occur. Dummy (placebo) pills are usually taken during the fourth week to allow the levels of oestrogen and progesterone to drop and menstruation to occur.

'Morning-after' pills

These are also known as **emergency hormonal contraception pills.** They often contain higher doses of the hormones (progesterone and oestrogen) found in standard oral contraceptive pills. They are taken by the woman after unprotected sexual intercourse to prevent implantation from occurring if fertilisation has taken place unintentionally. Ideally these pills should be taken as soon after unprotected sex as possible but may be effective up to about 72 hours.

'Mini pills'

These are also known as **progesterone-only pills** because they contain synthetic progesterone but not oestrogen. They work by causing thickening of cervical mucus, which reduces the viability of sperm and their access to the uterus.

Case Study | **Example of a physical contraceptive**

Although widely used to mean any system of birth control that prevents pregnancy, strictly speaking the word *contraception* means prevention of conception. If this definition is adhered to strictly, then an intra-uterine device (IUD) has to be described as a contragestic device since it prevents the *gestation* of an already conceived embryo.

In the UK, the term **intra-uterine device (IUD)** refers to a T-shaped plastic structure with copper wound around its outside (see Figure 11.8). It also has threads attached to it that can be used by a medical expert to gently remove the device when required. An IUD works in several ways:

- Its presence in the uterus stimulates the release of white blood cells and various substances that are hostile to sperm (and perhaps to the very early embryo).
- It impairs the mobility of sperm and prevents them reaching the egg.
- It irritates the lining of the uterus making it unreceptive to an embryo, which therefore fails to become implanted in the endometrium.

There is no denying that the IUD is an effective contraceptive device but it has tended to be less popular than other forms of contraception. This is probably due to the fact that it has had a history of causing **complications**, including inflammation of the uterus and oviducts and also ectopic pregnancies (those that result from the implantation of an embryo at a site other than the uterus).

In addition, many people feel uneasy about the **ethics** involved in preventing an already conceived embryo from becoming implanted in the endometrium. They prefer to use a method of contraception that stops sperm reaching an egg in the first place and therefore prevents any chance of conception taking place.

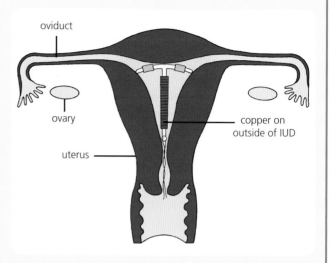

Figure 11.8 Intra-uterine device

Case Study   Example of a chemical contraceptive

Although 'mini pills' are also called progesterone-only pills, strictly speaking this is a misnomer because they contain a **synthetic** form of progesterone. In addition, this form of chemical contraception may be taken by the woman in the form of pills or be given as an **implant** the size of a matchstick. The latter is inserted under the skin on the inside of the upper arm and can give protection against pregnancy for up to 3 years.

The chemical's method of activity varies with dosage. At low doses it inhibits ovulation in only about 50% of cycles. Therefore the woman is largely dependent on progesterone's thickening effect on cervical mucus for protection against pregnancy. At higher doses, it inhibits ovulation in the majority of cycles and brings about thickening of cervical mucus.

Advantages and disadvantages
Some advantages of the 'mini pill' are as follows:

- It does not interfere with sexual spontaneity.
- It can be used during breast feeding and does not affect the milk supply.
- It can reduce the cramp and heavy bleeding suffered by some women during menstruation.
- It can be taken by women who cannot take oestrogen.
- It can be taken by women who suffer health problems such as high blood pressure.

Some of the disadvantages of the 'mini pill' are as follows:

- It must be taken at the same time every day.
- It can cause breast tenderness and mood swings.
- It can lead to irregular menstrual cycles.
- It can lead to weight gain.

This form of contraception is not necessarily suitable for all women. Individuals are advised to seek expert advice before using it.

Testing Your Knowledge

1 a) Distinguish clearly between the terms *continuous fertility* and *cyclical fertility* with reference to human beings. (2)

   b) i) For how long does a woman's period of fertility last during each menstrual cycle, on average?

       ii) Describe TWO signs that give an approximate indication of when this time occurs. (3)

2 a) What is *artificial insemination*? (1)

   b) Under what TWO sets of circumstances might artificial insemination be used as a means of treating infertility? (2)

3 a) The following list gives the steps in the procedure employed during *in vitro* fertilisation. Arrange them into the correct order. (1)

       A Incubation of fertilised eggs in nutrient medium.

       B Deep-freezing of unused fertilised eggs.

       C Stimulation of ovaries to bring about multiple ovulation.

       D Mixing of eggs with sperm in a dish.

       E Insertion of two or three fertilised eggs into the uterus.

       F Removal of eggs from the mother's body.

   b) What treatment is used in step **C**? (1)

   c) What is the purpose of carrying out steps **A** and **B**? (2)

4 Rewrite the following sentences about contraception using the correct answer from each underlined choice. (5)

   a) The use of a cervical cap is a <u>barrier/cyclical</u> method of contraception.

   b) The 'morning after' pill is a <u>physical/chemical</u> means of preventing pregnancy.

   c) An intrauterine device prevents implantation of an embryo in the <u>oviduct/endometrium</u>.

   d) The 'mini pill' contains synthetic <u>oestrogen/progesterone</u>, which works by making cervical mucus thicker.

   e) Sterilisation in a woman involves a procedure called <u>tubal ligation/vasectomy</u>.

12 Ante- and postnatal screening

The health of a pregnant woman and her developing fetus can be monitored using a variety of techniques and tests. Several methods of **antenatal (prenatal) screening** can be employed to identify the risk of the fetus inheriting a genetic disorder or chromosomal abnormality. Further tests can then be carried out if necessary.

Antenatal care

With her consent, the mother's blood pressure is monitored, her blood type identified and general health checks such as blood and urine tests carried out.

Ultrasound imaging

When an **ultrasound scanner** is held against a pregnant woman's abdomen, it picks up high-frequency sounds that have bounced off the fetus. These are converted to an ultrasound image on a computer screen.

Dating scan

Ultrasound imaging is carried out at 8–14 weeks to produce a **dating scan** (see Figures 12.1 and 12.2). This

Figure 12.2 The bad old days before ultrasound scanning

scan is used to determine the stage of the pregnancy (gestational age assessment) and to calculate the date when the baby is due to be born. Dating scans are used in conjunction with biochemical tests for marker chemicals (see below).

Anomaly scan

Further ultrasound imaging is performed at 18–20 weeks to produce an **anomaly scan**. This allows a check to be made for the presence of any serious physical abnormalities in the fetus.

Biochemical tests

A woman's body undergoes many physiological changes during pregnancy. This is the normal course of events. Many of these changes, such as the concentration of human chorionic gonadotrophin (HCG), can be monitored by **biochemical tests**. (The detection of HCG in blood and urine following implantation of an embryo is also the basis of early pregnancy tests.)

At 16–18 weeks, the pregnant woman is offered a series of biochemical tests that check for three **markers**.

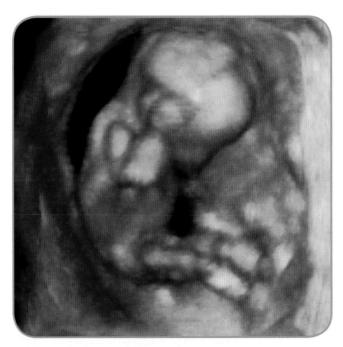

Figure 12.1 Ultrasound image at 14 weeks

(One of these markers is AFP. See Related Topic – Assessing the risk of Down's syndrome using screening tests.) The results of these tests, used in conjunction with the mother's age, enable medical experts to assess the risk of chromosomal abnormalities being present in the fetus.

Normally other routine tests are also carried out to check the health of the pregnant woman and the developing fetus by monitoring altered renal, liver and thyroid functions and other biochemical changes. (See Related Activity – Examining data on altered biochemistry during pre-eclampsia.)

False positives and false negatives

Some medical conditions are indicated by the presence of certain **marker chemicals** in blood and urine. However, these marker chemicals vary during pregnancy. At one stage the presence of a high (or low) concentration of a certain marker may indicate the presence of a genetic disorder in the fetus, whereas at another stage in the pregnancy, a high (or low) level of the marker may be of no significance. For example, in a normal pregnancy, the level of human chorionic gonadotrophin (HCG) increases during weeks 6–10 and then decreases to a steady low level later in gestation, as shown in Figure 12.3. However, it remains at a high level if the fetus has Down's syndrome. Risk assessment based on a result at 10 weeks would be meaningless since both a normal pregnancy and a Down's syndrome one would show an elevated result.

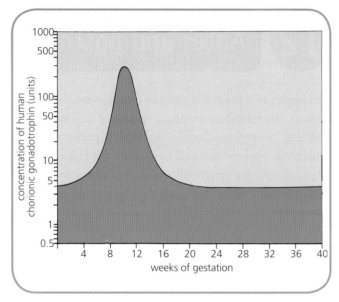

Figure 12.3 HCG levels during normal pregnancy

If a marker chemical was measured at an inappropriate point in the timescale and found to be high, significance could be attributed to it mistakenly. This would lead to a **false positive** result where the test would show the fetus to have the condition when, in fact, it does not have it. Similarly, if the test for the marker was carried out and found to be low at a time when the normal value is also low, this could give a **false negative** result. It would suggest that the fetus does not have the condition when in fact it might really have it (see Appendix 4). It is for this reason that the times chosen for biochemical tests are closely synchronised with information deduced from ultrasound dating scans.

Related Activity

Examining data on altered biochemistry during pre-eclampsia

Pre-eclampsia is a medical condition that affects a minority of pregnant women. It is regarded as the most common cause of several dangerous complications that can arise during pregnancy. A woman suffering pre-eclampsia displays some or all of the following symptoms:

- high blood pressure (hypertension)
- excess protein in blood plasma
- changes to blood biochemistry caused by factors such as altered liver or renal function.

The data in Table 12.1 refer to a series of studies carried out on a large population of women. Some were non-pregnant (NP), some pregnant but not suffering pre-eclampsia (P) and some pregnant and suffering pre-eclampsia (PE).

From the table it is concluded that the concentration of **urea** in blood plasma is significantly **higher** for the PE women than for the NP and P women. Also, the concentration of **calcium** in urine is significantly **lower** for the PE women than for the NP and P women. This latter difference becomes even more apparent when the data are presented as the bar chart shown in Figure 12.4 and drawn to include **error bars** (see Appendix 3).

Group	Urea in blood plasma (mg l⁻¹)	Calcium in urine (mg l⁻¹)
NP	189.3 ± 13.7	163.4 ± 24.8
P	187.0 ± 14.1	177.5 ± 39.1
PE	228.7 ± 20.2	91.8 ± 21.2

Table 12.1 Results of biochemical tests

The differences are thought to be the result of a decrease in renal blood flow and glomerular filtrate rate in PE women causing them to retain higher concentrations of urea and calcium in their blood. Altered levels of these chemicals are just two of many possible indicators of pre-eclampsia. At present there is no cure for the condition and in serious cases the baby may be induced early or be delivered by Caesarean section to avoid the mother's life being put in danger and the baby suffering long-term adverse effects.

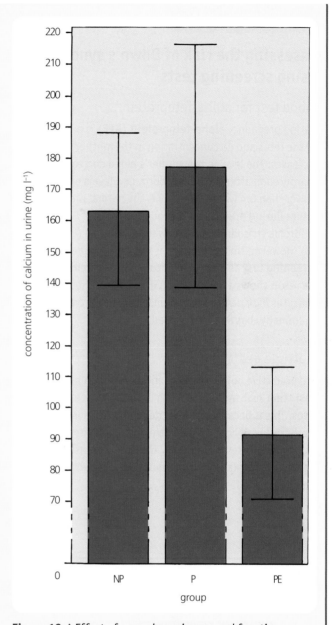

Figure 12.4 Effect of pre-eclampsia on renal function

Diagnostic testing

A **screening test** is one that is used to detect signs and symptoms associated with a certain condition or disorder. If the signs are found, the probability that the individual is suffering the condition can be assessed as a **degree of risk**. (See Related Topic – Assessing the risk of Down's syndrome using screening tests.)

A **diagnostic test**, on the other hand, is a definitive test that produces results that can be used to establish without doubt whether or not the person is suffering a specific condition or disorder. Certain diagnostic tests may be offered to a pregnant woman if:

- evidence of a potential problem has already emerged from the results of earlier routine screening tests
- there is a history of a harmful genetic disorder in her family
- she is already known to belong to a high-risk category (e.g. women over the age of 35). (See Related Topic – Risks associated with Down's syndrome testing.)

Assessing the risk of Down's syndrome using screening tests

Blood test for alpha-fetoprotein

During pregnancy **alpha-fetoprotein (AFP)** is produced by the fetus and its concentration in the mother's blood increases. The level in the mother's blood decreases sharply soon after the baby is born. Low levels of AFP (lower than 0.4 where 0.5–2.49 is the normal range of values during pregnancy) are found in cases of Down's syndrome (trisomy 21) and Edwards syndrome (trisomy 18). However, this test marker is part of a **biochemical screening test** and not a diagnostic one. Therefore, even if a result shows a low level of AFP to be present and indicates *high risk*, it does not mean that a chromosomal abnormality has been **diagnosed**.

Multiple gestation

The predictive power of biochemical tests for a **multiple gestation** such as twins is much lower than for a single fetus. This is because the tests depend on analysis of the mother's blood and may still give normal results even if one of the fetuses is abnormal. The nuchal translucency ultrasound test (see below) is much more reliable since it examines each fetus individually.

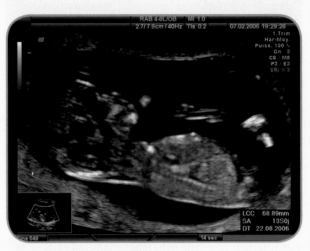

Figure 12.5 Nuchal translucency scan

Nuchal translucency scan (NT)

A **nuchal translucency scan** helps experts to estimate more accurately the risk of a woman having a Down's syndrome baby than by considering her age alone (as shown in Table 12.2 on page 147). The test is carried out at 11–14 weeks (the time found to give the most reliable results). It enables an assessment to be made of the thickness of the fluid in the tissue at the nape of the fetus's neck by viewing it as a **nuchal translucency** (see Figure 12.5). If the nuchal translucency exceeds the normal value, there is a risk of a chromosomal abnormality in the fetus. However, as before, this is not a diagnostic test.

Risks associated with Down's syndrome testing

There are two different aspects of risk associated with Down's syndrome. The first relates to the **mother's age** and the associated risk of her having a baby with this condition. In older women the germline cells that produce eggs are found to be more prone to a type of mutation that leads to eggs being formed that contain an extra copy of chromosome 21. Therefore, the older the woman, the higher the risk that she will have a baby with Down's syndrome, as shown in Table 12.2. When a combination of maternal age, biochemical tests and thickness of nuchal translucency all indicate a high risk of Down's syndrome, the woman may be advised to have an **amniocentesis** or a **chorionic villus sampling** test, both of which are **diagnostic** but invasive.

The second aspect of risk relates to the **tests themselves**. Amniocentesis (carried out at a later stage in gestation) slightly increases the risk of miscarriage, whereas chorionic villus sampling (carried out at an earlier stage in gestation) runs a much higher risk of losing the baby. Therefore when making a choice, the pregnant woman has to weigh up the risk of a miscarriage against the risk of wanting to seek a termination fairly late in the pregnancy.

Maternal age at full term of gestation period (years)	Risk of Down's syndrome
20	1:1450
22	1:1450
24	1:1400
26	1:1300
28	1:1150
30	1:940
32	1:700
34	1:460
36	1:270
38	1:150
40	1:85
42	1:55
44	1:40
46	1:30

Table 12.2 Risk of Down's syndrome with increasing maternal age

Use of karyotype

A person's **karyotype** is a visual display of their complete chromosome complement, with the chromosomes arranged as pairs showing their size, form and number. The two diagnostic tests described below depend on fetal material being obtained to allow a karyotype to be prepared for examination.

Amniocentesis

Amniocentesis is carried out at about 14–16 weeks. It involves the withdrawal of a little amniotic fluid containing fetal cells (see Figure 1.15 on page 12). These are cultured, stained and examined under the microscope. A full chromosome complement is photographed and the chromosomes arranged into pairs to form the karyotype, as shown in Figure 12.6. This technique, which takes about 2 weeks, allows

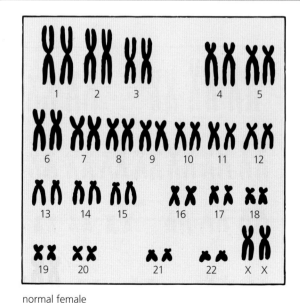

normal female

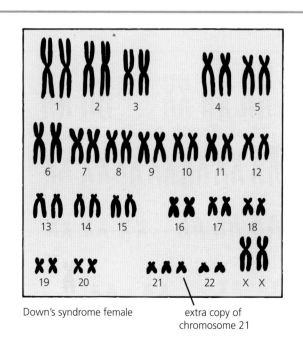

Down's syndrome female   extra copy of chromosome 21

Figure 12.6 Normal and Down's syndrome karyotypes

chromosomal abnormalities to be detected. A karyotype containing an extra copy of chromosome 21, for example, indicates Down's syndrome. The parents may then elect to have the pregnancy terminated.

Chorionic villus sampling

Chorionic villus sampling (CVS) involves taking a tiny sample of placental cells using a fine tube inserted into the mother's reproductive tract (see Figure 12.7). The cells are then cultured and used for karyotyping as before. One benefit of CVS is that it can be carried out as early as 8 weeks into the pregnancy. The prospect of a termination at this stage is much less traumatic for many would-be parents than at 16 or more weeks following amniocentesis. However, CVS causes a higher incidence of miscarriages than amniocentesis.

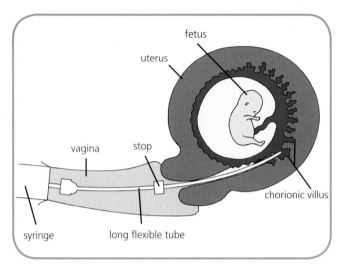

Figure 12.7 Chorionic villus sampling

Karyotypes indicating genetic disorders

Figure 12.6 on page 147 compares the karyotypes of a normal female with a female with **Down's syndrome (trisomy 21)**. This condition, caused by the presence of an extra copy of chromosome 21, is characterised by learning difficulties and distinctive physical features. It occurs in 1 in 800 live births.

Figure 12.8 shows the karyotype of a person with **Turner's syndrome**, caused by the lack of one of the two X chromosomes. It occurs with a frequency of about 1 in 2500 female live births. Individuals affected in this way are always female and short in stature. Since their ovaries do not develop, they are infertile and fail to develop secondary sexual characteristics at puberty.

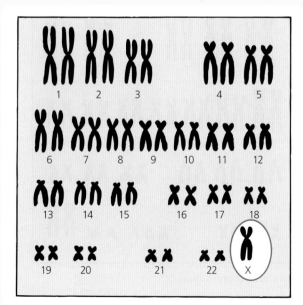

Figure 12.8 Turner's syndrome karyotype

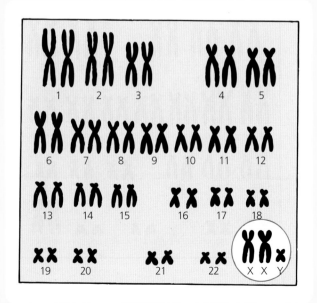

Figure 12.9 Klinefelter's syndrome karyotype

Autosomal dominant inheritance

Figure 12.18 shows a family history of Huntington's disease (chorea).

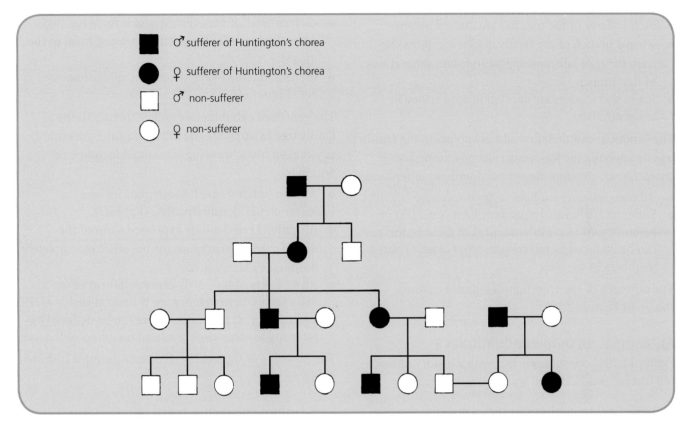

Figure 12.18 Autosomal dominant inheritance

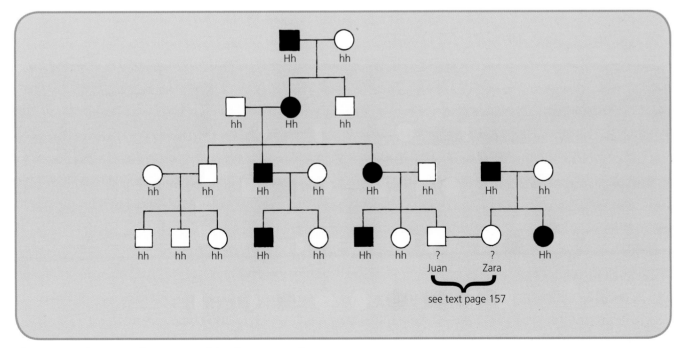

Figure 12.19 Genotypes for autosomal dominant example

The geneticist recognises that such a trait shows a typical **autosomal dominant** pattern of inheritance because:

- the trait appears in every generation
- each sufferer of the trait has an affected parent
- when a branch of the family does not express the trait, the trait fails to reappear in future generations of that branch
- males and females are affected in approximately equal numbers.

The geneticist can therefore add genotypes to the family tree by applying the following rules governing any characteristic showing autosomal dominant inheritance:

- All non-sufferers are homozygous recessive (e.g. hh).
- Sufferers are homozygous dominant (e.g. HH) or heterozygous (e.g. Hh) and most of these genotypes can be deduced by referring to other closely related members of the tree.

The outcome of the Huntington's disease example is shown in Figure 12.19.

Autosomal incomplete dominance

Figure 12.20 shows a family tree with a history of sickle-cell disease and sickle-cell trait (see also page 55).

The geneticist recognises that such a disorder shows a typical **autosomal incompletely dominant** pattern of inheritance because:

- the fully expressed form of the disorder occurs relatively rarely
- the partly expressed form occurs much more frequently
- each sufferer of the fully expressed form has two parents who suffer the partly expressed form of the disorder
- males and females are affected in approximately equal numbers.

The geneticist can therefore add genotypes to the family tree by applying the following rules governing any characteristic showing autosomal incomplete dominance:

- All non-sufferers are homozygous for one incompletely dominant allele (e.g. HH).
- All sufferers of the fully expressed form of the disorder are homozygous for the other incompletely dominant allele (e.g. SS).
- All sufferers of the partly expressed form of the disorder are heterozygous for the two alleles (e.g. HS) and most or all of the genotypes can be deduced by referring to other closely related members of the tree.

The outcome of the sickle-cell disease example is shown in Figure 12.21.

Sex-linked recessive trait

Figure 12.22 shows a family with a history of haemophilia. (Further information about sex-linked inheritance is given in the Related Topic on page 156.)

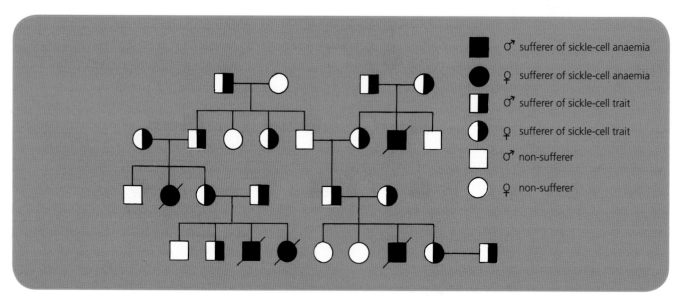

Figure 12.20 Autosomal incomplete dominance inheritance

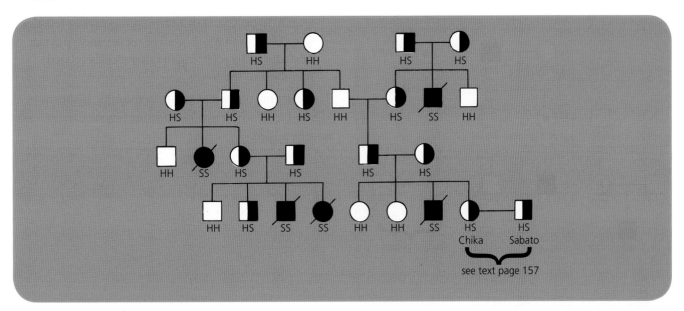

Figure 12.21 Genotypes for autosomal incomplete dominance example

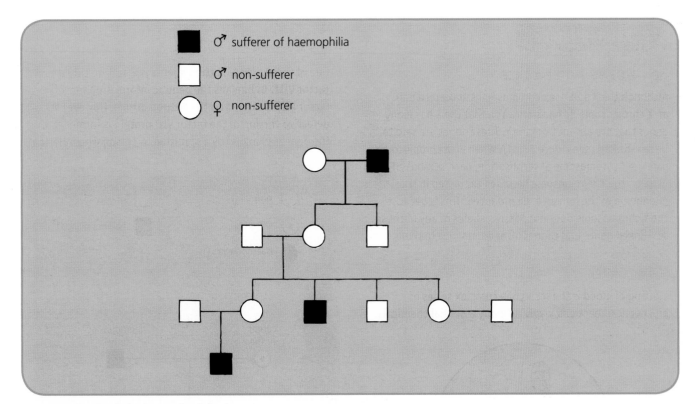

Figure 12.22 Sex-linked recessive inheritance

The geneticist recognises that such a trait shows a typical **sex-linked recessive** pattern of inheritance because:

- many more males are affected than females (if any)
- none of the sons of an affected male shows the trait
- some grandsons of an affected male do show the trait.

The geneticist can therefore add genotypes to the family tree by applying the following rules governing any characteristic showing sex-linked recessive inheritance:

- All sufferers of the trait are 'homozygous' recessive (normally male, e.g. X^hY, and very rarely female X^hX^h).
- Non-sufferers are 'homozygous' dominant

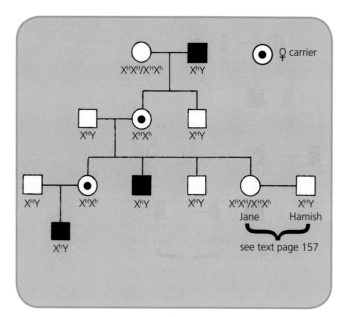

Figure 12.23 Genotypes of sex-linked recessive example

(e.g. X^HY or X^HX^H) or heterozygous carrier females (e.g. X^HX^h) and most or all of these genotypes can be deduced by referring to other closely related members of the tree.

The outcome of the haemophilia example is shown in Figure 12.23, where symbols for the female carriers can now be added.

Related Topic

Sex-linked inheritance

Although the X and Y sex chromosomes make up a pair, an X chromosome differs from a Y in that the X has many genes that are absent from the Y. These genes are said to be **sex-linked** (see Figure 12.24). When an X chromosome meets a Y chromosome at fertilisation, each sex-linked gene on the X chromosome becomes expressed in the phenotype of the human male produced. This is because the Y chromosome does not possess alleles of any of these sex-linked genes and cannot offer dominance to them.

Haemophilia

Clotting of blood is the result of a complex series of biochemical reactions involving many essential chemicals.

One of these blood-clotting agents is a protein called **factor VIII**. In humans the genetic information for factor VIII is carried on the X chromosome. However, a defective version of the factor VIII protein is formed if the gene is changed by a mutation. A person who inherits

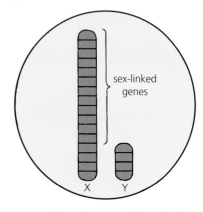

Figure 12.24 Sex-linked genes

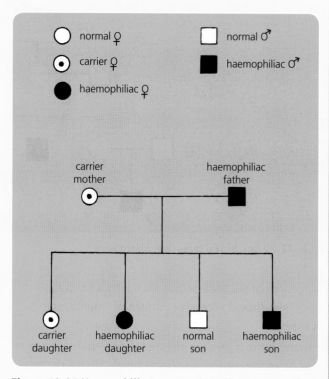

Figure 12.25 Haemophilia A cross using alternative symbols

the altered genetic material suffers a condition called **haemophilia A** (see page 183). The sufferer's blood takes a very long time (or even fails) to clot, resulting in prolonged bleeding from even the tiniest wound. Internal bleeding may occur and continue unchecked, leading to serious consequences.

Since haemophilia A is caused by a recessive allele carried on the X but not the Y chromosome, it is a **sex-linked** condition. The genotypes of individuals in crosses involving haemophilia A are normally represented by the following symbols: X^H (normal blood-clotting allele), X^h (haemophilia) and Y (no allele for this gene). Figure 12.25 shows the outcome of a cross between a carrier woman and a haemophiliac man using standardised human pedigree symbols. Figure 12.26 shows the same cross using symbols where the sex chromosomes are represented by X and Y and the alleles of the sex-linked gene by appropriate superscripts.

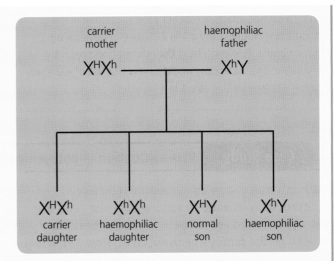

Figure 12.26 Haemophilia A cross using superscript symbols

Related Activity

Assessing the risk

Once the genetic counsellor has constructed the pedigree chart and established as many genotypes as possible, they are in a position to assess risk and state probabilities.

Autosomal recessive

Returning to the **cystic fibrosis** example shown in Figure 12.17, consider the situation that Sandra and Ian find themselves in. Cystic fibrosis is known to exist in Sandra's family but not that of Ian. The genetic counsellor would work out from the family tree that Sandra has a 2 in 3 chance of being a carrier. The counsellor would already know that the frequency among the British population of an individual being heterozygous for the cystic fibrosis allele is 1 in 25, and would therefore regard this as the risk of Ian being a carrier. Combining all these probabilities, they would conclude that the risk of Sandra and Ian having a child with cystic fibrosis is fairly low.

Autosomal dominant

Returning to the **Huntington's disease** example shown in Figure 12.19, consider the situation that Juan and Zara find themselves in. Unlike their siblings, both are still too young to know whether or not they have received the harmful allele from an affected parent. At present it is a 1 in 2 chance that each is heterozygous

for the condition and destined to develop it later in life. From the information available, the genetic counsellor would conclude that there is a high risk that each of their children would suffer this debilitating disease. For example, if both Juan and Zara turn out to be Hh then 3 in 4 of their children on average would suffer Huntington's disease.

Autosomal incompletely dominant

Returning to the **sickle-cell disease** example shown in Figure 12.21, consider the situation that Chika and Sabato find themselves in. The genetic counsellor would explain to them that each of their children would have a 1 in 4 chance of suffering the fully expressed condition of the disease, a 1 in 2 chance of suffering the milder condition and a 1 in 4 chance of being unaffected.

Sex-linked recessive

Returning to the **haemophilia** example shown in Figure 12.23, consider the position that Jane and Hamish find themselves in. Jane is anxious to know if she could pass the trait on to her sons. There is no history of the disorder in Hamish's family. From the information in the family tree, the genetic counsellor would note that Jane's brother and nephew have developed this sex-linked disorder. This shows that Jane's mother and sister are carriers. They would conclude therefore that there is a 1 in 2 chance of Jane being a carrier. If she does turn out to be a carrier, then each of her sons (but none of her daughters) would stand a 1 in 2 chance of developing haemophilia. However, from the information presently available, the counsellor would assess the risk of each son being a haemophiliac as a 1 in 4 chance.

Case Study   Risk evaluation in polygenic (multiple gene) inheritance

When a trait shows an autosomal recessive pattern of inheritance involving only one gene, a straightforward assessment of risk can be made based on the family's pedigree chart. However, when a genetic defect or disorder shows a **polygenic** pattern of inheritance, an evaluation of risk is much more difficult to make with any degree of accuracy.

This is because the condition is determined partly by the effects of **multiple genes** and partly by **environmental** influences.

Risk evaluation in such cases is usually **empirical**. This means that it is based on data derived from many real case histories rather than on genetic theory alone. A few examples are shown in Table 12.3.

Disorder	Incidence (per 100)	Risk of normal parents having a second affected child (per 100)	Risk of affected parent having:	
			an affected child (per 100)	a second affected child (per 100)
Asthma	4.0	10	26	*
Cleft palate	0.1	2	7	15
Congenital heart disease	0.5	1–4	1–4	10
Epilepsy	0.5	5	5	10
Manic-depressive psychosis	0.4	10–15	10–15	*
Schizophrenia	1.0	10	16	*

Table 12.3 Empirical risk for multiple-gene disorders (* = data unavailable)

Testing Your Knowledge 2

1 a) By what means is postnatal screening for phenylketonuria (PKU) carried out? (1)
   b) Why is PKU described as an inborn error of metabolism? (Hint: see chapter 4, page 56.) (2)
   c) What treatment is given to sufferers of PKU? (1)
2 a) Name TWO characteristics of a family's pedigree chart that would enable a geneticist to recognise that it showed a pattern of autosomal recessive inheritance. (2)
   b) Give ONE rule that the geneticist would apply when adding genotypes to a family tree showing a pattern of autosomal recessive inheritance. (1)
3 a) Name TWO characteristics of a family's pedigree chart that would enable a geneticist to recognise that it showed a pattern of sex-linked recessive inheritance. (2)
   b) Give ONE rule that the geneticist would apply when adding genotypes to a family tree showing a pattern of sex-linked recessive inheritance. (1)

What You Should Know Chapters 9–12

amniocentesis	follicle-stimulating	ovaries
anomaly		oviducts
antigens	genetic	ovulation
anti-Rhesus	germline	pedigree
avoidance	implantation	phase
barriers	injection	postnatal
biological	insemination	progesterone
chemical	interstitial	prostate
chorionic	karyotype	seminal
chromosomal	luteinising	sperm
continuous	luteum	sterilisation
cyclical	marker	stimulate
dating	menstrual	testes
diagnostic	miscarriage	testosterone
endometrium	mucus	vascularisation
follicle	negative	vitro
	oestrogen	

Table 12.4 Word bank for chapters 9–12

1 The _____ of the human male produce sperm from germline cells in seminiferous tubules and make testosterone in _____ cells.

2 The mobility and viability of sperm are maintained by fluids secreted by the _____ gland and _____ vesicles.

3 The _____ of the human female contain _____ cells that produce ova (eggs) each surrounded by a protective _____. Hormones made by the ovary are oestrogen and _____.

4 The pituitary gland releases _____ hormone (FSH) and interstitial-cell-stimulating hormone (ICSH)/_____ hormone (LH).

5 In men, FSH stimulates sperm production and ICSH promotes _____ production. The concentration of testosterone is maintained at a steady level by _____ feedback control.

6 In women, FSH stimulates the development of a follicle containing an ovum (egg) and the secretion of _____. LH triggers _____ and brings about the development of the corpus _____ which secretes progesterone.

7 Oestrogen stimulates the proliferation of the _____ and progesterone promotes its further development and _____.

8 The _____ cycle lasts for about 28 days and involves a follicular _____ and a luteal phase.

9 Fertility in men is _____; fertility in women is _____, being restricted to the 1–2 days following ovulation in each monthly cycle.

10 Infertility may be caused by failure to ovulate, blockage of _____ or failure of _____ in women, and low _____ count in men.

11 Methods of treatment of infertility include the use of drugs that _____ ovulation, artificial _____, in-_____ fertilisation (IVF) and intracytoplasmic sperm _____ (ICSI). Pre-implantation _____ diagnosis may be used during IVF to check an embryo for chromosomal defects before implantation.

12 Some methods of contraception are based on _____ knowledge of the menstrual cycle and the _____ of fertile periods. Other physical methods depend on _____, intra-uterine devices or _____. Some _____ methods prevent follicles from being stimulated and eggs from being released. Others cause thickening of cervical _____.

13 During antenatal care, a _____ scan is made by ultrasound imaging to determine the stage the pregnancy has reached. An _____ scan is used to detect physical problems in the fetus.

14 Signs of medical conditions suffered by pregnant women can be detected using screening tests for _____ chemicals. These allow risk of genetic disorders in the fetus to be assessed and may be followed up by _____ tests.

15 A _____ is a display of a complement of chromosomes arranged in pairs to show their form, size and number.

16 During _____, a sample of amniotic fluid is taken to obtain cells for karyotyping to check for _____ abnormalities. During _____ villus sampling, cells for the same purpose are obtained from the placenta. This procedure carries a higher risk of _____ than amniocentesis.

17 A Rhesus-negative mother is given _____ antibodies after the birth of a Rhesus-positive baby to destroy any Rhesus _____ before her immune system has time to respond to them.

18 _____ screening is carried out on newborn babies to check for metabolic disorders such as PKU. Information about a particular characteristic can be collected from the members of a family and be used to construct a _____ chart. Single gene disorders show different patterns of inheritance, such as autosomal recessive.

13 The structure and function of arteries, capillaries and veins

Cardiovascular system

In the human body, substances need to be exchanged continuously between the different structures that make up the body's internal environment. In addition, exchanges must be made between the organism as a whole and the external environment. These requirements are met by the **cardiovascular system**, which contains a fluid connective tissue (**blood**) confined to tubes (**vessels**). The smallest of these vessels transport materials to within rapid diffusion distance of every living cell. A muscular pump (**heart**)

Related Topic

Circulation of blood

Figure 13.1 shows the route taken by blood as it passes from the heart to a region of the body and then back to the heart. **Arterial** branches of the aorta supply **oxygenated** blood to all parts of the body. **Deoxygenated** blood leaves the organs in **veins**. These unite to form the vena cava, which returns blood to the heart.

Pulmonary system

This is the route by which blood is circulated from the heart to the lungs and back to the heart. It should be noted that as a rule arteries carry oxygenated blood and veins carry deoxygenated blood. However, the pulmonary system is exceptional in that the artery carries deoxygenated blood and the vein carries oxygenated blood.

Hepatic portal vein

Whereas veins normally carry blood from a capillary bed in an organ directly back to the heart, the hepatic portal vein is exceptional in that it carries blood from the capillary bed of one organ (the intestine) to the capillary bed of a second organ (the liver). This means that the liver has *three* blood vessels associated with it.

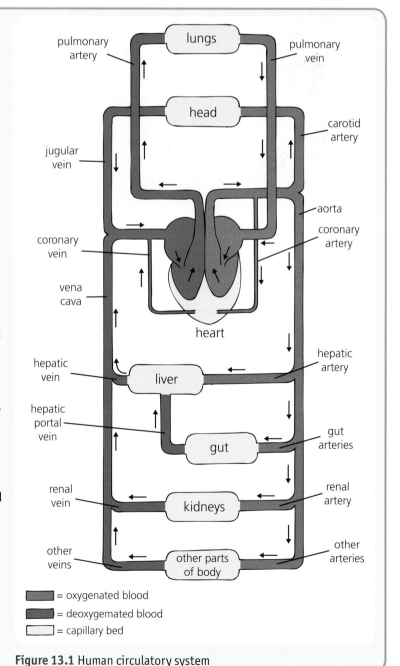

= oxygenated blood

= deoxygemated blood

= capillary bed

Figure 13.1 Human circulatory system

continuously circulates blood round the system. (Remarkably there are approximately 60 000 miles of vessels in a human adult!)

The distribution of blood is under efficient control at all times, allowing the cardiovascular system to work in close harmony with the digestive, respiratory, excretory, locomotor and endocrine systems. (See Related Topic – Circulation of blood.)

Structure and function of blood vessels

The lumen (central cavity) of a blood vessel is lined with a thin layer of epithelial cells called the **endothelium**. The composition of the vessel wall surrounding the endothelium is found to be different in an artery compared with a vein, as shown in Figure 13.2.

Arteries

Arteries carry blood away from the heart. The wall of an artery possesses a thick middle layer composed of smooth muscle and elastic fibres surrounded by an outer layer of connective tissue containing more elastic fibres. The elastic fibres enable the wall of an artery to **pulsate** (stretch and recoil with a rhythmical beat) thereby accommodating the surge of blood received after each heartbeat.

Vasoconstriction and vasodilation

The smooth muscle in the walls of **arterioles** (small arteries) can contract (or become relaxed) depending on the body's requirements. This process allows the changing demands of different tissues to be met by finely tuned adjustments in the local distribution of blood. For example, during strenuous exercise, arterioles leading to working muscles undergo **vasodilation** (see Figure 13.3). This allows an increase in blood flow to the skeletal muscles involved in the strenuous exercise. At the same time the arterioles leading to abdominal organs such as the small intestine undergo **vasoconstriction**, which reduces blood flow to these parts.

Capillaries

Blood is transported from **arterioles** to **venules** (small veins) by passing through a dense network of tiny microscopic vessels called **capillaries** (see Figure 13.4). Capillaries are the most numerous type of blood vessel in the body. They are referred to as the **exchange vessels** since all exchanges of substances between blood and living tissues take place through their thin walls. These are composed of epithelium and are only one cell thick.

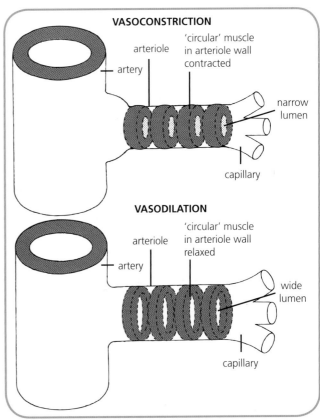

Figure 13.3 Simplified version of vasoconstriction and vasodilation

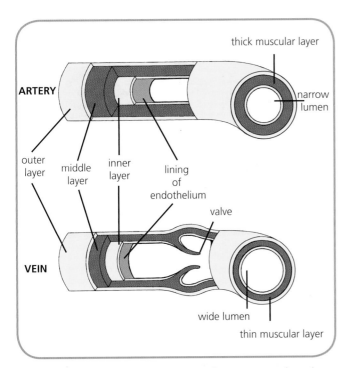

Figure 13.2 Comparison of structure of an artery and a vein

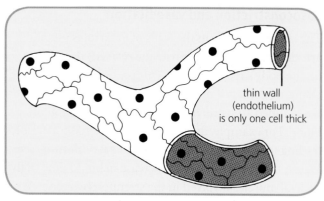

Figure 13.4 Capillary

thin wall
(endothelium)
is only one cell thick

Veins

Veins carry blood back to the heart. The muscular layer and the layers of elastic fibres in the wall of a vein are thinner than those in an artery because blood flows along a vein at low pressure. Compared with an artery, the lumen of a vein is **wider**. This reduces resistance to flow of blood to a minimum. **Valves** are present in veins (but not in arteries) to prevent backflow of blood. (See Related Activity – Demonstrating the presence of valves in veins.)

Related Activity

Demonstrating the presence of valves in veins

The presence of a valve in a vein in the arm can be demonstrated by the method shown in Figure 13.5.

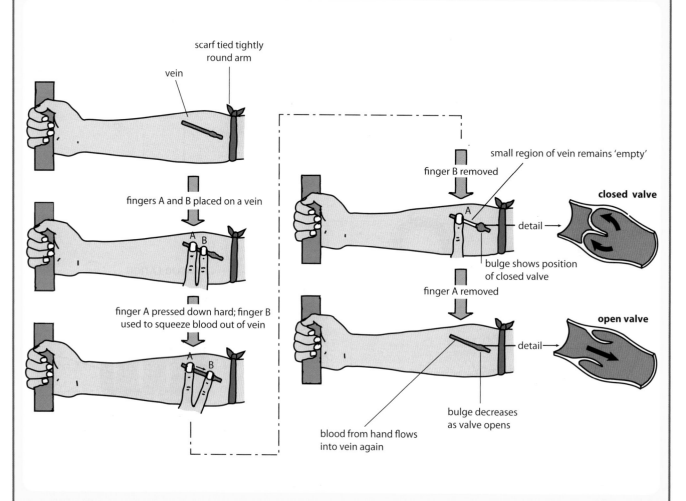

Figure 13.5 Demonstrating the presence of valves in a vein

Related Activity

Measuring the degree of stretching in arteries and veins

Figure 13.6 shows the apparatus set up and ready for use. The rings of artery and vein used in this experiment are cut from the aorta and vena cava of a cow or sheep. The length of a ring of artery with a mass carrier attached to it is measured. This is regarded as the 'original length' for all calculations. A 10 g mass is added to the carrier. The new length of the ring of artery is recorded and the percentage change in length (compared with the original length) calculated. The procedure is continued using additional 10 g masses up to 50 g and then repeated using a ring of vein.

A greater percentage increase in length is obtained for arteries than for veins in response to increasing mass added. This shows that arteries are able to stretch more than veins and it is explained by the fact that arteries contain more elastic fibres in their walls than veins.

When the experiment is extended to measure percentage change in length on *removing* each applied mass, arteries are found to return to their original length more readily than veins, showing that they are capable of more elastic recoil.

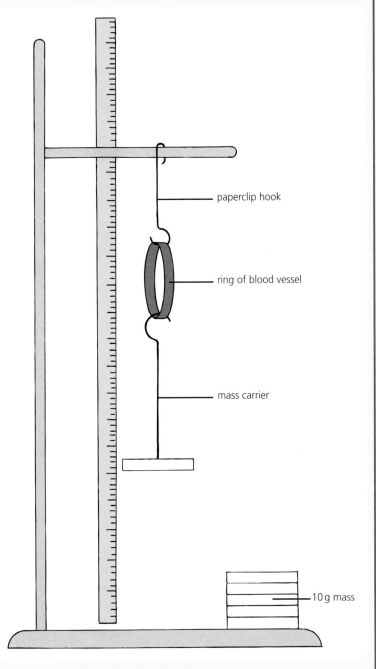

paperclip hook

ring of blood vessel

mass carrier

10 g mass

Figure 13.6 Measuring the stretching of a blood vessel

Tissue fluid

Blood consists of red blood cells, white cells and platelets bathed in plasma. **Plasma** is a watery yellow fluid that contains many dissolved substances such as glucose, amino acids, respiratory gases, plasma proteins and useful ions.

Blood arriving at the arteriole side of a capillary bed (see Figure 13.7) is at a higher pressure than blood in the capillaries. As blood is forced into these narrow exchange vessels, it undergoes a form of **pressure filtration** and much of the plasma (containing small dissolved molecules) is squeezed out through the thin walls. This liquid is called **tissue fluid**. It differs from blood plasma in that it contains little or no protein.

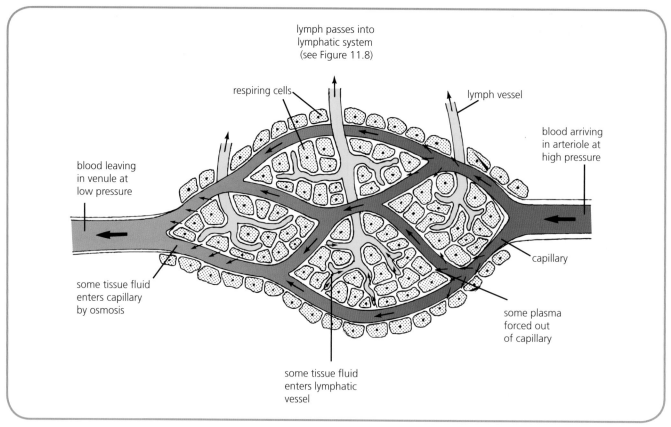

lymph passes into
lymphatic system
(see Figure 11.8)

respiring cells

lymph vessel

blood arriving
in arteriole at
high pressure

blood leaving
in venule at
low pressure

some tissue fluid
enters capillary
by osmosis

capillary

some plasma
forced out
of capillary

some tissue fluid
enters lymphatic
vessel

Figure 13.7 Exchange of materials in a capillary bed

Exchange of materials

The network of capillaries in a capillary bed is so dense that every living cell in the body is located close to a blood capillary and is constantly bathed in tissue fluid. Since tissue fluid contains a high concentration of soluble food molecules, dissolved oxygen and useful ions, these diffuse down a concentration gradient into nearby cells, supplying them with their requirements. At the same time, carbon dioxide and other metabolic wastes diffuse out of the cells into the tissue fluid to be excreted.

Osmotic return of tissue fluid

Much of the tissue fluid returns to the blood in the capillaries at the venule side of the capillary bed. This process is brought about by **osmosis**, with water passing from a region of higher water concentration (tissue fluid lacking plasma proteins) to a region of lower water concentration (blood plasma rich in soluble proteins) down a water concentration gradient. Carbon dioxide and metabolic wastes enter the bloodstream by diffusion.

Lymphatic return of tissue fluid

Some of the tissue fluid does not return to the blood in the capillaries. Instead this excess tissue fluid is absorbed by thin-walled lymphatic vessels, which have blind ends and are located in the connective tissue among the living cells (see Figure 13.7). Once in a lymphatic vessel, the tissue fluid is called **lymph**.

Lymphatic system

Lymph is collected by a vast network of tiny lymph vessels, which unite to form larger vessels, a few of which are shown in Figure 13.8. Flow of lymph through the lymphatic system is brought about mainly by the vessels being **periodically compressed** when muscles contract during breathing, locomotion and other body movements. Backflow of lymph is prevented by the presence of **valves** in the larger vessels. These vessels eventually return their contents via two lymphatic **ducts**, which enter the veins coming from the arms (see Figure 13.8).

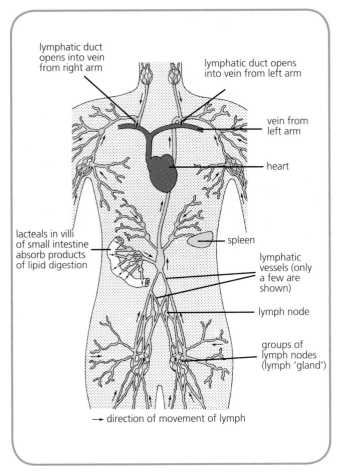

lymphatic duct opens into vein from right arm

lymphatic duct opens into vein from left arm

vein from left arm

heart

lacteals in villi of small intestine absorb products of lipid digestion

spleen

lymphatic vessels (only a few are shown)

lymph node

groups of lymph nodes (lymph 'gland')

→ direction of movement of lymph

Figure 13.8 Lymphatic system

The lymphatic system is regarded as a specialised part of the cardiovascular system because it consists of:

- lymph fluid that is derived from blood
- a system of vessels that returns lymph to the bloodstream.

Case Study — Disorders of the lymphatic system

Oedema

Oedema is the name given to the condition where tissue fluid accumulates in the spaces between cells and blood capillaries, causing tissues to swell up. Oedema can be caused by several factors, including the following.

Blood pressure

High blood pressure can result in tissue fluid being produced at a rate faster than it is drained away by the lymphatic system.

Malnutrition

A prolonged dietary deficiency of protein can result in the plasma protein level in the blood being so low that it is similar in concentration to that of the tissue fluid. Under these circumstances, little or no tissue fluid is returned osmotically to the blood. The lymphatic system is unable to remove the extra volume of fluid, which tends to gather in the abdominal region. An abdomen swollen in this way is a symptom of **kwashiorkor** (see Figure 13.9). This is a severe form of malnutrition suffered by young children (especially infants weaned off breast milk by a mother with a newborn baby to feed).

Parasites

The tiny larvae of one type of parasitic worm are transmitted by mosquitoes. Once inside the body, they invade the lymphatic system. When they mature into adult worms, they take up residence in, and block, lymph vessels, especially those in the legs. This obstruction along with excessive growth of neighbouring tissues in the infected area results in **elephantiasis** (see Figure 13.10), an enormous enlargement of the affected extremity.

some parts of Africa  USA

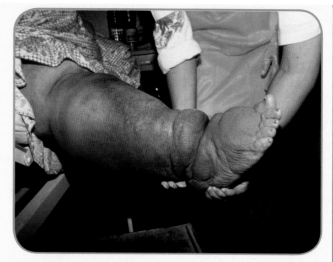

Figure 13.10 Elephantiasis

Figure 13.9 Kwashiorkor

Testing Your Knowledge

1 Decide whether each of the following statements is true or false and then write T or F to indicate your choice. Where a statement is false, give the word that should have been used in place of the word in bold print. (6)
   a) The **carotid** vein returns deoxygenated blood from the head to the vena cava.
   b) The **renal** artery carries blood to the kidneys to be purified.
   c) The pulmonary **vein** carries deoxygenated blood from the heart to the lungs.
   d) The **hepatic portal** vein carries deoxygenated blood from the gut to the liver.
   e) The **coronary** vein carries oxygenated blood from the lungs to the heart.
   f) The hepatic **artery** carries deoxygenated blood from the liver to the vena cava.

2 a) Give TWO structural differences between an artery and a vein. (2)
   b) Give ONE functional difference between an artery and a vein. (1)

3 a) What is *tissue fluid*? (2)
   b) Name a substance that passes from body cells into tissue fluid. (1)
   c) Briefly describe TWO methods by which tissue fluid returns to the bloodstream. (2)

4 a) Briefly describe the means by which lymph in a lymph vessel is forced along through the lymphatic system. (1)
   b) What structures prevent backflow of lymph in the lymphatic system? (1)
   c) Which structures enable lymph to return to the blood circulatory system? (1)

14 Structure and function of the heart

Structure of the heart

The continuous circulation of blood round the body is maintained by a powerful muscular pump, the **heart**. This organ is divided into four chambers, two **atria** and two **ventricles** (see Figure 14.1). The right atrium receives deoxygenated blood from all parts of the body via two main veins called the **venae cavae**. This deoxygenated blood passes into the right ventricle and then leaves the heart by the **pulmonary artery** which divides into two branches each leading to a lung.

Following oxygenation in the lungs, blood returns to the heart by the **pulmonary veins** and enters the left atrium. It flows from the left atrium into the left ventricle and leaves the heart by the **aorta**, the largest artery in the body.

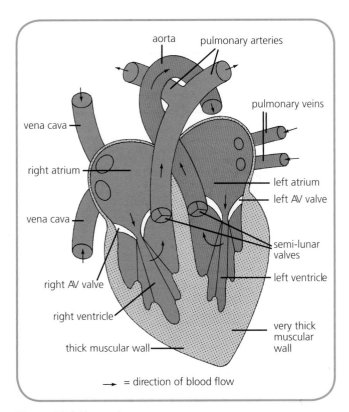

Figure 14.1 Human heart

Thickness of ventricle walls

The wall of the left ventricle is particularly thick and muscular since it is required to pump blood all round the body. The wall of the right ventricle is less thick since it only pumps blood to the lungs.

Valves

Figure 14.1 shows the four heart valves. Two of these, situated between the atria and the ventricles, are called the **atrio-ventricular** (**AV**) valves. They allow blood to flow from atria to ventricles but prevent backflow from ventricles to atria.

The other two heart valves, situated at the origins of the pulmonary artery and aorta, are called **semi-lunar** (**SL**) valves. These valves open during ventricular contraction allowing blood to flow into the arteries. When arterial pressure exceeds ventricular pressure, they close, preventing backflow. The presence of the valves ensures that blood is only able to flow in **one direction** through the heart.

Cardiac function

At each contraction of the heart, the right ventricle pumps the same volume of blood through the pulmonary artery (and round to the lungs) as the left ventricle pumps through the aorta (and round the body).

Heart rate (pulse) is the number of heartbeats that occurs per minute. (See Related Activity – Measuring pulse rate using a pulsometer.)

Stroke volume is the volume of blood expelled by each ventricle on contraction. The stronger the contraction, the greater the stroke volume.

Cardiac output is the volume of blood pumped out of a ventricle per minute. Thus cardiac output (CO) = heart rate (HR) × stroke volume (SV).

Table 14.1 shows an example of the effect of exercise on cardiac output for a human adult.

State of body	Heart rate (beats/min)	Stroke volume (ml)	Cardiac output by each ventricle (l/min)
At rest	60	60	3.6
During exercise	120	70	8.4
During strenuous exercise	180	80	14.4

Table 14.1 Effect of exercise on cardiac output

Related Activity

Measuring pulse rate using a pulsometer

Pulse rate can be measured by using a **pulsometer**, as shown in Figure 14.2. **Resting pulse rate** is a measure of pulse rate when the body is at rest and has not been exercising for some time. On average, resting pulse rate for men is approximately 75 beats/minute and slightly higher for women, although any value between 60 and 90 is regarded as being within the normal range.

Pulse as a health indicator

If a person is fit, the relative quantity of cardiac muscle present in their heart wall is greater and more efficient than that of an unfit person. A very fit person tends to have a **lower pulse rate** than an unfit person because the fit person's heart is larger and stronger. Therefore it does not need to contract as often to pump an equal volume of blood round the body. (In other words, the stroke volume is greater.)

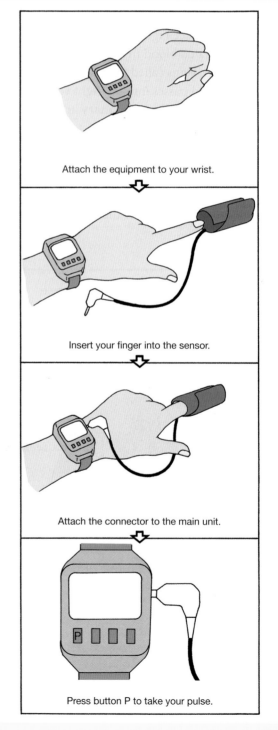

Attach the equipment to your wrist.

Insert your finger into the sensor.

Attach the connector to the main unit.

Press button P to take your pulse.

Figure 14.2 Using a pulsometer

Cardiac cycle

The term **cardiac cycle** refers to the pattern of contraction (**systole**) and relaxation (**diastole**) shown by the heart during one complete heartbeat. On average, the length of one cardiac cycle is 0.8 seconds, as shown in Figure 14.3 which is based on a heart rate of 75 beats per minute.

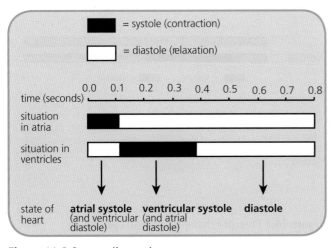

Figure 14.3 One cardiac cycle

During **diastole**, the return of blood via the venae cavae and pulmonary veins to the atria causes the volume of blood in the atria to increase. Eventually atrial pressure exceeds that in the ventricles, the AV valves are pushed open and blood starts to enter the ventricles.

During **atrial systole** (see Figure 14.4) the two atria contract simultaneously and send the remainder of the blood down into the ventricles through the open AV valves. The ventricles (still in the relaxed state of ventricular diastole) fill up with blood and the SL valves remain closed.

Atrial systole is followed about 0.1 seconds later by **ventricular systole**. This stage involves the contraction of the ventricles and the closure of the AV valves. The pressure exerted on the blood in the ventricles (as the cardiac muscle contracts) soon exceeds the blood pressure in the arteries. The SL valves are pushed open and blood is pumped out of the heart and into the aorta and pulmonary arteries.

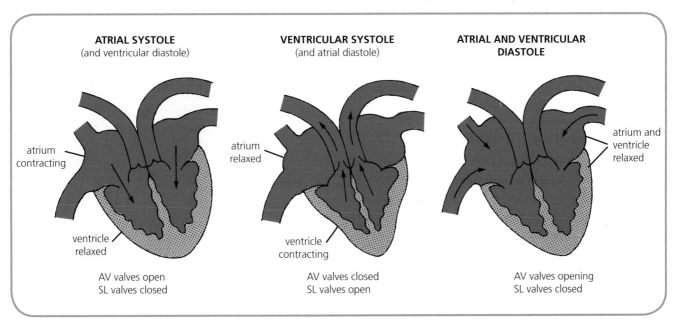

Figure 14.4 Systole and diastole

Valves and heart sounds

Figure 14.5 refers to some of the changes that occur during the cardiac cycle. At point W in the graph (which refers to the left side of the heart only), ventricular pressure exceeds atrial pressure forcing the AV valve to close. This produces the first heart sound ('**lubb**') which can be heard using a **stethoscope**. It can also be detected as a pattern shown on a **phonocardiogram** (see Figure 14.5).

At point X, ventricular pressure exceeds aortic pressure forcing open the SL valve. At point Y, ventricular pressure falls below aortic pressure causing the SL valve to close. This produces the second heart sound ('**dupp**') heard through a stethoscope. At point Z, ventricular pressure falls below atrial pressure and the AV valve opens.

The heart sound 'lubb' is heard at the start of ventricular systole and 'dupp' at the start of ventricular diastole. Abnormal heart sounds produced by abnormal patterns of cardiac blood flow are called **heart murmurs**. These are often caused by defective valves, which fail to open or close fully. This type of condition can be inherited or result from diseases such as rheumatic fever.

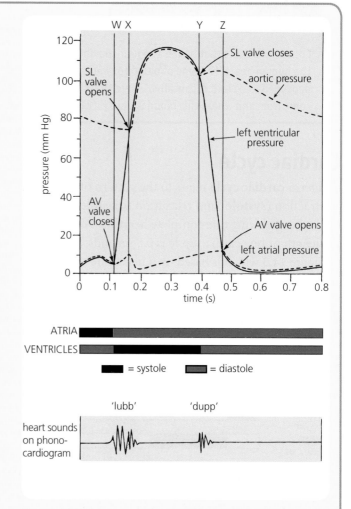

Figure 14.5 Pressure changes and heart sounds

During **diastole** the higher pressure of blood in the arteries closes the SL valves again and the next cardiac cycle begins. The closing of the AV and SL valves are responsible for the sounds that can be heard during each cardiac cycle. (See Related Topic – Valves and heart sounds.)

Cardiac conducting system

The sequence of events that occurs during each heartbeat is brought about by the activity of the **pacemaker** and the **conducting system** of the heart (see Figure 14.6). The pacemaker, also known as the **sino-atrial node (SAN)**, is located in the wall of the right atrium. It is a small region of specialised tissue (composed of autorhythmic cells) that exhibits **spontaneous excitation**. This means that the pacemaker initiates electrical impulses that make cardiac muscle cells contract at a certain rate. This rate can be regulated

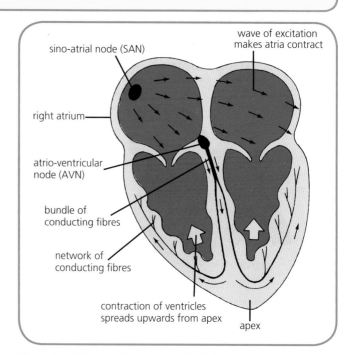

Figure 14.6 Conducting system of the heart

by other factors to suit the body's requirements (see below). The pacemaker works automatically and would continue to function even in the absence of nerve connections from the rest of the body.

A wave of excitation originating in the SAN spreads through the muscle cells in the wall of the two atria making them contract simultaneously (**atrial systole**). The impulse is then picked up by the **atrio-ventricular node (AVN)** located centrally near the base of the atria.

The impulse passes from the AVN into a bundle of **conducting fibres,** which divides into left and right branches. Each of these branches is continuous with a dense network of tiny conducting fibres in the ventricular walls. Stimulation of these fibres causes simultaneous contraction of the two ventricles (**ventricular systole**) starting from the heart apex and spreading upwards.

Such coordination of heartbeat ensures that each type of systole involves the combined effect of many muscle cells contracting and that ventricular systole occurs slightly later than atrial systole, allowing time for the ventricles to fill completely before they contract.

Regulation

The pacemaker tissue alone initiates each heartbeat. However, heart rate is not fixed as it is altered by **nervous** and **hormonal** activity.

Autonomic nervous control

The heart is supplied with branches of the opposing parts of the **autonomic nervous system** (see page 207). Control centres located in the medulla of the brain regulate heart rate (see Figure 14.7). The **cardio-accelerator centre** sends its nerve impulses via a sympathetic nerve to the heart; the **cardio-inhibitor centre** sends its information via a parasympathetic nerve.

The two pathways are **antagonistic** to one another in that they have opposite effects on heart rate. An increase in the relative number of nerve impulses arriving at the SAN (pacemaker) via the **sympathetic** nerve results in an **increase** in heart rate. An increase in the relative number of impulses arriving at the SAN via the **parasympathetic** nerve results in a **decrease** in heart rate. The actual rate at which the heart beats is determined by which system exerts the greater influence over the heart at any given moment.

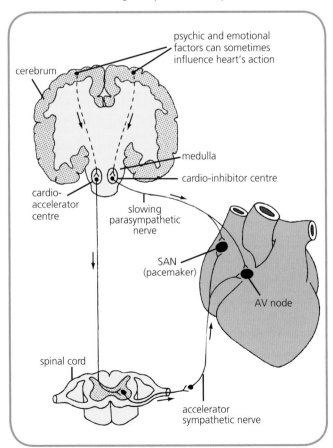

Figure 14.7 Autonomic nervous control of heart rate

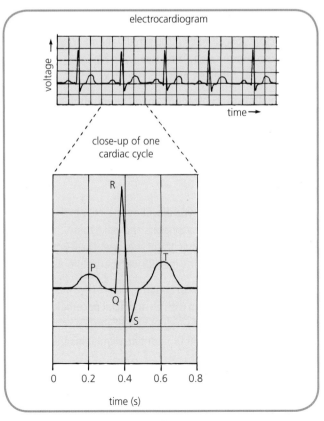

Figure 14.8 Normal electrocardiogram

Neurotransmitter substances released by these nerves influence the SAN. Sympathetic accelerator nerves release the neurotransmitter **norepinephrine** (**noradrenaline**), whereas slowing parasympathetic nerves release **acetylcholine** (see page 256).

Hormonal control

Under certain circumstances (such as exercise or stress), the sympathetic nervous system acts on the adrenal glands, making them release the hormone **epinephrine** (**adrenaline**) into the bloodstream. On reaching the SAN, this hormone makes the pacemaker generate cardiac impulses at a higher rate and bring about an **increase** in heart rate.

Electrocardiogram

The electrical activity of the heart generates tiny currents that can be picked up by electrodes placed on the skin surface. The electrical signals, once amplified and displayed on an oscilloscope screen, produce a pattern called an **electrocardiogram** (**ECG**).

The normal ECG pattern is shown in Figure 14.8. It consists of three distinct waves normally referred to as P, QRS and T. The **P wave** corresponds to the wave of electrical excitation spreading over the atria from the SAN. The **QRS complex** represents the wave of excitation passing through the ventricles. The **T wave** corresponds to the electrical recovery of the ventricles occurring towards the end of ventricular systole.

Related Activity

Examining abnormal ECGs

Abnormal heart rhythms and some forms of heart disease can be detected and diagnosed using **ECGs** because these produce unusual but identifiable patterns. Some examples are shown in Figure 14.9.

When extremely rapid rates of electrical excitation occur, these lead to an increase in rate of contraction of either atria or ventricles. In an **atrial flutter**, for example, the contractions occur much more rapidly than normal but do remain coordinated. The example in Figure 14.9 shows three P waves for every one QRS complex.

In a **fibrillation**, contractions of different groups of heart muscle cells occur at different times, making it impossible for coordinated pumping of the heart chambers to take place. Ventricular fibrillation, for example, produces an ECG with an irregular pattern. This condition is lethal if it is not corrected.

During **ventricular tachycardia**, abnormal cells in the ventricle walls act like pacemakers and make these chambers beat rapidly and independently of the atria. The P (atrial) waves are absent and the wide QRS waves are abnormal.

Relief for some sufferers of abnormal heart rhythms can be provided by fitting them with an **artificial pacemaker**. This acts as a stimulator and sends out small electric impulses to the heart, making it beat in a normal, regular manner.

Emergency

CPR (cardiopulmonary resuscitation) is an emergency procedure involving chest compressions administered to

atrial flutter

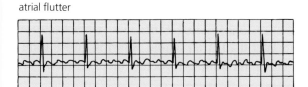

ventricular fibrillation

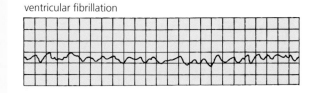

ventricular tachycardia

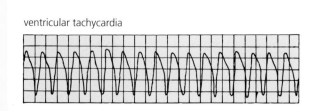

Figure 14.9 Abnormal ECGs

a person who has suffered a cardiac arrest. If it is followed soon after by **defibrillation** (the administration of an electric shock to the subject's heart by trained staff), the person's chance of survival is increased by up to 30%. Defibrillation is only effective for certain abnormal heart rhythms such as fibrillation and ventricular tachycardia.

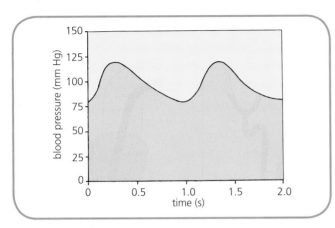

Figure 14.10 Blood pressure trace

Blood pressure

Blood pressure is the force exerted by blood against the walls of the blood vessels. It is measured in millimetres of mercury (mm Hg) as described below.

Blood pressure is generated by the contraction of the ventricles and it is therefore highest in the large elastic arteries (aorta and pulmonary artery). As the heart goes through systole and diastole during each cardiac cycle, the **arterial pressure rises and falls**. For example, during ventricular **systole**, the pressure of blood in the aorta rises to a maximum (for example, 120 mm Hg); during ventricular **diastole**, it drops to a minimum (for example, 80 mm Hg). Figure 14.10 shows the blood pressure trace for a normal 18-year-old at rest.

Decreasing blood pressure during circulation

Although the pumping action of the heart causes fluctuations in aortic blood pressure, the average pressure in the aorta remains fairly constant. Figure 14.11 shows how a progressive **decrease in pressure** occurs as blood travels round the circulatory system dropping to almost zero by the time it reaches the right atrium again.

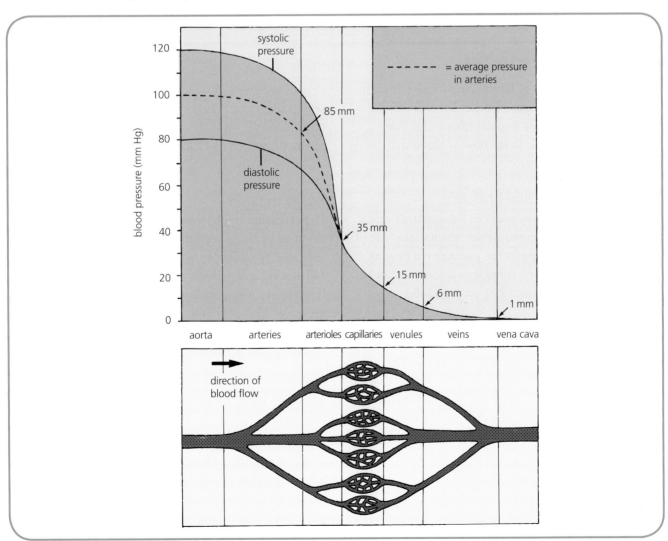

Figure 14.11 Decrease in blood pressure

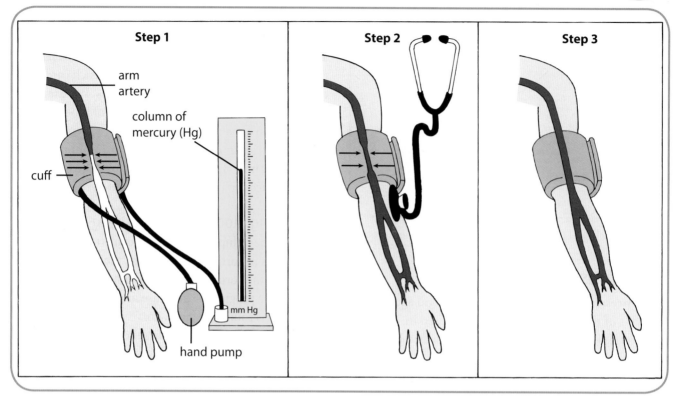

Figure 14.12 Measuring blood pressure

As blood flows through a narrow blood vessel, friction occurs between the blood and the vessel wall making the wall resist blood flow to some extent. It is this resistance by vessel walls to the flow of blood that causes the decrease in its pressure.

Measurement of blood pressure

Systolic and diastolic pressures can be measured using an inflatable **sphygmomanometer**, which makes use of a column of mercury to give the pressure readings as shown in Figure 14.12. (A sphygmomanometer can be digital, as shown in Figure 19.21 on page 268.)

- In step 1, the cuff is inflated until the pressure that it exerts stops blood flowing through the arm artery.
- In step 2, the cuff is allowed to deflate gradually until the pressure of blood in the artery exceeds the pressure in the cuff. A pulse can now be felt and blood can be heard to spurt through the arm artery again using a stethoscope. The pressure at which this first occurs is a measure of **systolic pressure**.
- In step 3, more air is released from the cuff until the sound of spurting blood disappears and a pulse is no longer detected. The pressure at which this first occurs is a measure of **diastolic pressure**.

Blood pressure is found to vary considerably from person to person. A typical set of values for a healthy young adult would be a systolic pressure of 120 mm Hg and a diastolic pressure of 70 mm Hg. These values are normally written as 120/70 mm Hg and referred to as '120 over 70'.

Hypertension

Hypertension (high blood pressure) is the prolonged elevation of blood pressure when at rest. It is normally indicated by values of systolic pressure greater than 140 mm Hg and diastolic pressure greater than 90 mm Hg. It is rare in young people but fairly common in adults over the age of 35. It is a major risk factor for many diseases that have a relatively high incidence later in life such as **coronary heart disease** and **strokes**.

Hypertension is commonly found in people with an unhealthy lifestyle that includes some or all of the following:

- being overweight
- not taking enough exercise
- eating a diet excessively rich in fatty food especially animal fat
- consuming too much salt
- drinking alcohol to excess regularly
- being under continuous stress.

Testing Your Knowledge 2

1 a) Distinguish between the terms *systole* and *diastole*. (2)

   b) Construct a table to compare atrial systole and ventricular systole with reference to state of atrial wall, state of ventricular wall, state of AV valves and state of SL valves. (4)

2 a) i) By what other name is the heart's pacemaker known?

      ii) Briefly describe the function performed by the pacemaker. (2)

   b) i) What heart structure is represented by the letters AVN?

      ii) This structure passes impulses on to the conducting fibres. In which region of the heart are these fibres located?

      iii) Which stage of the cardiac cycle occurs as a direct result of the conducting fibres passing on the impulses? (3)

3 a) What is an *electrocardiogram*? (1)

   b) i) Of how many waves does a normal ECG consist?

      ii) How many of these represent waves of electrical excitation affecting regions of the heart? (2)

4 a) i) Is the pressure of blood in the aorta at its maximum during ventricular systole or ventricular diastole?

      ii) Explain your answer. (2)

   b) What name is given to an instrument used to measure blood pressure? (1)

   c) Give ONE reason why prolonged hypertension is dangerous. (1)

What You Should Know Chapters 13–14

aorta	endothelium	semi-lunar
arterioles	fibres	sino-atrial
atria	heartbeats	sounds
atrio-ventricular	high	sphygmomano-meter
autorhythmic	hormonal	
backflow	hypertension	stretch
capillary	low	stroke
cardiovascular	lumen	systole
chambers	lymph	tissue fluid
circulatory	nerves	valves
closing	osmosis	vasoconstriction
contraction	output	veins
decrease	protein	venae cavae
diastole	pulmonary arteries	ventricles
elastic		
electro-cardiogram	pulmonary veins	
	pulse	

Table 14.1 Word bank for chapters 13–14

1 The _____ is the inner cellular layer of a blood vessel's wall that lines the central cavity (_____).

2 Arteries carry blood away from the heart at _____ pressure and their walls are thicker, more muscular and more _____ than those of _____ which carry blood back to the heart at _____ pressure.

3 The elasticity of arterial walls enables them to _____ and recoil in response to the surge of blood that arrives after each _____ of the heart. Veins have _____ to prevent backflow of blood.

4 Flow of blood to particular body parts can be controlled by _____ and vasodilation of _____.

5 When blood is forced through a _____ bed, some plasma passes out through the vessel walls. This liquid, which bathes the cells, is called _____. It differs from plasma in that it contains little or no plasma _____.

6 Some tissue fluid returns to blood capillaries by _____; the remainder is absorbed by tiny lymphatic vessels and becomes _____.

7 The heart has two upper _____ called atria and two lower chambers called _____. Deoxygenated blood returns to the heart from the body by the _____; it is pumped by the heart to the lungs via the _____. Oxygenated blood returns to the heart from the lungs by the _____; it is pumped by the heart to the body via the _____.

8 The atrio-ventricular (AV) valves in the heart prevent _____ of blood from the ventricles to the _____. The _____ (SL) valves prevent backflow from the large arteries to the ventricles.

9 Heart rate (_____) is the number of _____ that occurs per minute. _____ volume is the volume of blood expelled by each ventricle on contraction. Cardiac _____ is the volume of blood pumped out of a ventricle per minute.

10 A cardiac cycle consists of a period of contraction called _____ and a period of relaxation called _____. During a cardiac cycle two separate heart _____ can be heard; each indicates the _____ of a set of valves.

11 Heartbeat is initiated in the heart itself by the _____ cells of the _____ node (pacemaker) which set it at a certain rate. This rate of heartbeat is then regulated by autonomic _____ and _____ control.

12 Impulses from the SAN spread through the atria and are picked up by the _____ node and passed via conducting _____ to the ventricular walls which respond by contracting.

13 The electrical activity of the heart can be displayed on a screen as an _____.

14 Blood pressure shows a progressive _____ as blood travels round the _____ system.

15 Arterial blood pressure can be measured using a _____. High blood pressure (_____) is a major risk factor for _____ disease.

15 Pathology of cardiovascular disease

Pathology of cardiovascular disease (CVD)

In the UK, **cardiovascular diseases** are responsible for a high proportion of deaths annually, as shown in Figure 15.1.

Atherosclerosis

Atherosclerosis is the formation of plaques called **atheromas** beneath the inner lining (endothelium) in the wall of an artery. Initially plaques are composed largely of fatty material (mainly **cholesterol** – see page 185), but as the years go by they become enlarged by the addition of fibrous material, calcium and more cholesterol (see Figure 15.2).

The presence of these larger atheromas leads to:
- a significant reduction in the diameter of the affected artery's lumen
- the restriction of blood flow to the capillary bed served by that artery
- an increase in blood pressure.

In addition, large plaques hardened by deposits of calcium cause arterial walls to become thicker and lose their elasticity. This process, which occurs as a direct result of atherosclerosis, is often called **hardening of the arteries**. Symptoms of atherosclerosis normally remain absent until later in life when problems can arise. Then the condition can lead to the development of various cardiovascular diseases such as **coronary heart disease** (including angina), **strokes** and **heart attacks** (myocardial infarctions). Atherosclerosis is also the root cause of peripheral vascular disease (see page 183).

Women

- injuries and poisoning (3%)
- respiratory disease (15%)
- other cancers (14%)
- breast cancer (4%)
- lung cancer (5%)
- bowel cancer (2%)
- diabetes (1%)
- other CVD including heart attacks (12%)
- coronary heart disease including angina (13%)
- stroke (8%)
- all other causes (23%)

Men

- injuries and poisoning (5%)
- respiratory disease (13%)
- other cancers (20%)
- lung cancer (7%)
- bowel cancer (3%)
- diabetes (1%)
- other CVD including heart attacks (8%)
- coronary heart disease including angina (18%)
- stroke (7%)
- all other causes (18%)

Figure 15.1 Causes of death in the UK

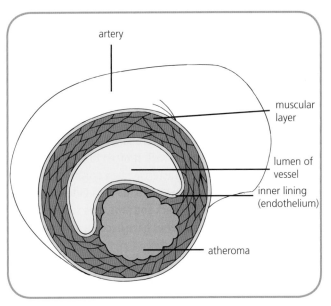

Figure 15.2 Atheroma in an artery

Coronary heart disease

Coronary arteries

The first two branches of the aorta are the left and right **coronary arteries** (see Figure 15.3). These vessels spread out over the surface of the heart and divide into an enormous number of tiny branches leading to a dense network of capillaries among the cardiac muscle cells that make up the wall of the heart.

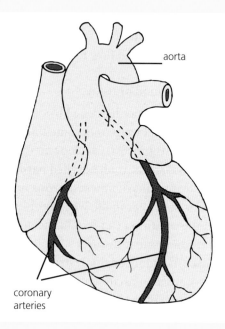

Figure 15.3 Coronary arteries

Each cardiac muscle cell is within 10 μm of a capillary, compared with the average distance of 60–80 μm in other organs. This close proximity to exchange vessels allows very rapid diffusion of oxygen and food into the actively respiring cardiac muscle cells.

Coronary heart disease (CHD)

This general term refers to any disease that results in **restriction** or **blockage** of the coronary blood supply to part of the heart's muscular wall. It often takes the form of **angina**, a condition characterised by a crushing pain in the centre of the chest, which tends to radiate out into the left arm and up to the neck and jaws. It is suffered by people whose coronary arteries have become narrowed by atherosclerosis. Coronary arteries obstructed to such an extent that their diameter is reduced by about 70% allow sufficient blood to flow to the cardiac muscle only when the person is at rest.

However, during exercise or stress, the heart beats faster and the demand for oxygen by cardiac muscle cells increases accordingly. This demand cannot be met because of the reduced blood flow through the narrowed coronary arteries. Therefore a sudden pain occurs in the chest, often accompanied by feelings of suffocation.

Examining league tables

CHD is the most common cause of premature death in many developed countries. Table 15.1 contains data from a European survey. It compares death rates from CHD per 100 000 population in men (all ages) for several European countries for two different years.

From the data it can be seen that the countries occupying the six positions at the 'undesirable' top of the 'league table' have remained unchanged over the 8-year period but that all of these countries have shown a decrease in CHD deaths except Hungary, which remains the undisputed leader.

The data also show that the four positions at the 'desirable' bottom end of the table have also remained unchanged and that all four countries have shown a decrease in CHD deaths. The four countries in positions 7–10 in the league have also shown an improvement over the 8-year period and the UK has moved to a slightly better position (though its new position still leaves plenty of room for improvement).

Table 15.2 contains data collected in the UK. It compares death rates from CHD per 100 000 population from the four parts of the UK over a 10-year period. From the data it can be concluded that the number of deaths caused by CHD is decreasing in all parts of the UK with time. It can also be seen that death rate per 100 000 population is always higher for men than for women in any year in any part of the UK. In addition, the **death rate in Scotland** for both men and women is always higher than that in any other part of the UK in any given year.

Year			
2000		**2008**	
Country	Deaths per 100 000 population	Country	Deaths per 100 000 population
Hungary	302	Hungary	303
Czech Republic	256	Czech Republic	239
Finland	255	Finland	193
Bulgaria	247	Bulgaria	190
Ireland	234	Ireland	155
Poland	205	Poland	151
UK	200	Austria	142
Austria	183	Iceland	138
Sweden	176	UK	132
Iceland	166	Sweden	130
Norway	164	Norway	103
Switzerland	129	Switzerland	93
Netherlands	125	Netherlands	73
France	76	France	55

Table 15.1 Deaths rates from coronary heart disease for European men

		Year									
		1999	2000	2001	2002	2003	2004	2005	2006	2007	2008
Women aged 35–74	UK mean	39	36	35	32	30	27	24	22	21	20
	England	37	34	32	30	28	25	23	21	19	19
	Wales	41	40	39	36	33	32	27	25	23	21
	Scotland	54	50	45	41	41	37	35	33	30	28
	N. Ireland	46	41	38	36	30	30	28	27	23	22
Men aged 35–74	UK mean	112	104	98	92	87	80	74	69	65	61
	England	108	100	94	88	83	77	71	66	62	59
	Wales	128	113	109	104	95	83	82	75	69	65
	Scotland	146	133	120	113	112	101	98	88	89	81
	N. Ireland	133	115	105	99	90	89	80	76	74	65

Table 15.2 Deaths from coronary heart disease for British men and women per 100 000 population

Clotting of blood

Blood clotting is a protective device triggered by damage to cells. Normally it occurs to prevent loss of blood at a wound. The presence of damaged cells leads to the release of blood **clotting factors** that activate the cascade of reactions shown in Figure 15.4. The enzyme **prothrombin**, which is always present in blood plasma but inactive, now becomes converted to its active form called **thrombin**. Thrombin promotes the conversion of molecules of **fibrinogen** (a soluble plasma protein) into threads of **fibrin** (an insoluble protein). These fibrin threads become interwoven into a framework to which platelets adhere, forming a **blood clot**. By this means the wound is sealed and a scaffold is produced upon which scar tissue can be formed.

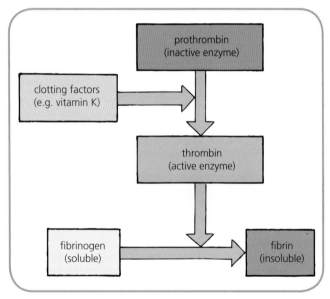

Figure 15.4 Chemical reactions resulting in fibrin formation

Thrombosis

Atheromas on the inside lining of an artery make the surface uneven and disturb the smooth flow of blood. As an atheroma gradually becomes enlarged it may eventually burst through the endothelium and damage it (see Figure 15.5). Under these circumstances, **thrombosis** may occur. Thrombosis is the formation of a blood clot (**thrombus**) in a vessel.

Embolus

The presence of a thrombus in an artery causes further blockage in addition to that caused by atheromas. If a thrombus breaks loose, it is known as an **embolus**.

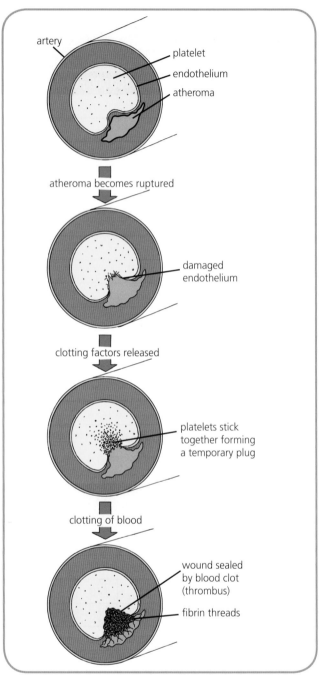

Figure 15.5 Formation of a thrombus in a blood vessel

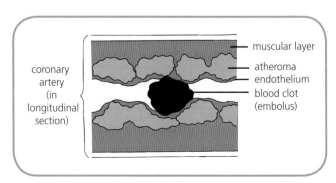

Figure 15.6 Coronary thrombosis

An embolus is carried along by the blood until it blocks some narrow vessel and causes blood flow to be severely restricted or even brought to a complete halt. Blockage of a coronary artery by this type of thrombus is called **coronary thrombosis** (see Figure 15.6). It deprives part of the heart muscle of oxygen and may lead to a **myocardial infarction** (heart attack). A thrombus that causes a blockage in an artery in the brain may lead to a **stroke**. This normally results in the death of some of the tissues served by that artery because they are deprived of oxygen.

Research Topic | **Use of thrombolytic medications**

Background

The main constituent of a thrombus is fibrin. Under normal circumstances, once a blood clot has served its purpose, the clot is broken down and removed. For this to happen, **plasminogen**, the inactive form of the necessary enzyme, must first be converted to **plasmin**, the active form. This conversion is brought about by a further enzyme called **tissue plasminogen activator (tPA)** present on the endothelial cells lining blood vessels. The series of reactions involved is shown in Figure 15.7.

Thrombolysis

The formation of a blood clot in a blood vessel and its subsequent movement through the circulatory system as an embolus that finally blocks a narrow blood vessel is the root cause of several serious conditions such as myocardial infarction, stroke and deep vein thrombosis (see page 183). **Thrombolysis** is the process by which such a clot is broken down using a special medication in order to limit the damage caused by the blockage. Examples of thrombolytic drugs are **streptokinase** (originally extracted from bacteria) and **tissue plasminogen activator** (produced by recombinant DNA technology).

Both of the above drugs work by converting inactive plasminogen to active plasmin and they are given

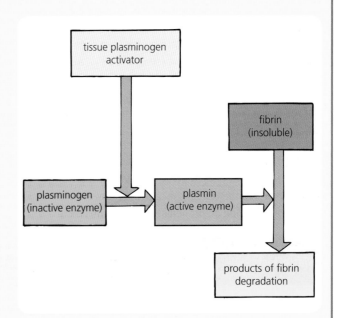

Figure 15.7 Action of tissue plasminogen activator

intravenously as soon as possible after the onset of the heart attack or stroke. However, great care must be taken to ensure that their use is appropriate. For example, they are suitable for treatment of a stroke caused by an embolus (blood clot) but not for a stroke caused by a haemorrhage (bleeding from a ruptured blood vessel).

Related Topic

Comparison of use of antiplatelet and anticoagulant therapies

Both of these forms of medication are used to prevent the formation of clots in the circulatory system.

Antiplatelet drugs

An **antiplatelet** drug is a form of pharmaceutical medication that interferes with the formation of a blood clot by inhibiting the sticking together of platelets. Therefore antiplatelet drugs are used to prevent the formation of a thrombus that could cause a coronary thrombosis or a stroke in people who run a significant risk of developing such a condition.

Aspirin is an example of a relatively mild antiplatelet drug. It inhibits the action of an enzyme essential for the production of a chemical that makes platelets

181

stick together. However, daily use of aspirin, even in low-to-moderate doses, is accompanied by the risk of gastrointestinal bleeding in some individuals. **Glycoprotein inhibitors** are a more potent form of antiplatelet drug. They also cause gastrointestinal bleeding in some patients and their use is restricted to a hospital setting because they can cause complications such as low blood pressure.

Anticoagulants

Anticoagulants are used in the treatment of thrombotic disorders such as deep vein thrombosis (see page 183) and pulmonary embolism (see page 184). An anticoagulant differs from an antiplatelet drug in that it reduces or prevents blood clotting by interfering with a stage in the biochemical pathway that leads to fibrin production (see Figure 15.4).

In the UK, **warfarin** is the most commonly used anticoagulant. It works by preventing **vitamin K** from carrying out its function in the pathway. However, excessive depletion of vitamin K increases the risk of

calcification (hardening) of the arteries. **Heparin** is another anticoagulant. It works by preventing thrombin from playing its role in the pathway.

Side effects

Like antiplatelet drugs, the most serious side effect of anticoagulants is bleeding. This may take the form of prolonged nosebleeds, gastrointestinal bleeding or increased bleeding during menstruation.

Clinical investigation

Table 15.3 shows the results of a randomised trial comparing the two types of treatment on a large population of patients recovering from the implantation of metallic stents (supports) in their coronary arteries. From these results it was concluded that in this trial, three of the side effects occurred significantly less frequently in antiplatelet therapy compared with anticoagulation therapy. However, there is insufficient evidence available at present to enable experts to favour conclusively one type of therapy over the other.

Side effect	Percentage of patients in sample affected by side effect	
	Antiplatelet therapy	**Anticoagulation therapy**
Swelling caused by partially clotted blood	25	34
Discoloured 'bruised' areas of skin	16	38
Prolonged bleeding during surgical repair	1	2
All bleeding complications	33	48

Table 15.3 Side effects of two therapies

| **Research Topic** | **Bleeding disorders** |

Von Willebrand disease

Von Willebrand disease (vWD) is caused by deficiency of a protein called **von Willebrand factor (vWF)** needed to make platelets stick together during blood clotting. There are several forms of this inherited condition. In the most common one the person is **heterozygous** for the defective gene and their level of vWF is significantly lower than normal. However, their blood still clots normally and they are unaware of the condition except during unusual

circumstances such as dental surgery or nosebleeds and during menstruation when bleeding may be prolonged.

In a rarer form of vWD, affected individuals are **homozygous** for the defective gene. They lack vWF and are more seriously affected. They can even suffer bleeding of the joints. Nevertheless, individuals with vWD normally do not require regular treatment. However, they may be given a synthetic form of vWF in exceptional circumstances, for example immediately prior to undergoing major surgery.

Haemophilia A, B and C

Haemophilia is a group of inherited disorders, each of which impairs the body's blood clotting mechanism and leads to prolonged bleeding from even the tiniest wound. The three forms of the condition are summarised in Table 15.4.

Haemophilia A and B occur in approximately 1 in 5000 males (and very rarely in females). Of these males about 85% suffer haemophilia A and about 15% haemophilia B.

The less severe form, haemophilia C, affects about 1 in 100 000 of the population (male and female equally).

Up until the 1950s, people with severe haemophilia rarely lived beyond the age of 10–11 years. Since the 1960s, when effective treatment first became available, the life expectancy of haemophiliacs has been extended to near normal levels. Modern treatments make use of versions of Factors VIII, IX and XI produced by recombinant DNA technology.

Haemophilia	Genetics	Cause of the problem	Symptoms	Treatment
A	Mutant allele of Factor VIII gene which is recessive and sex-linked	Lack of sufficient clotting Factor VIII	External and internal bleeding post-operatively and spontaneously into soft tissues and joints	Regular intravenous infusion of Factor VIII
B	Mutant allele of Factor IX gene which is recessive and sex-linked	Lack of sufficient clotting Factor IX		Regular intravenous infusion of Factor IX
C	Mutant allele of Factor XI gene which is incompletely dominant and autosomal	Lack of sufficient clotting Factor XI	Mild form of condition characterised by some prolonged bleeding	Normally no treatment required except in advance of surgery when Factor XI is given

Table 15.4 Types of haemophilia

Peripheral vascular disorders

The **peripheral** arteries are those *other than* the aorta and coronary and carotid arteries. When any of the peripheral arteries are affected by atherosclerosis, their central cavity becomes narrower (see Figure 15.2). This leads to **peripheral vascular disease**, which most commonly affects the leg arteries. When these blood vessels suffer an obstruction of this type, blood flow is restricted and pain is felt in the leg muscles because they are receiving an inadequate supply of oxygen.

Deep vein thrombosis

Deep vein thrombosis is the formation of a thrombus (blood clot) in a vein, most commonly one in the calf of the lower leg. Normally this causes the affected extremity to become painful and swell up (see Figure 15.8). In addition, veins close to the skin surface can become engorged with blood.

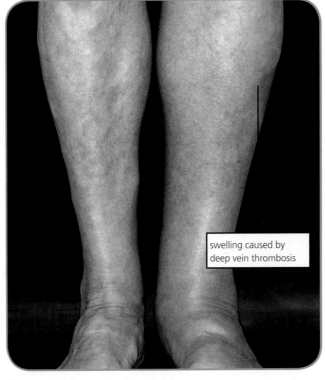

swelling caused by deep vein thrombosis

Figure 15.8 Deep vein thrombosis

Pulmonary embolism

If a thrombus in a vein breaks free, a serious complication may arise. The clot (now an embolus) is transported via the vena cava and heart chambers to the pulmonary artery where it may block a small arterial branch. This serious situation is called a **pulmonary embolism** (see Figure 15.9) and is characterised by symptoms such as chest pains, breathing difficulties and palpitations. Treatment takes the form of anticoagulant drugs or, in severe cases, thrombolytic drugs. If untreated, a pulmonary embolism can lead to collapse and sudden death.

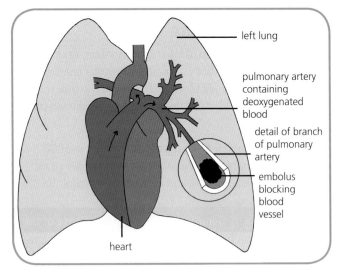

Figure 15.9 Pulmonary embolism

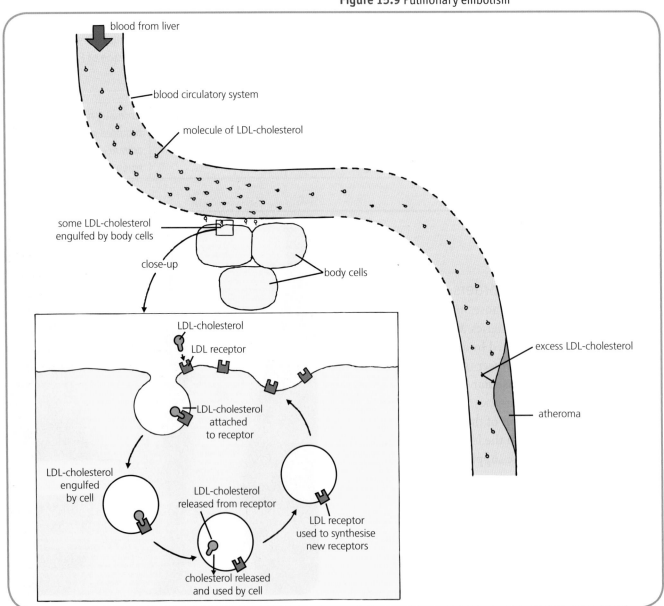

Figure 15.10 Transport of cholesterol by LDL

Cholesterol

The term **lipid** refers to a diverse group of organic compounds which includes simple lipids such as fats (saturated and unsaturated) and more complex substances such as steroids. **Cholesterol** is an important substance because it is a precursor for the synthesis of steroids (such as sex hormones) and it is a basic component of cell membranes. Therefore its presence in the bloodstream at an appropriate concentration is essential to the health and well-being of the human body. It is produced in liver cells from saturated fats present in a normal balanced diet.

Transport of cholesterol

Lipoproteins are molecules containing a combination of lipid and protein. They are present in blood plasma and transport lipids from one part of the body to another.

Low-density lipoprotein (LDL)

Cholesterol is transported to body cells by **low-density lipoproteins (LDL)** produced by the liver. Most body cells synthesise LDL **receptors** which then become inserted in their cell membrane. When a molecule of LDL carrying cholesterol (LDL-cholesterol) becomes attached to a receptor, the cell engulfs the LDL-cholesterol and the cholesterol is released for use by the cell (see Figure 15.10).

Atherosclerosis

Once a body cell contains an adequate supply of cholesterol, a negative feedback system is triggered which inhibits the synthesis of new LDL receptors. Therefore less of the LDL-cholesterol circulating in the bloodstream is absorbed by body cells. Instead, some of it is taken up by endothelial cells lining the inside of an artery. The cholesterol is then deposited in an atheroma in the wall of the artery. This process is very likely to occur if the person:

- eats a diet rich in saturated fat throughout their life
- suffers an inherited condition called familial hypercholesterolaemia (see page 187).

High-density lipoprotein (HDL)

Some excess cholesterol is transported by **high-density lipoproteins (HDL)** from body cells to the liver for elimination. Under normal circumstances, this process prevents a high level of cholesterol accumulating in the bloodstream. In addition, HDL-cholesterol is not taken into artery walls and therefore does not contribute to atherosclerosis. However, these benefits are dependent on a **healthy balance** existing between the HDL-cholesterol molecules (sometimes called 'good cholesterol') and the LDL-cholesterol molecules (sometimes called 'bad cholesterol'). (It should be noted that the cholesterol attached to a molecule of LDL is identical to that attached to HDL.)

Normally HDL molecules carry about 20–30% of blood cholesterol and LDL molecules carry about 60–70%. A higher ratio of HDL to LDL results in a decrease in blood cholesterol and a reduced chance of atherosclerosis and cardiovascular disease. The reverse is true of a lower ratio of HDL to LDL.

The concentration of HDL-cholesterol in the bloodstream is normally higher and the risk of CVD lower in people who exercise regularly. In addition, evidence suggests that HDL levels may also be raised by the replacement of some saturated fat with unsaturated fat in the diet and the consumption of less total fat.

Statins

The level of cholesterol in the blood can be reduced by medication. Drugs such as **statins** bring about this effect by inhibiting an enzyme essential for the synthesis of cholesterol by liver cells. (See Research Topic – Action of cholesterol-reducing drugs.)

Research Topic | Action of cholesterol-reducing drugs

HMG-CoA reductase is an enzyme found in liver cells that promotes the first step in a long chain of biochemical reactions that leads to the production of cholesterol (see Figure 15.11). **Statins** are the most widely used cholesterol-reducing drugs. They act by competing with the enzyme's substrate (**HMG-CoA**) for the active sites on the molecules of the enzyme. This process of **competitive inhibition** reduces the rate at which the enzyme is able to produce intermediate 1 and subsequently reduces the rate at which the cell is able to make cholesterol.

Since more cholesterol originates from internal manufacture in liver cells than from the diet, a drop in cholesterol production by the liver soon results in a drop in the level of cholesterol in the blood. Statins also act by increasing the number of LDL receptors made by liver cells. These receptors bring about a reduction in the level of 'bad cholesterol' by removing some molecules of LDL-cholesterol from the bloodstream.

Night or day

Normally cholesterol synthesis occurs mainly at night. Therefore statins that only act over a short time span are more effective if taken at bedtime rather than in the morning.

Clinical trials

Several clinical trials have shown the benefits of statins in reducing death rates among patients suffering cardiovascular disease (CVD). In recent years, further clinical trials have been carried out to investigate if statins reduce coronary and cerebrovascular events in people who are **at risk of**, but are not suffering, CVD. Table 15.5 shows the results from one of these trials. It was carried out over a period of 4 years on many thousands of participants of average age 63 years. About 25% were suffering diabetes but none of them was suffering CVD at the start of the trial.

The results from this trial and many others have been collated. From the vast quantity of data produced, it has been concluded that statins, by reducing cholesterol levels in patients at risk of CVD, bring about a significant reduction in the risk of major cardiovascular events and a significant decrease in mortality caused by CVD.

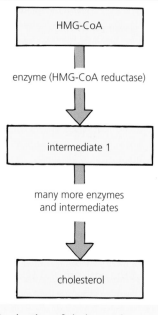

Figure 15.11 Production of cholesterol

Condition	Percentage of group affected after 4 years		Percentage risk reduction
	Control group	Statin group	
Major coronary event	5.4	4.1	24
Major cerebrovascular event	2.3	1.9	17
All-cause mortality	5.7	5.1	11

Table 15.5 Results of clinical trial investigating action of statin

Investigating current views on the use of statins

In the UK, advice to NHS staff includes the following guidelines:

- Statin therapy is recommended for adults with **clinical evidence** of CVD.
- Statin therapy is recommended as part of the management strategy for primary prevention of CVD for adults who have a **20% or greater 10-year risk** of developing CVD. (Risk assessment is based on factors such as whether the individual is elderly, suffers diabetes or belongs to a high-risk ethnic group.)
- Any decision to initiate therapy should only be made following an **informed discussion** between the doctor and the patient about the risks and benefits of statin treatment.

Therefore the current view in the UK is that the use of statins should be restricted to the treatment of patients who are **at risk of or are already suffering CVD**. Despite concerted efforts by pharmaceutical companies, there is no suggestion at present that people who have elevated LDL-cholesterol levels but are *not* at risk of CVD should be taking statins.

Although most people find the guidelines helpful, some people find them unsatisfactory. They are concerned that the '20% 10-year risk of developing CVD' is difficult to define quantitatively and is open to various interpretations. In their opinion many doctors, under pressure from pharmaceutical companies and from increased public awareness of a 'potential health problem', may be disposed to prescribe statins for people whose LDL-cholesterol is above a notional target level but who are otherwise perfectly healthy and not at risk of CVD. The critics add that these people would be undergoing **many years of unnecessary drug therapy**. They stress that it is important that each person makes an informed decision before embarking on a course of statin therapy.

Familial hypercholesterolaemia (FH)

Familial hypercholesterolaemia is an inherited disorder that shows an autosomal dominant pattern of inheritance. (See Related Activity – Analysing an FH pedigree.)

Under normal circumstances, molecules of LDL-cholesterol bind to LDL receptors on cell membranes. However, in sufferers of FH, the mutated gene causes a decrease in the number of LDL receptors present in the cell membranes or a change in their structure that renders them non-functional. Either way, molecules of LDL-cholesterol are unable to unload their cholesterol in cells. It is for this reason that FH is characterised by the possession of **very high LDL-cholesterol levels** in the bloodstream.

If the condition is left untreated, large quantities of cholesterol are deposited in the walls of arteries from an early age. Therefore affected individuals suffer cardiovascular problems such as coronary artery disease at a much younger age than the members of the population as a whole. People whose family has a history of FH can take a genetic test to determine if they have inherited an FH allele. The condition can be treated by the individual adopting a **modified lifestyle** and taking **medication** such as statins. (See Related Topic – Investigating treatments for FH.)

Related Activity

Analysing an FH pedigree

Figure 15.12 shows the pedigree of Laura, a sufferer of familial hypercholesterolaemia (FH). Members of Laura's generation and Laura's father's generation affected by this inherited disorder have all received treatment and are still alive. However, affected members of Laura's grandmother's generation (and earlier generations) did not receive treatment and died prematurely compared with the non-sufferers of their own generation.

In this family tree, the mutant allele for FH is dominant and each affected person is a heterozygote. Each heterozygote has married an unaffected homozygote.

Therefore, on average, there is a 1 in 2 (50%) chance that each of their children will be affected.

FH shows typical features of **autosomal dominant inheritance**:

● The trait appears in every generation.
● Each sufferer of the trait has an affected parent.
● When a branch of the family does not express the trait, the trait fails to reappear in future generations of that branch.
● Males and females are affected in approximately equal numbers.

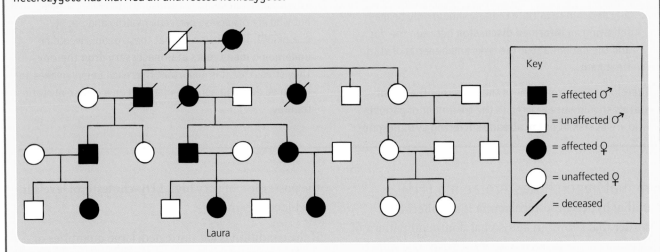

Figure 15.12 Pedigree of familial hypercholesterolaemia

Related Topic

Investigating treatments for familial hypercholesterolaemia (FH)

Change in lifestyle

For the treatment to be successful, the sufferer has to follow a diet that reduces total fat intake to less than 30% of their total intake of kilojoules. In addition, this diet concentrates on a reduction in saturated fats by the affected individual:

● decreasing the quantities of beef, pork, lamb and chicken that they eat
● replacing full-fat dairy products with low-fat dairy and soy products
● eliminating foods rich in oils such as palm and coconut.

Sufferers are also encouraged to take regular exercise because this is thought to lower levels of LDL-cholesterol.

Medication

If a dietary change does not succeed in lowering LDL-cholesterol level sufficiently, the sufferer of FH is given medication. Some drugs such as statins work by lowering the level of LDL-cholesterol ('bad cholesterol'); others such as nicotinic acid (a form of vitamin B_3) may help to increase the level of HDL-cholesterol ('good cholesterol').

Filtration

Sufferers of very extreme forms of FH are treated by having their blood filtered to remove excess LDL-cholesterol.

Testing Your Knowledge

1 a) What is meant by the term *atherosclerosis*? (2)
   b) Identify TWO ways in which the structure of an artery is altered by atherosclerosis. (2)

2 a) Arrange the following four stages that occur during thrombosis into the correct order. (1)
      A clotting factors are released
      B soluble fibrinogen changes into threads of insoluble fibrin
      C an atheroma bursts and damages the endothelium
      D inactive prothrombin becomes active thrombin
   b) What might be the result of a thrombus blocking:
      i) an artery in the brain
      ii) a coronary artery? (2)

3 a) Give ONE reason why all cells need a supply of cholesterol. (1)
   b) How is cholesterol transported:
      i) from the liver to body cells
      ii) from body cells to the liver? (2)

   c) By what means do drugs such as statins reduce blood cholesterol level? (1)

4 Decide whether each of the following statements is true or false and then use T or F to indicate your choice. Where a statement is false, give the word that should have been used in place of the word in bold print. (5)
   a) Atherosclerosis causes the lumen of an affected artery to become **wider**.
   b) A thrombus that has broken loose and is free to travel through the bloodstream is called an **angina**.
   c) The development of an atheroma in the wall of an artery results in an **increase** in blood pressure.
   d) Peripheral vascular disease is characterised by the narrowing of **capillaries** in body extremities.
   e) If a small clot formed by deep vein thrombosis travels through the bloodstream it may cause a **pulmonary** embolism.

16 Blood glucose levels and obesity

Blood glucose levels concentration

Normally blood plasma contains glucose at a concentration of around 5 millimoles per litre (mmol/l) with this glucose concentration varying slightly depending on demand by respiring tissues. However, if a person is suffering untreated **diabetes** (see page 191) their level of blood glucose may become **elevated** to an abnormal and chronic level such as 30 mmol/l. Under these circumstances endothelial cells lining blood vessels absorb far more glucose than normal. This process causes damage to blood vessels and may lead to peripheral vascular disease, CVD or stroke.

Microvascular (small vessel) disease

When the endothelial cells lining a small blood vessel such as an arteriole (see Figure 16.1) take in more glucose than normal, their basement membrane becomes thicker but weaker. Therefore, as the walls of an affected blood vessel become abnormally thick, they lose their strength and may **burst** and **bleed** (haemorrhage) into the surrounding tissues. This leakage reduces rate of flow of blood through the body. A tissue may be affected either by being flooded with leaked blood or by not receiving an adequate oxygen supply. Microvascular disease can cause damage to the **retina**, affecting vision, and to the **kidneys**, causing renal failure. In addition it affects nerves in the body's extremities, which suffer **peripheral nerve dysfunction**. (See Related Activity – Investigating symptoms associated with microvascular disease.)

Regulation of blood glucose level

All living cells in the human body need a continuous supply of energy released from the breakdown of **glucose** during tissue respiration. However, the body obtains supplies of glucose only on those occasions when food is eaten. To guarantee that a regular supply of glucose is present in the bloodstream and available for use by cells regardless of when and how often food is consumed, the body employs a system of **negative feedback control**.

Liver as a storehouse

About one hundred grams of glucose are stored as **glycogen** in the liver. Glucose can be added to or removed from this reservoir of stored carbohydrate depending on changes in supply or demand.

Insulin and glucagon

A rise in blood glucose concentration to above its normal level (for example, following a meal) is detected by receptor cells in the **pancreas**. These cells produce **insulin**. This hormone is transported in the bloodstream to the liver where it is picked up by insulin receptors. Excess glucose is absorbed by the liver cells and an enzyme is activated that catalyses the reaction:

$$glucose \rightarrow glycogen$$

This process brings about a decrease in blood glucose concentration to around its normal level. If the blood glucose drops below its normal level (for example,

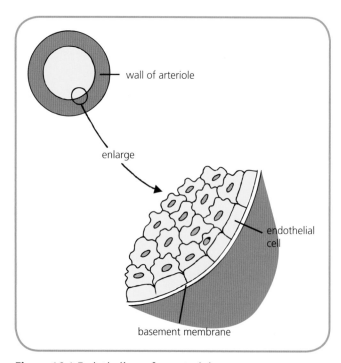

Figure 16.1 Endothelium of an arteriole

Related Activity

Investigating symptoms associated with microvascular disease

Whereas **macrovascular** (large vessel) disease is caused by an obstruction blocking a main artery, **microvascular** disease results from damage to small blood vessels. Symptoms of a macrovascular problem such as coronary heart disease tend to be readily apparent, severe and often life-threatening. Symptoms of a microvascular problem such as peripheral nerve dysfunction tend to lack specific expression for a long time but to become severe eventually. They tend to be restricted to localised areas and to respond to treatment if an associated condition such as elevated blood glucose level is controlled in time.

Microvascular complications are common among long-term sufferers of poorly controlled diabetes. **Diabetic retinopathy** resulting from prolonged high blood glucose levels, for example, causes small vessels in the retina to haemorrhage and leak into the back of the eye. This early sign of diabetes is normally picked up during a routine eye test. In its early stages, diabetic retinopathy can occur without obvious symptoms or pain. As the condition becomes more serious, vision is affected and, if left untreated, it eventually causes blindness.

Microvascular disease that causes damage to kidney arterioles can lead to **renal failure**. This occurs because the kidneys are eventually no longer able to perform their function of filtering and purifying blood. This form of kidney disease is common among people with diabetes who do not control their condition strictly. Its symptoms include:

- swelling of the ankles, feet and lower legs
- production of darker urine, containing traces of blood
- shortage of breath when climbing stairs.

In order to have kidney disease diagnosed and treated before it reaches an extreme stage, people with diabetes are advised to be screened annually for kidney complications.

Peripheral neuropathy (peripheral nerve dysfunction) is a condition resulting from microvascular disease of blood vessels closely associated with nerves that serve the body's extremities such as hands and feet. These nerves are damaged by prolonged exposure to elevated levels of blood glucose. Symptoms can take the form of numbness, tingling or pain in hands, arms, toes, feet or legs. Peripheral neuropathy is common among older, overweight diabetics who continue to exert poor control over their condition. If left untreated, it can lead to the development of ulcers and eventually to the amputation of the affected extremity.

---

between meals or during the night), different receptor cells in the pancreas detect this change and release **glucagon**. This second hormone is transported to the liver and activates a different enzyme, which catalyses the reaction:

$$\text{glycogen} \rightarrow \text{glucose}$$

Now glucose is released from liver cells and the blood glucose concentration rises to around its normal value. Figure 16.2 gives a summary of this **homeostatic system** and shows how insulin and glucagon act antagonistically.

Epinephrine

During exercise or 'fight or flight' reactions (see page 209) when the body needs additional supplies of glucose to provide energy quickly, the adrenal glands secrete an increased quantity of the hormone **epinephrine** (adrenaline) into the bloodstream. Epinephrine overrides the normal homeostatic control of blood glucose level by inhibiting the secretion of insulin and promoting the breakdown of glycogen to glucose. Once the crisis is over, secretion of epinephrine is reduced to a minimum and blood glucose level is returned to normal by the appropriate corrective mechanism involving insulin or glucagon.

Diabetes

People who suffer **diabetes** are unable to control their blood glucose level. If untreated, it can rise to 10–30 mmol/l compared with the normal blood glucose concentration of around 5 mmol/l. There are two types of diabetes and they are compared in Table 16.1.

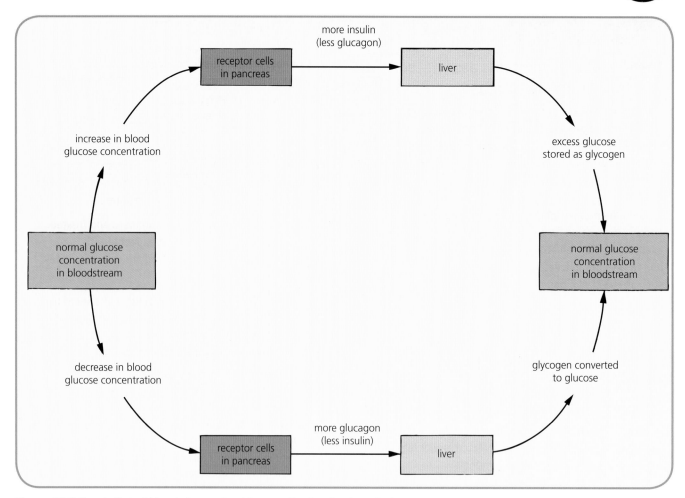

Figure 16.2 Regulation of blood glucose level by negative feedback control

	Type 1 diabetes	Type 2 diabetes
Percentage of all cases	5–10	90–95
Stage of life at which condition normally first occurs	Childhood or early teens (therefore often referred to as 'juvenile onset' diabetes)	Adulthood (therefore often referred to as 'adult onset' diabetes)
Typical body mass of sufferer	Normal or underweight	Overweight or obese
Ability of pancreatic cells to produce insulin	Absent	Present
Sensitivity of cells (e.g. liver and skeletal muscle) to insulin	Cells have the normal number of insulin receptors on their surfaces. They respond to the presence of insulin (if given as treatment) and bring about the opening of glucose channels into the cells.	Cells have a decreased number of insulin receptors on their surfaces, making them less sensitive (or even resistant) to insulin. Therefore few (or no) glucose channels are opened, much glucose fails to enter the cells and normal conversion of glucose to glycogen is prevented.
Treatment	Regular injections of insulin and a careful diet	Exercise, weight loss, diet control (and insulin in some cases)

Table 16.1 Comparison of two types of diabetes

Both types of diabetes, if untreated, result in a rapid increase in blood glucose level following a meal. The filtrate formed in the kidneys of an untreated diabetic is so rich in glucose that much of the glucose is not reabsorbed into the bloodstream but instead excreted in **urine**. Therefore, testing urine for glucose is often used as an **indicator** of diabetes.

Many vascular diseases are closely associated with chronic (long-lasting) complications of diabetes. Whereas it used to be a fatal disorder, diabetes can now be successfully treated provided that the affected individual is prepared to adopt a healthy lifestyle.

Glucose tolerance test

Glucose tolerance is the capacity of the body to deal with ingested glucose. This depends on the body being able to produce adequate quantities of insulin. Measurement of glucose tolerance is a clinical test used to **diagnose** diabetes. It investigates by indirect means whether (or not) insulin production is normal. After fasting for 8 hours, a person has their blood glucose level measured and then consumes a known mass of glucose to give a **glucose load**. Their blood glucose level is monitored over a period of 2½ hours and the results plotted to give a glucose tolerance curve. (See Related Activity – Analysing glucose tolerance curves.)

Analysing glucose tolerance curves

Figure 16.3 shows three different glucose tolerance curves resulting from glucose tolerance tests.

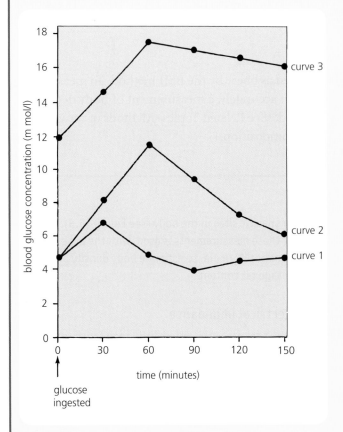

Figure 16.3 Glucose tolerance curves

Curve 1

The person's glucose concentration rises to a maximum at around 30 minutes and then quickly drops to its initial low level well within the 2½ hour period. This indicates that insulin production is normal. The increase in blood glucose concentration has triggered the sequence of events shown in Figure 16.2 and has been brought back to normal by negative feedback control. In this example, the process is so effective that the blood glucose level close to the end of the test dips below the initial fasting level for a short time.

Curve 2

The person's blood glucose concentration begins at the normal fasting level but continues to rise to a maximum at around 60 minutes (or even later) before beginning to decrease. The delay in insulin response to glucose load indicates a mild form of type 2 diabetes. This condition will probably respond to a careful diet and not require treatment with insulin.

Curve 3

After fasting, the person's blood glucose concentration is still at an abnormally high level. After ingestion of glucose, it continues to rise for 60 minutes (or more) and then shows a slight decrease but fails to drop even to its initial high level. This sequence of events indicates severe diabetes (probably type 1). The person is either producing no insulin or their cells are failing to make a normal insulin response to glucose load. Regular injections of insulin and a carefully controlled diet will probably be needed to control this condition.

Obesity

Obesity is a condition characterised by the accumulation of **excess body fat** in relation to lean tissue such as muscle. Being obese greatly increases the individual's risk of suffering a variety of health problems (see Figure 16.4).

Body mass index (BMI)

There is an ideal body mass (weight) for each person. This varies from one individual to another. A person's **body mass index (BMI)** is calculated using the formula:

$$BMI = \frac{body\ mass}{height^2}$$

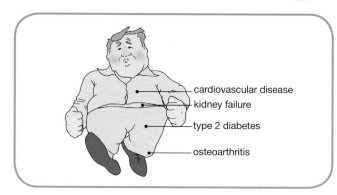

cardiovascular disease
kidney failure
type 2 diabetes
osteoarthritis

Figure 16.4 Obesity and health problems

(where body mass is measured in kg and height is measured in m). The currently accepted classification of BMI values is shown in Table 16.2.

BMI value	Opinion of experts	Risk of associated health problems
20–25	Ideal for height	Average
26–30	Overweight	Increased
31–40	Obese (very overweight)	Greatly increased
Over 40	Very obese (grossly overweight)	Very greatly increased

Table 16.2 BMI values

Limitation of BMI method

Some individuals such as body-builders, who have a relatively low percentage of body fat and an unusually high percentage of muscle bulk, would be wrongly classified as obese by the BMI method. To measure body fat accurately, a measurement of **body density** is required. (See Related Topic – Methods of measuring body composition.)

Related Topic

Methods of measuring body composition

Densitometry

This method assumes that the body's **fat mass** (all chemical fat in the body) and the **fat-free mass** (muscle, bone, etc.) have densities of 0.9 g/cm³ and 1.11 g/cm³ respectively. Measurements of the person's body weight in air and then under water provide information that can be used with the aid of a mathematical formula to estimate the percentage of total body mass made of fat.

Skin-fold thickness

A **skin-fold calliper** is used to measure the thickness of a fold of skin containing a layer of subcutaneous fat at four specific sites in the body (see Figure 16.5). The sum of these measurements is used, with the aid of a mathematical formula, to estimate body density and percentage fat content.

Bioelectrical impedance

The body's fat-free mass offers little **resistance** (or, more accurately, **impedance**) to the flow of a small electric current compared with the fat mass, which is an insulator and offers a high resistance. Therefore, an estimate of ratio of fat-free mass to fat mass can be made by applying a small electric current to parts of the body and measuring bioelectrical impedance.

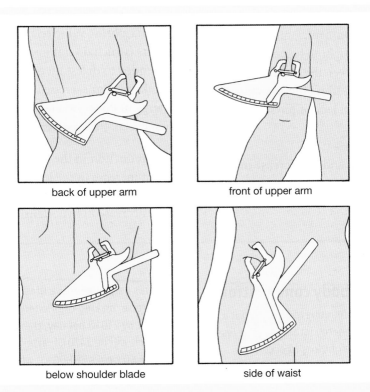

back of upper arm

front of upper arm

below shoulder blade

side of waist

Figure 16.5 Positions used for taking skin fold measurements

Waist/hip ratio

Although the total quantity of fat in the body is a very important factor when assessing health risk, the **distribution pattern** of the fat is even more important. 'Apple'-shaped people (see Figure 16.6) with excess abdominal fat are now known to be, on average, at a greater risk of type 2 diabetes and CVD than 'pear-shaped' people with excess fat round their hips. Women are considered to be at risk when their waist/hip ratio is greater than 0.8 and men are considered to be at risk when their waist/hip ratio is greater than 1.0.

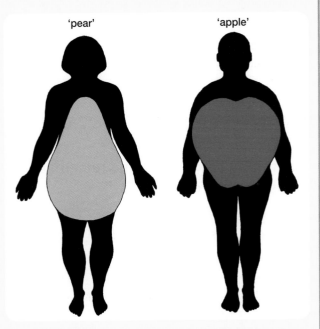

'pear'   'apple'

Figure 16.6 Two patterns of fat distribution

Causes of obesity

Genetic, psychological, environmental, metabolic and dietary factors are all thought to be possible contributors to obesity. However, the most common cause is the excessive consumption of food rich in fats and free sugars combined with lack of physical activity.

Treatment

A **reduction in energy intake** and **an increase in energy expenditure** are the mainstays of treatment for obesity. Affected individuals are advised to reduce energy intake by limiting their consumption of fat (which has a high calorific value relative to other foods) and free sugars (which do not need any metabolic energy to be expended to digest them). They are also encouraged to take regular exercise in order to expend some of the energy in their body fat while keeping their muscle tissue lean. Increased exercise also helps to bring about:

- a reduction in the risk factors associated with CVD
- an increase in the level of HDL-cholesterol ('good cholesterol') in the blood
- a decrease in hypertension and stress.

Related Topic

Effect of exercise on body composition

Exercise does not need to be strenuous to bring about a loss in weight. However, total energy intake must be less than total energy output for weight loss to occur.

Worked example

1 kg of body fat contains 29.4 megajoules (MJ) of energy.

1 MJ = 1000 kJ. Therefore, 29.4 MJ = 29 400 kJ.

Imagine an individual wants to lose 2 kg of body fat (equivalent to 58 800 kJ).

If this person adopts a lifestyle that includes some gentle exercise, making their energy output exceed their energy input by 210 kJ per day, then after 1 week their energy deficit will be 1470 kJ, after 10 weeks 14 700 kJ and after 40 weeks 58 800 kJ, which is equivalent to 2 kg of body fat.

On the other hand, if the person adopts a lifestyle that includes more vigorous exercise, making their energy output exceed their energy input by 840 kJ per day, then after 1 week their energy deficit will be 5880 kJ and after 10 weeks 58 800 kJ, which is equivalent to 2 kg of body fat.

Related Activity

Examining risk factors and remedial measures in treating CVD

Risk factors

The main CVD risk factors are:

- high LDL-cholesterol level
- obesity
- unhealthy diet high in saturated fat and cholesterol
- physical inactivity
- high blood pressure
- smoking
- diabetes
- drinking alcohol to excess
- very stressful lifestyle.

All of these risk factors can be reduced significantly by adopting a healthy lifestyle supported by medication where necessary.

Treatment of cardiovascular disease

Medication

People who are suffering cardiovascular disease or have suffered a heart attack may be given:

- **nitroglycerin** to improve blood flow to heart muscle by making coronary arteries become dilated
- **heparin** to act as an anticoagulant and reduce the tendency of blood to clot in coronary arteries
- **tissue plasminogen activator** to dissolve blood clots in vessels
- **beta-blockers** to reduce the heart's workload by decreasing heart rate and blood pressure.

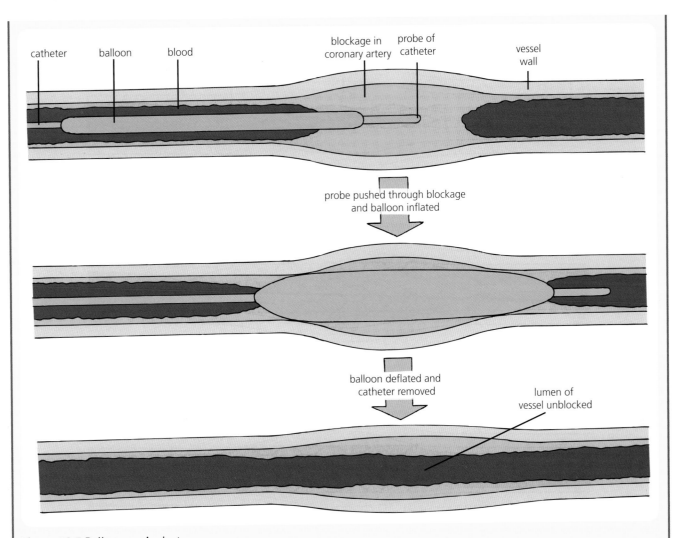

Figure 16.7 Balloon angioplasty

Medical intervention

Angioplasty

Where partial or almost complete blockage of a coronary artery is detected, **angioplasty** may be used. Figure 16.7 shows the tip of a balloon catheter reaching a blockage. When the balloon is inflated, it disperses the fatty deposits and opens up the vessel. The balloon is then deflated and removed.

Stenting

A **stent** (see Figure 16.8) is a supporting device that may be inserted during balloon angioplasty of a coronary artery to hold the vessel open.

Coronary bypass

In cases of extensive coronary artery disease, a **coronary bypass** operation may be performed. This is done by connecting the aorta to the coronary artery using a length of vein from the patient's leg (see Figure 16.9).

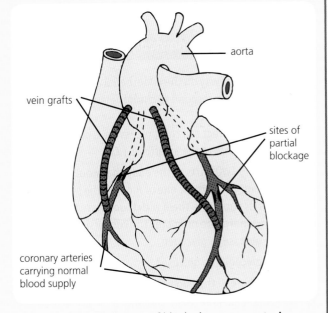

Figure 16.9 Double bypass of blocked coronary arteries

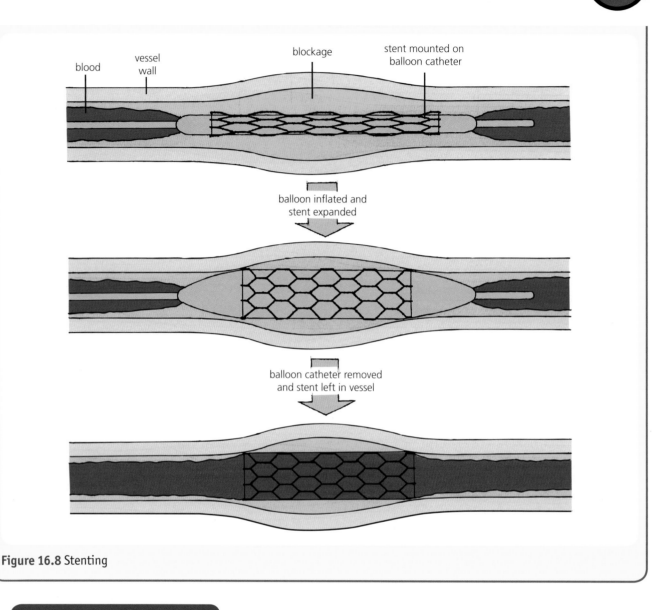

Figure 16.8 Stenting

Testing Your Knowledge

1 a) i) In what way do the endothelial cells lining an arteriole respond to chronic elevation of blood glucose level?

   ii) What effect can this process have on the structure of the arteriole? (2)

   b) Identify TWO parts of the body that would be seriously affected if one or more of the small blood vessels supplying them with blood were to burst. (2)

2 a) Which hormone promotes the conversion of:

   i) glucose to glycogen

   ii) glycogen to glucose?

   iii) Where in the body are these hormones produced?

   iv) In which organ do the conversions promoted by these hormones occur? (4)

   b) Construct a table to show THREE differences between type 1 and type 2 diabetes. (3)

   c) i) What is meant by the term *glucose tolerance*?

   ii) What is the purpose of a glucose tolerance test? (2)

3 a) What is meant by the term *obesity*? (1)

   b) Give the formula used to calculate body mass index (BMI). (1)

   c) What are the TWO main ways in which an obese person can reduce the risk factors for CVD? (2)

What You Should Know Chapters 15–16

activity	embolism	lean
arterioles	embolus	liver
atheromas	endothelium	low-density
atherosclerosis	exercise	membranes
attack	familial	negative
burst	fibrin	obesity
cardiac	glucagon	peripheral
cardiovascular	glycogen	receptors
cholesterol	high	restricted
clotting	high-density	statins
coronary	hormone	stroke
deep	indicator	thrombosis
diabetes	insulin	tissues
diagnosed	kidneys	tolerance

Table 16.3 Word bank for chapters 15–16

1 Atherosclerosis is the accumulation of plaques (_____) under the _____ in the walls of arteries. The diameter of an affected artery becomes reduced and blood flow is _____. The condition is the root cause of many cardiovascular diseases.

2 _____ is the formation of a blood clot (thrombus). Cells damaged by the rupture of an atheroma release _____ factors that activate a series of chemical reactions. These result in the formation of threads of _____ that clot blood and seal the wound.

3 A thrombus that travels through the bloodstream and eventually blocks a blood vessel is called an _____. If it blocks the _____ artery, it can cause a heart _____. If it blocks an artery to the brain, it can cause a _____.

4 The narrowing of arteries other than those supplying parts of the body core, such as the _____ muscle and brain, causes _____ vascular disease.

5 The formation of a blood clot in a large peripheral vein is called _____ vein thrombosis. If the clot travels to a blood vessel supplying the lungs, it can cause a pulmonary _____.

6 Cholesterol is a fatty substance made in the _____ and needed as a component of cell _____. It is transported to body cells by _____ lipoproteins (LDL) and back to the liver by _____ lipoproteins (HDL).

7 Excess LDL-cholesterol in the blood can result in _____ being deposited in atheromas causing _____. Drugs such as _____ reduce blood cholesterol levels.

8 _____ hypercholesterolaemia is an inherited disorder that causes sufferers to develop abnormally _____ levels of cholesterol.

9 Very high levels of blood glucose can cause damage to _____ and lead to vascular disease. Damage to small blood vessels can make them _____ and leak their contents into nearby _____.

10 Blood glucose level is regulated by _____ feedback control. When the glucose level is too high, the _____ insulin promotes the conversion of glucose to _____; when it is too low, the hormone _____ promotes the conversion of glycogen to glucose.

11 People with _____ are unable to control their blood glucose level effectively.

12 Those with type 1 diabetes are unable to make _____. Those with type 2 diabetes are able to make insulin but their cells lack an adequate number of insulin _____ and fail to allow enough glucose to enter.

13 The fact that, in both types of diabetes, the _____ cannot cope with the elevated blood glucose level and release glucose in urine, is used as an _____ of the condition.

14 Diabetes is _____ using the glucose _____ test.

15 _____ is characterised by the accumulation of excess body fat relative to _____ body tissue. Obesity is closely associated with lack of physical _____ and a diet rich in fat.

16 Risk factors for _____ disease are reduced by taking _____ and keeping body weight at an optimum level.

Applying Your Knowledge and Skills

Chapters 9–16

1 The male reproductive system is shown in Figure 16.10.
   a) Using only the appropriate letters, indicate the route taken by sperm from site of production to point of exit from the male body. (1)
   b) Copy and complete Table 16.4. (8)

2 Figure 16.11 represents some of the events that occurred during a complete menstrual cycle in a woman.
   a) i) On which dates would sexual intercourse have been most likely to result in fertilisation in this woman?
      ii) Give TWO reasons to explain how you arrived at your answer. (3)
   b) Give a further physical sign unrelated to the graph that might have helped the woman to identify her most fertile time. (1)
   c) What evidence from the graph tells you that fertilisation did not occur during the cycle shown? (1)

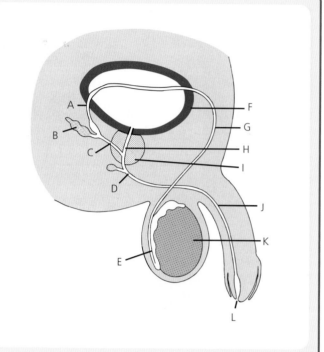

Figure 16.10

Letter in Figure 13.22 indicating accessory gland	Name of this accessory gland	Example of substance secreted by accessory gland that contributes to fertilisation	Way in which named substance contributes to fertilisation

Table 16.4

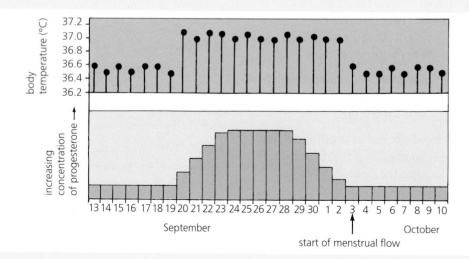

Figure 16.11

d) i) Assuming that the cycle remains the same length, on which dates in October will sexual intercourse be most likely to lead to conception in this woman?

ii) If fertilisation were to occur, in what way would the progesterone curve differ from the one in Figure 16.11? (2)

e) i) Instead of trying to avoid the fertile phase, the woman decided to use a hormonal method of contraception containing progesterone. Explain how this prevents fertilisation.

ii) Why do women 'on the pill' take pills lacking sex hormones 1 week in every four?

iii) Suggest why women are recommended to take the placebo pills rather than no pills at all during that week. (4)

3 The graph in Figure 16.12 shows the relationship between incidence of Down's syndrome and maternal age.

a) Use the graph to estimate the chance of a woman at each of the following ages having a Down's syndrome baby: **i)** 20 years, **ii)** 30 years, **iii)** 40 years. (3)

b) Account for the trend shown by the graph. (1)

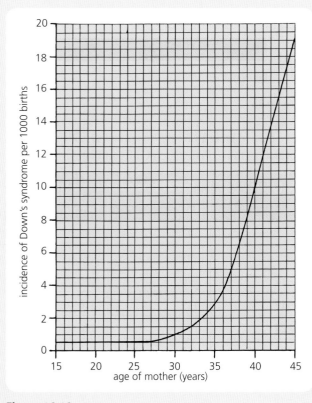

Figure 16.12

4 Figure 16.13 shows a pedigree chart for the inherited trait of deaf-mutism.

a) i) Which of the following patterns of inheritance is shown by this trait?
  A autosomal recessive
  B autosomal dominant
  C sex-linked recessive

ii) Give ONE reason to support your choice.

iii) Give ONE reason for deciding against each of the other choices. (4)

b) Copy the pedigree chart and, using symbols of your choice, attempt to supply the genotype of each person (giving both possibilities in uncertain cases). (9)

c) If X marries a man who is homozygous for the normal allele, what is the percentage chance of each child of this union being:
  i) a deaf mute
  ii) a carrier of the deaf-mute allele? (2)

d) If Y marries a woman who is homozygous for the normal allele, what is the chance of each child of this union being:
  i) a deaf mute
  ii) a carrier of the deaf-mute allele? (2)

e) If X marries Y, what is the chance of each child being:
  i) a deaf mute
  ii) a carrier of the deaf-mute allele? (2)

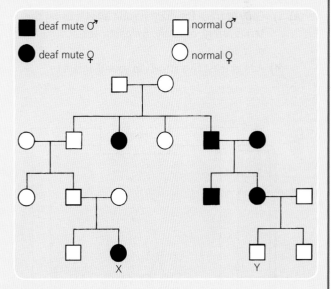

Figure 16.13

5 Table 16.5 shows a set of results for the experiment shown in Figure 13.6 on page 163 using a 2 mm-thick ring of artery and a 2 mm-thick ring of vein from a cow.

Mass added (g)	Change in length (%)	
	Vein	Artery
0	0.0	0.0
10	72.5	75.0
20	79.0	84.5
30	91.5	95.0
40	95.0	102.0
50	94.5	105.5

Table 16.5

a) Plot the data on the same sheet of graph paper and draw curves of best fit. (4)
b) i) Draw TWO conclusions from the data.
   ii) Relate your answer to i) to the structure of arteries and veins. (3)
c) Identify TWO points of procedure relating to the animal material being used that were carried out to eliminate additional variable factors from the experiment. (2)
d) State TWO ways in which the experiment could be improved to increase the reliability of the results. (2)

6 Figure 16.14 shows a simplified version of the circulatory system.
   a) i) Copy or trace the diagram and label it using the terms *artery, capillary, lymphatic vessel, lymphatic duct* and *vein*.
      ii) Add at least five arrows to your diagram to show the direction of flow of the liquids in the vessels. (7)
   b) Which letter in the diagram indicates the presence of: i) blood plasma, ii) lymph, iii) tissue fluid? (3)

7 Figure 16.15 represents the repeated series of events that occurs during the human heartbeat.
   a) i) Which lasts longer, atrial or ventricular systole?
      ii) Explain why this difference is necessary. (2)
   b) i) Name the stages of the cardiac cycle represented by X and Y.
      ii) In what state is cardiac muscle in the atria during stage X? (3)
   c) i) How many complete heartbeats are represented by the diagram?
      ii) Express the person's pulse rate in beats per minute. (2)
   d) Redraw part of the diagram to represent one complete heartbeat and then add the letters L and D and arrows to indicate when the two heart sounds would be heard. (3)

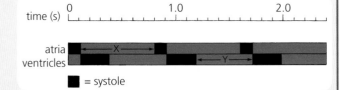

Figure 16.15

8 The data in Table 16.6 refer to normal systolic blood pressures found in humans at different ages.

Age (years)	Systolic blood pressure (mm Hg)	
	Male	Female
0–4	93	93
5–9	97	97
10–14	106	106
15–19	119	116
20–24	123	116
25–29	125	117
30–34	126	120
35–39	127	124
40–44	129	127
45–49	130	131

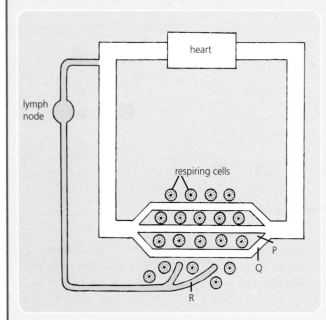

Figure 16.14

50–54	135	137
55–59	138	139
60–64	142	144
65–69	143	154
70–74	145	159
75–79	146	158

Table 16.6

a) Present the data as a two-tone bar chart. (4)

b) i) Make a generalisation about the effect of age on blood pressure.

  ii) Give a possible explanation (related to lifestyle) to account for this trend. (2)

c) It has been suggested that female sex hormones may in some way offer protection against high blood pressure. What information from the table seems to support this theory? (1)

9 Figure 16.16 shows box plots of body weight for three populations of British men of average height at three different times in recent history.

a) What percentage of data is contained in the box in a box plot? (See Appendix 2 for help.) (1)

b) i) By what means is the median value of the data indicated in a box plot?

  ii) State the median value for each of the populations. (2)

c) i) Which box set shows the widest distribution of values between its median and its upper quartile?

  ii) Which box set shows the smallest overall distribution of values? (2)

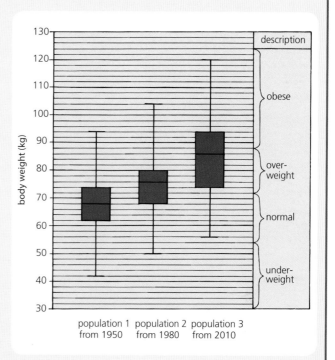

Figure 16.16

d) Does a whisker represent a 95% level of confidence or an actual value? (1)

e) What was the lowest value recorded? (1)

f) If the data were plotted as a graph of number of individuals against body weight, which population would give a symmetrical, bell-shaped curve? (1)

g) i) What trend is shown by the three box plots?

  ii) Suggest a reason for this trend that does not refer to food intake. (2)

10 Give an account of thrombosis following the rupture of an atheroma. (9)

Note: Since this group of questions does not include examples of every type of question found in SQA exams, it is recommended that students also make use of past exam papers to aid learning and revision.

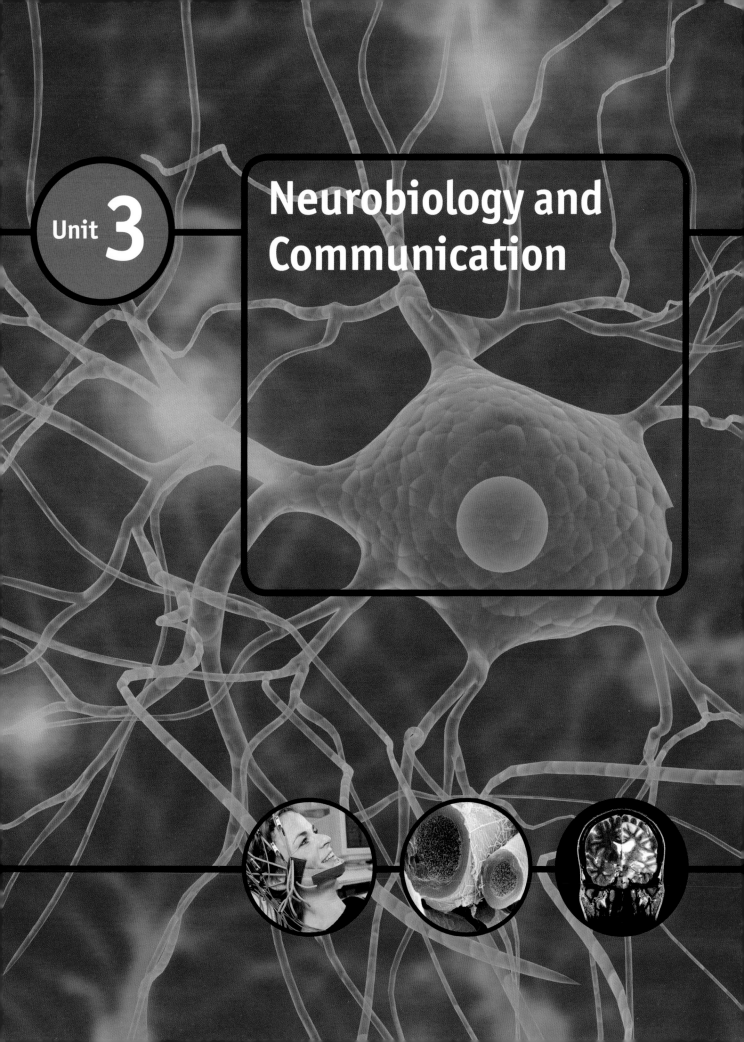

Unit 3

Neurobiology and Communication

17 Divisions of the nervous system and parts of the brain

Sensory information from receptors in contact with the external environment and the internal environment (the inside of the body) is collected by the **nervous system** and analysed. Some of this information is stored for possible future use. Meanwhile, appropriate voluntary and involuntary motor responses are initiated which lead to muscular contractions or glandular secretions.

Divisions of the nervous system

Based on **structure** and **location** of component parts, the nervous system can be divided as shown in Figure 17.1. Figure 17.2 shows, in a simple way, where these parts are located in the human body.

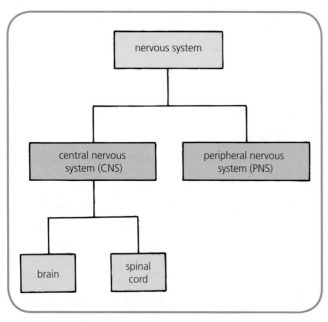

Figure 17.1 Structural division of the nervous system

Sensory and motor pathways

Peripheral nerves contain a **sensory pathway** consisting of sensory nerve cells and/or a **motor pathway** consisting of motor nerve cells. Sensory pathways carry nerve impulses to the CNS from **receptors**. Some receptors are located in external sense organs (such as the skin, eye retina and ear cochlea); others are found in internal sense organs (such as CO_2 receptors in

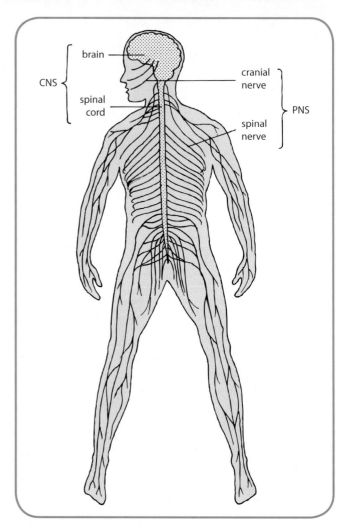

Figure 17.2 Location of CNS and PNS

carotid arteries and thermoreceptors in the cerebellum – see page 213). Sensory pathways keep the brain in touch with what is going on in the body's external and internal environments.

The brain analyses, interprets, processes and stores some of this constant stream of information, which is based on stimuli such as sounds, sights, colours, tastes, temperature of skin and blood and CO_2 concentration and water concentration of blood. The brain's association centres (see page 216) may act on this information by sending nerve impulses via the motor pathways to **effectors** (e.g. muscles and glands). These then bring about the appropriate **response**, such

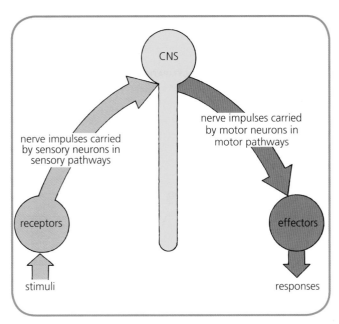

Figure 17.3 Flow of information through the nervous system

as muscular contraction or enzyme secretion. This relationship is summarised in Figure 17.3.

Functional division

A further method of dividing up the nervous system is based on the different **functions** performed by the two separate branches of the peripheral nervous system, as shown in Figure 17.4.

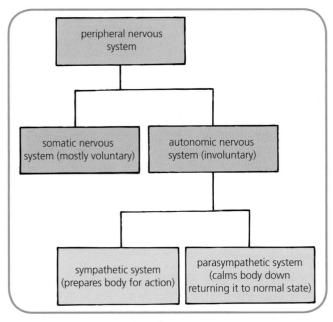

Figure 17.4 Functional divisions of the nervous system

Somatic nervous system

The **somatic nervous system** (**SNS**), which includes the spinal nerves, controls the body's skeletal muscles. This involves sensory and motor pathways, as outlined in Figure 17.3. The somatic nervous system is responsible for bringing about certain **involuntary reflex** actions (for example, limb withdrawal), but most of the control that it exerts is over **voluntary movements** of skeletal muscles.

Imagine, for example, that you are invited to select your four favourite chocolates from a large box of a familiar make. The displayed chocolates act as visual stimuli. Nerve impulses pass from each retina via sensory nerve cells in the optic nerve to the brain. There, association centres process the information and compare it with previous experiences. Decisions are taken and nerve impulses pass to the brain's motor area (see page 216). This in turn sends impulses via motor nerve cells to the appropriate skeletal muscles of the arm and hand allowing the voluntary responses needed to pick out the four sweets. This series of events involves the somatic nervous system.

Autonomic nervous system

The **autonomic nervous system** (**ANS**) (see Figure 17.5) regulates the **internal** environment by controlling structures and organs such as the heart, blood vessels, bronchioles and alimentary canal. This control is generally **involuntary** because it normally works **automatically** without the person's conscious control being involved (although, under exceptional circumstances, some people are able to heighten or suppress certain autonomic responses intentionally).

The nerves that comprise the autonomic nervous system arise from nerve cells in the brain and emerge at various points down the spinal cord to reach the effectors (cardiac muscle, smooth muscle and glands) that they stimulate with nerve impulses.

Antagonistic nature of autonomic nervous system

The **sympathetic** and **parasympathetic** systems that make up the autonomic nervous system are described as being **antagonistic**. This means that they affect many of the same structures but exert opposite effects on them. Figure 17.5 shows only a few of the many tissues and organs controlled in this way.

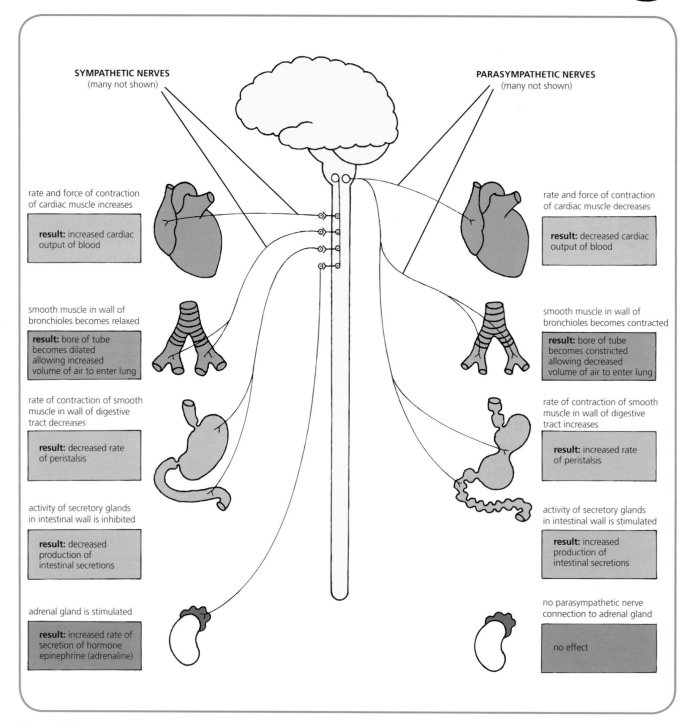

Figure 17.5 Autonomic nervous system

Harmonious balance

The autonomic nervous system is concerned with maintaining a **stable internal environment** by playing its part in the process of **homeostasis** (see Related Topic – Homeostasis). Under normal circumstances the sympathetic and parasympathetic systems are constantly working in an equal but opposite manner, with neither gaining the upper hand. The activity of a tissue or organ under their control is the result of the two **opposing influences**. It is therefore normally in a state somewhere between the extremes of hypo- and hyperactivity.

Finely tuned control

The stimulation of an effector by both sympathetic and parasympathetic nerves provides a **fine degree of control** over the effector. The system works like a vehicle equipped with both an **accelerator** and a **brake**. If the car had an accelerator but no brake then the process of reducing speed would depend solely on decreasing the pressure on the accelerator. This method would require too much time to elapse before the car responded and slowed down. Use of a brake, which can be applied in addition to decreasing the pressure on the accelerator, allows for a much more rapid and effective means of regulating the car's speed.

There are exceptions to the rule of dual innervation of an effector by both parasympathetic and sympathetic nerves. For example, the adrenal gland, which secretes the hormone **epinephrine (adrenaline)**, receives a supply of sympathetic nerves only (see Figure 17.5).

Fight or flight

On being stimulated and briefly gaining the upper hand, the **sympathetic** system arouses the body in preparation for action and the expenditure of energy during '**fight or flight**'. The heart rate and blood pressure increase. Blood supplies are diverted to the skeletal muscles (in great need of an increased supply of oxygen) and away from the gut and skin (which require minimal servicing during the crisis). Rate of nervous perspiration also increases. Hence a thudding heart, a face white with fear and a clammy sensation in localised areas of the body that are in a 'cold sweat' (e.g. armpits and palms of hands) are all characteristic responses to a crisis.

The hormone epinephrine helps to sustain the arousal effects until the emergency is dealt with. This might involve taking a determined and defensive stand, perhaps involving a fight or cutting your losses and running away. In either case the vast amount of energy required by the skeletal muscles is supplied by their increased blood flow.

Rest and digest

When the excitement is over, the **parasympathetic** system takes over for a brief spell, calming the body down and returning it to normal for '**rest and digest**'. Heart rate and blood pressure drop to normal. Rate of peristaltic contractions in the digestive tract increases. Blood is diverted to the intestines where it can resume its job of absorbing the end products of digestion now that the crisis is over. The effects brought about by the parasympathetic nerves help the body to **conserve resources** and **store energy**.

Related Topic

Homeostasis

Homeostasis is the maintenance of the body's internal environment within certain tolerable limits despite changes in the body's external environment (or changes in the body's rate of activity).

Principle of negative feedback control

When some factor affecting the body's internal environment deviates from its normal optimum level (called the **norm** or **set point**), this change in the factor is detected by **receptors**. These send out nerve or hormonal messages which are received by **effectors**. The effectors then bring about certain responses that counteract the original deviation from the norm and return the system to its set point. This corrective homeostatic mechanism is called **negative feedback control** (see Figure 17.6). It provides the stable environmental conditions needed by the body's community of living cells to function efficiently and survive.

Effect of exercise on respiratory system

Rate and **depth** of breathing increase during exercise. This results in increased ventilation of the lungs which promotes the uptake of oxygen and the removal of carbon dioxide.

Carbon dioxide as the stimulus

The graph in Figure 17.7 shows the results from an experiment to compare the effect on breathing rate of inhaling normal air, 'abnormal' air type 1 and 'abnormal' air type 2. Only the 'abnormal' air type 2 is found to cause breathing rate to increase sharply. It is concluded that it is the high level of carbon dioxide in the 'abnormal' air type 2 (and not the low level of oxygen in 'abnormal' air

$\rightarrow$

209

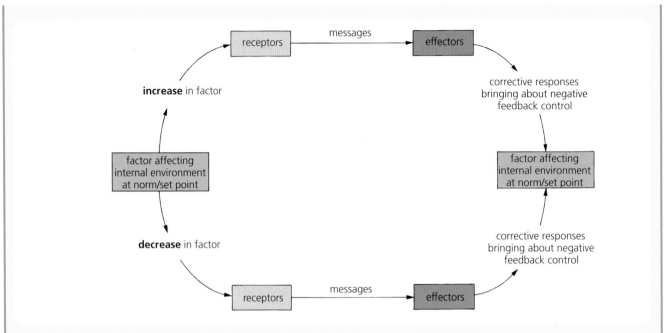

Figure 17.6 Principle of negative feedback control

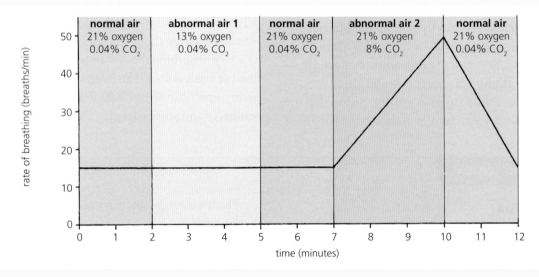

Figure 17.7 Effect of gases on breathing rate

type 1) that acts as the stimulus triggering increased rate of breathing.

Further experiments show that the depth of breathing also increases in response to inhalation of air rich in carbon dioxide. Similarly in the body of a person undergoing vigorous exercise, it is the increased level of carbon dioxide in the bloodstream that acts as the main stimulus bringing about an increase in rate and depth of breathing. However, severe lack of oxygen eventually also causes increased rate and depth of breathing.

Homeostatic control

Chemoreceptors in the carotid arteries and aorta (see Figure 17.8) are sensitive to the concentration of carbon dioxide present in the bloodstream. A rise in carbon dioxide level during vigorous exercise causes these sensory cells to send an increased number of nerve impulses via sensory neurons to the **respiratory control centre** in the **medulla**. This region of the brain responds by sending a greater number of nerve impulses via motor neurons to the intercostal muscles and diaphragm. The subsequent increased activity of these structures brings about an **increase in rate and depth of breathing**. Excess carbon dioxide is removed and the internal environment is kept within tolerable limits. This homeostatic pathway is summarised in Figure 17.9.

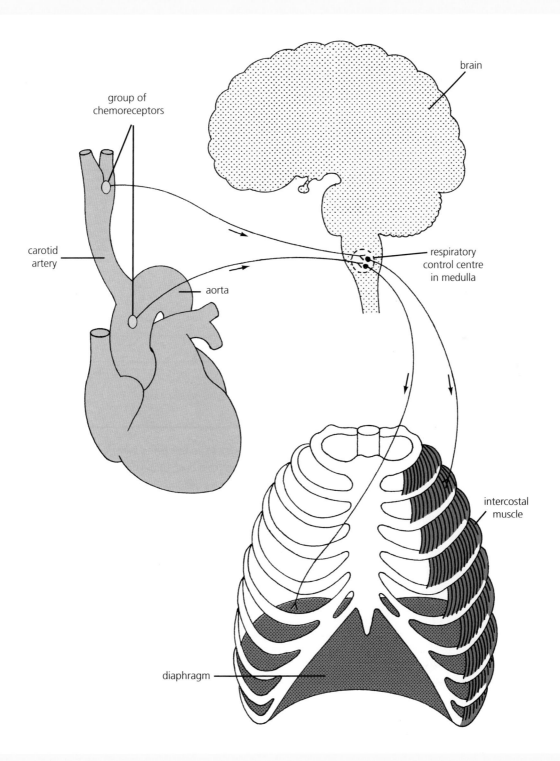

Figure 17.8 Nerve pathway triggered by chemoreceptors

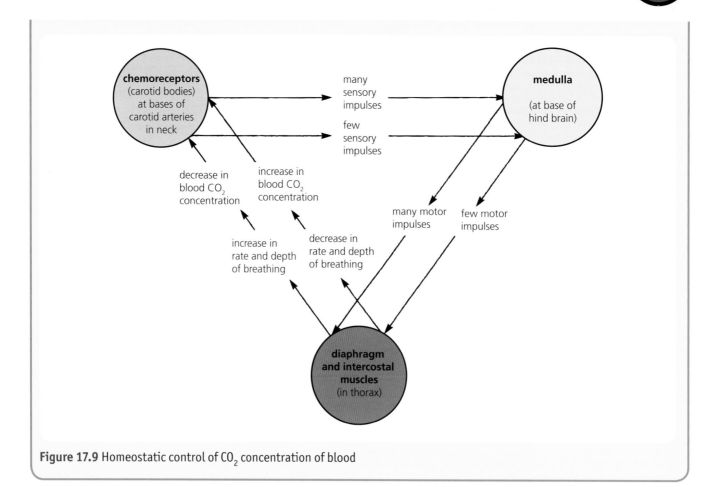

Figure 17.9 Homeostatic control of CO_2 concentration of blood

Testing Your Knowledge 1

1 a) Name the TWO components of the central nervous system (CNS). (2)

   b) What collective name is given to all the nerves excluding the central nervous system? (1)

2 Differentiate between the following pairs of terms:

   a) *sensory* and *motor* pathways (2)

   b) *receptors* and *effectors*. (2)

3 a) Name the TWO branches of the autonomic nervous system. (2)

   b) State the effect of each branch of the autonomic nervous system on:

      i) cardiac output

      ii) width of bore of the bronchioles. (4)

4 Rewrite the following sentences, choosing the correct word from each underlined choice.

   The body becomes aroused ready for 'fight or flight' by the action of the <u>sympathetic/parasympathetic</u> nerves. As a result, rate of blood flow to the heart <u>increases/decreases</u>, rate of peristalsis <u>increases/decreases</u> and rate of breathing <u>increases/decreases</u>. (4)

Brain

The **brain** (see Figure 17.10) is a complex organ composed of three interconnected, concentric layers:

- the central core
- the limbic system
- the cerebral cortex (outer layer of cerebrum).

Central core

The **central core** contains several important regions of the brain. One of these is the **medulla**, which controls many essential processes of life such as breathing, heart rate, sleep and arousal (the state of being awake and aware of the external environment). The **cerebellum** is also part of the brain's central core. It controls balance and muscular coordination, enabling the body to move and to adopt appropriate posture.

Limbic system

The **limbic system** is a composite region of the brain illustrated in a simple way in Figure 17.10. It also possesses structures (not shown in the diagram) that project deep into the cerebral cortex. The structures that make up the limbic system are involved in a variety of functions such as:

- processing information needed to form long-term memories
- regulating emotional states (e.g. anxiety, fear and aggression)
- influencing biological motivation (e.g. hunger, thirst and sex drive).

Hypothalamus

The limbic system also contains the **hypothalamus** (see Figure 17.10). This is a physically small but functionally significant region of the brain.

Pituitary gland

The hypothalamus is connected to the **pituitary gland**, as shown in Figure 17.11, and therefore acts as a link between the nervous system and the hormonal (endocrine) system.

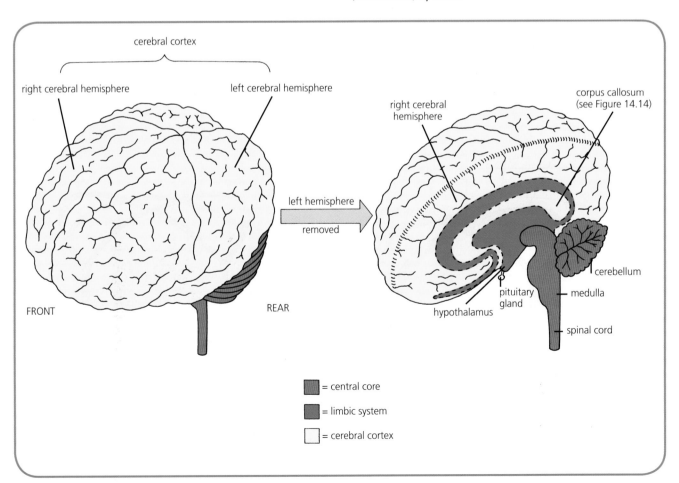

Figure 17.10 Human brain

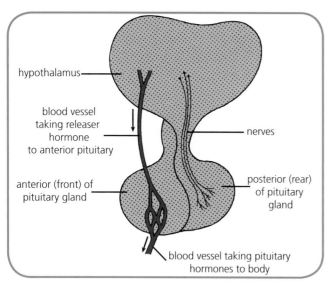

Figure 17.11 Pituitary gland

The hypothalamus contains secretory cells that produce **releaser hormones**. These are transported in the bloodstream to the anterior pituitary gland where they trigger the release of pituitary hormones such as:

- growth hormone – essential for the normal growth of the body and development of long bones
- thyroid stimulating hormone – needed to make the thyroid gland produce thyroxin, which is essential for the control of metabolism

- gonadotrophic hormones – first released during puberty to stimulate reproductive organs to produce and release gametes (see chapter 10).

Contraction of smooth muscle

Axons of neurons whose **cell bodies** lie in the hypothalamus extend to the sympathetic and parasympathetic centres in the brain's central core. This enables the hypothalamus to **regulate autonomic activities** such as the contraction and relaxation of smooth (involuntary) muscle involved in homeostatic mechanisms. These include vasoconstriction and vasodilation of blood vessels during control of body temperature.

Control of body temperature

The hypothalamus contains the **thermoreceptors**, which are sensitive to changes in temperature of blood that reflect changes in the temperature of the body core. The thermoregulatory centre in the hypothalamus responds to this information by sending appropriate nerve impulses to effectors such as sweat glands and the blood vessels involved in vasodilation and vasoconstriction (see Figure 17.12). These trigger **corrective feedback mechanisms** and return the body temperature to its normal level.

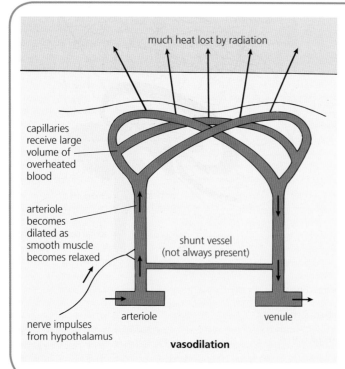

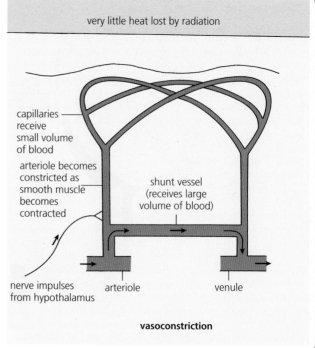

Figure 17.12 Vasodilation and vasoconstriction of skin arterioles

Control of water balance

If the water concentration of the blood decreases due to increased sweating or consumption of salty food, then **osmoreceptors** in the hypothalamus are stimulated. These trigger the secretion of antidiuretic hormone (ADH) by the posterior pituitary gland into the bloodstream. On arriving in the kidneys, ADH increases the permeability of the tubules and collecting ducts to water. As a result, more water is reabsorbed into the bloodstream and its water concentration is returned to normal.

Cerebral cortex

The **cerebral cortex** is the outer layer of the cerebrum, which forms the largest and most complex part of the brain. The cerebrum is split by a deep cleft into two halves called **cerebral hemispheres**. The left hemisphere processes information from the right visual field and controls the right side of the body. The reverse is true of the right hemisphere.

The two hemispheres are not completely separated but are connected by a large bundle of nerve fibres called

---

Case Study   Role of the limbic system

Researchers have developed partial maps of the regions of the brain responsible for **emotions**. Much of this information has been obtained from studies of people who have suffered some form of brain damage or disorder. For example, some people whose limbic system undergoes excessive activity as a result of a disorder such as an epileptic seizure are often found to exhibit one or more of the following 'emotional' behaviours:

- unaccountable fear responses
- unreasonable, aggressive impulses and actions
- uncontrollable laughter
- strong sexual arousal
- feelings of bliss and 'at oneness' with the universe.

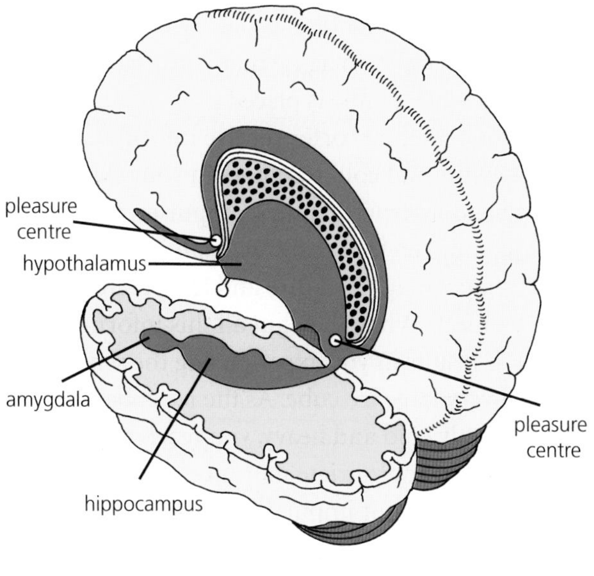

Figure 17.13 Details of the limbic system

On the other hand, people whose limbic system has been seriously damaged by, for example, a tumour, are often found to suffer an absence of certain emotions and desires. Figure 17.13 shows further details of the limbic system.

Amygdala

This region of the limbic system is closely associated with the ability to experience and express emotions such as fear, anger and aggression, each of which plays its part in the survival of a human in a crisis. Lesioning of the amygdala on each side of the limbic system is found to make mammals emotionally unresponsive and unaggressive. For example, monkeys who had previously responded to the presence of snakes with terror were found to show no fear of the same snakes following lesioning of their amygdalae.

Among humans, some people suffer **epileptic seizures** that cause abnormally high levels of activity of certain neurons in the amygdala. Evidence suggests that this results in feelings of hyper-aggressiveness, leading to violent behaviour. Removal of part of each amygdala by neurosurgery has been found to change the person's behaviour and eliminate rage attacks. The amygdala is also thought to be critical in the formation of memories associated with fear of dangerous objects and situations.

Pleasure centres

Several areas of the limbic system are found to be associated with feelings of **pleasure**. Two of these are shown in Figure 17.13. (See also Figure 19.22 on page 268.)

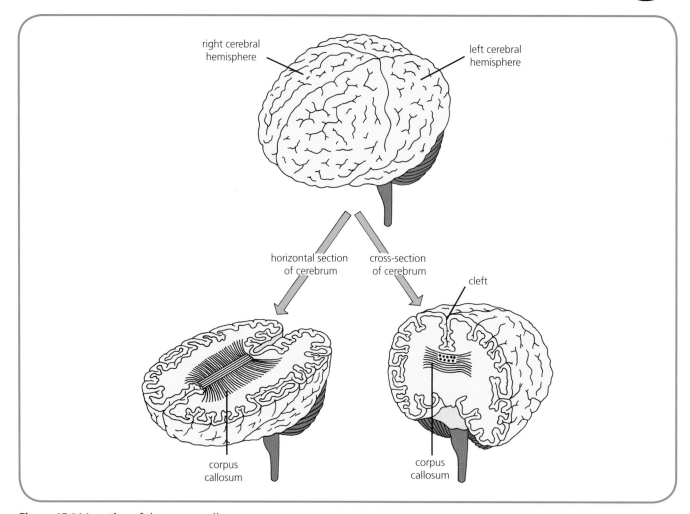

right cerebral hemisphere

left cerebral hemisphere

horizontal section of cerebrum

cross-section of cerebrum

cleft

corpus callosum

corpus callosum

Figure 17.14 Location of the corpus callosum

the **corpus callosum** (see Figure 17.14). This fibrous link between the two hemispheres is important to allow information to be transferred from one to the other. Whatever happens in one side of the brain is quickly communicated to the other side, thereby coordinating brain functions and enabling the brain to act as an integrated whole.

Functional areas of the cerebral cortex

The cerebral cortex contains three types of functional area: **sensory**, **association** and **motor**. Each area performs its own particular function distinct from the others. The sensory areas **receive information** as sensory impulses from the body's receptors (e.g. touch receptors in the skin and thermoreceptors in the hypothalamus). The association areas **analyse** and **interpret** these impulses, 'make sense' of them and 'take decisions' if necessary. The motor areas receive information from the association areas and 'carry out orders' by **sending motor impulses** to the appropriate

effectors (such as muscles). By this means, coordination of voluntary movement is effected.

As a simple example, imagine that you are blindfolded and a very large ice cube is placed in your hand. Sensory areas in your cerebral cortex receive information from touch, pressure and cold receptors in your skin. By analysing and interpreting these impulses, association centres in your cerebral cortex gain an impression of the size, shape, weight, texture and temperature of the 'mystery object'. When they put all this information together, it results in you experiencing the sensation of holding a very large ice cube. As the ice cube becomes uncomfortably cold and heavy, you reach a decision to let it go. The appropriate motor centre carries out orders by sending out impulses to the muscles that operate the hand, causing your grip on the ice cube to relax.

Localisation of function

A cerebral hemisphere consists of several **distinct regions**, as shown in Figure 17.15, which refers to the left side of the cerebrum. Each of these areas has a particular function to perform, as described below (see Related Topic – Functions of cerebral areas).

Every area shown in Figure 17.15 is **duplicated** (in mirror image) in the right cerebral hemisphere with the exception of the speech motor area. Each person has only one such region; it is situated on the left cerebral hemisphere in 90% of the population. Of particular importance for normal sensations are the **somatosensory**, **visual** and **auditory** areas. These localised regions receive separate sets of sensory impulses and register sensations that are evaluated in their association centres.

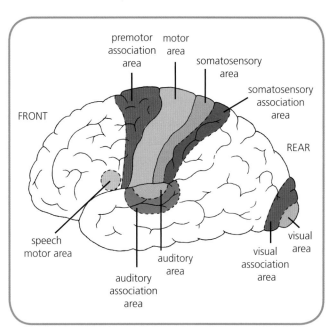

Figure 17.15 Detail of the left cerebral hemisphere

Related Topic

Functions of cerebral areas

The different functions of the various areas of the cerebral cortex are summarised in Table 14.1.

Area of cerebral cortex	Function
Somatosensory	Receives impulses from receptors in skin, organs and muscles
Somatosensory association	Receives impulses from somatosensory area, which it analyses, interprets and may 'act on'; also stores memories of previous experiences
Visual	Receives impulses from retina and interprets these as images involving shape, colour and so on
Visual association	Receives impulses from visual area and brings about recognition of observed objects by referring to memories of previous visual experiences
Auditory	Receives impulses from cochlea and interprets these as sounds involving pitch, volume and so on
Auditory association	Receives impulses from auditory area and brings about conversion of sounds into recognisable patterns, which can be understood depending on previous experience
Premotor association	Receives messages from other association areas and stores memories of past motor experiences, enabling it to plan and control complex sequences of motor activities such as writing, running down stairs and so on, in the light of past experience
Motor	Receives messages from premotor association area and obeys orders by sending motor impulses to appropriate skeletal muscles
Speech motor	Receives messages from other parts of cerebrum; translates thoughts into speech by sending impulses to the appropriate parts of the motor area which in turn send motor impulses to muscles controlling lips, tongue and vocal chords

Table 17.1 Functions of cerebral cortex

Interconnections

The cerebral cortex is the centre of conscious thought and no part of it works in complete isolation. **Interconnections** in the form of tiny nerve fibres link up the different areas and messages are constantly passing between them from sensory areas to motor areas via association areas. This allows for sophisticated perception of a situation involving several types of sensory impulse and the ability to make an **integrated response**. The cerebrum is also able to recall stored memories and then alter future behaviour in the light of past experience. In addition there are areas of the cerebral cortex that are responsible for higher mental processes such as intelligence, personality, creativity, imagination and conscience.

Related Topic

Evidence from brain injuries

Some forms of tumour and disease can injure certain areas of the brain. Similarly, specific regions of the brain can be damaged by accidents. Careful study of the effects of such **injuries** sometimes allows experts to infer the role played by a particular part of the brain.

Damaged frontal lobe

In 1848, an accident led to an inch-thick rod being driven into the head of a young American railroad worker. The rod entered beneath his left eye and exited through the top of his head (see Figure 17.16). Amazingly the man survived and was able to speak, think, remember and eventually return to work. However, he had changed. From having been mild-mannered and dependable, he had become ill-tempered, unreliable and no longer able to stick to a plan. This case and others involving damage to the **frontal lobes** of the brain show that they are involved in planning, goal setting and personality.

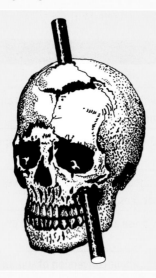

Figure 17.16 This brain damage was not fatal

Wife or hat?

In another case, a musician of great ability developed a problem in later life. He no longer recognised people or objects and failed to remember the past visually. He would chat to pieces of furniture thinking that they were people. On one occasion he reached out, took hold of his wife's head and tried to lift it to put it on, thinking that it was his hat! The problem was due to damage to his **visual association centres**.

Lesions

Lesions are small regions of damage. The location of the brain's **language areas** is verified by the fact that lesions in these regions give rise to speech defects. For example, extensive damage to the speech motor areas results in the person being unable to articulate words despite the fact that they fully understand the words that they hear.

Electroencephalograms (EEGs)

An **EEG** is a record of the cerebrum's **electrical activity** over a short period of time (e.g. 20–30 minutes). It is made using information from impulses picked up by electrodes placed on different regions of the scalp (see Figure 17.17). Different brain wave patterns indicate different levels of mental activity, as shown in Figure 17.18.

Figure 17.17 Preparing for an EEG

Compare the long, rolling wave pattern typical of sleep with the concentrated non-rolling pattern obtained when the subject is awake and concentrating. The more densely packed the 'spikes' in the pattern, the higher the level of electrical activity present in the brain. An extreme version of this is seen in the EEG of the person suffering an epileptic attack. In addition, patients who suffer personality problems accompanied by extreme changes of behaviour are often found to produce abnormal EEGs.

EEGs are useful but not very precise since they reflect the simultaneous activity of many cells all over the brain. Although an EEG may show an abnormal pattern indicating a possible problem, it fails to pinpoint the particular region of the brain responsible. However, reliable use can be made of EEGs in the diagnosis of comas and brain death.

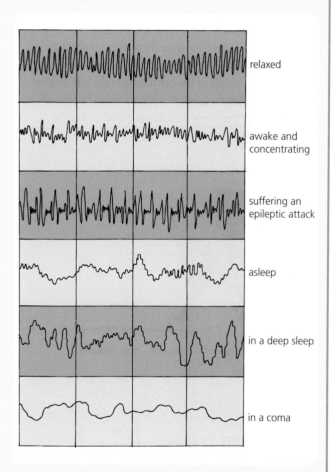

Figure 17.18 EEG wave patterns

Split-brain studies

Visual pathways

Normal situation

Each cerebral hemisphere controls the opposite side of the body. When the two eyes are focused on the centre of a two-tone field of view, each eye receives light from both sides of the field of view, as shown in Figure 17.19. However, each cerebral hemisphere only receives information from about half of this visual field. Everything to the left of the central line (the left field of view – blue in the diagram) is represented in the visual area of the right cerebral hemisphere; everything to the right of the central line (the right field of view –

orange in the diagram) is represented in the left cerebral hemisphere.

This occurs because half of the nerve fibres in each optic nerve **cross over** to the opposite side of the brain at the optic chiasma. Normally each side of the cerebrum quickly communicates its share of the information with the other side via the **corpus callosum**. As a result, both hemispheres perceive the whole field of view.

Abnormal situation

A person whose corpus callosum has been cut (for example, during an operation to try to relieve intractable epilepsy) is described as a '**split-brain**' patient. In such a person the exchange of information between the cerebral

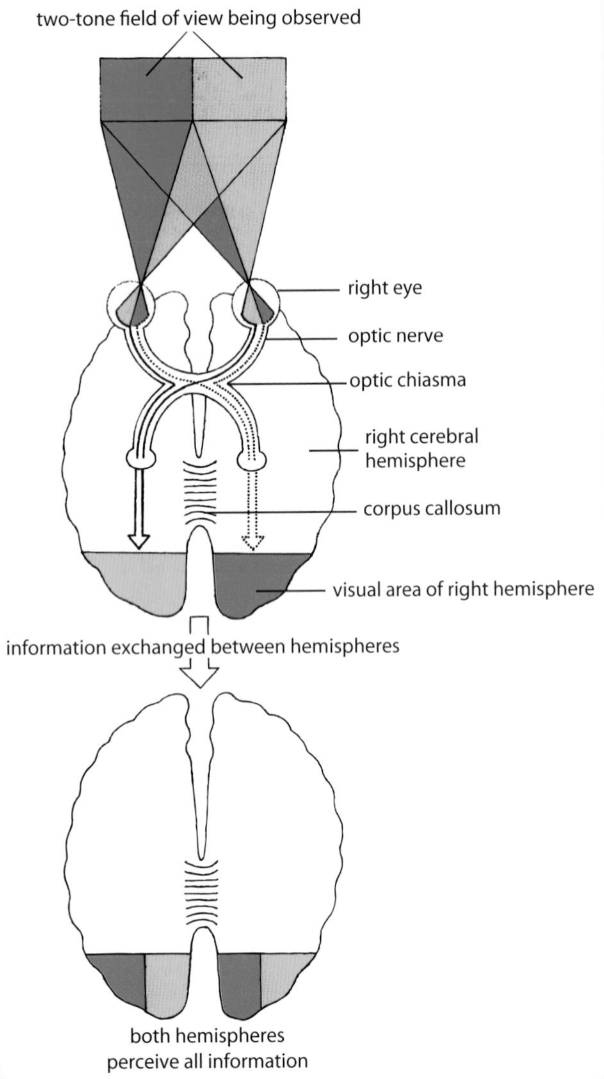

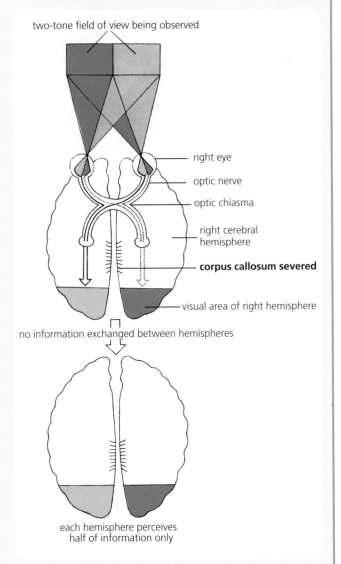

two-tone field of view being observed

right eye

optic nerve

optic chiasma

right cerebral hemisphere

corpus callosum

visual area of right hemisphere

information exchanged between hemispheres

both hemispheres perceive all information

Figure 17.19 Normal visual pathway

two-tone field of view being observed

right eye

optic nerve

optic chiasma

right cerebral hemisphere

corpus callosum severed

visual area of right hemisphere

no information exchanged between hemispheres

each hemisphere perceives half of information only

Figure 17.20 Abnormal visual pathway

hemispheres as described above cannot take place and each hemisphere receives only half of the information about the field of view (see Figure 17.20).

When asked to **say** what they see, the split-brain patient describes a field of view that is completely orange. This is further evidence of the localisation of brain function. It indicates that the motor speech area is in the left cerebral hemisphere only and that this is the left side of the brain 'talking'. However, when asked to **point with the left hand** to what they saw from a selection of possible fields of view, the person chooses a field of view that is completely blue since the left hand is controlled by the right side of the brain, which is now indicating its version of events.

Related Topic

Imaging the brain

Several techniques are now available that enable medical experts to create clear, visual **images of inside the brain** without involving surgery. These are used to study the normal workings of the brain and to diagnose various disorders. Two such techniques are described below.

PET (positron-emission tomography)

This type of **brain scan** reveals the location of active areas of the brain, showing **high metabolic activity** (an increased demand for glucose and oxygen). In preparation for a **PET** scan, the person receives an injection of glucose (or an appropriate alternative) labelled with a harmless isotope that shows up at the most active areas of the brain. These areas are detected by a PET camera and converted to a colourised image by a computer.

PET scans are used to diagnose abnormalities such as brain tumours (areas of high metabolic activity) and areas of neuron damage (low metabolic activity) that may lead to forms of dementia such as Alzheimer's disease (AD). For example, the PET scan in Figure 17.21 shows lower metabolic rates and lower levels of brain activity in a person who is affected by AD.

PET scans can also be used to identify those parts of the brain that show highest metabolic activity during particular **actions** and **emotions**. For example, a scan will show which area of the brain is most active when the person is, say, listening to music or stroking a furry object or suddenly feeling angry. These findings provide convincing evidence for the **localisation** of brain functions.

Language areas

The process of speech is found to involve several specific regions of the brain. These different **'language'** areas also show up on brain scans as regions of high metabolic activity (see Figure 17.22). When the information from several scans is put together, it gives a map of the brain's language areas, as shown in Figure 17.23.

fMRI (functional magnetic resonance imaging)

This technique employs radio waves and magnetic fields to produce images of the brain. It also depends on properties of haemoglobin present in blood that enable

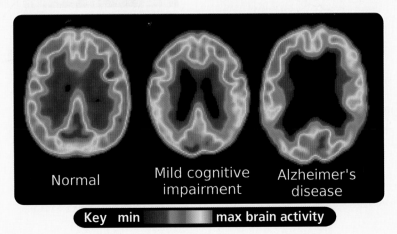

Figure 17.21 PET scans

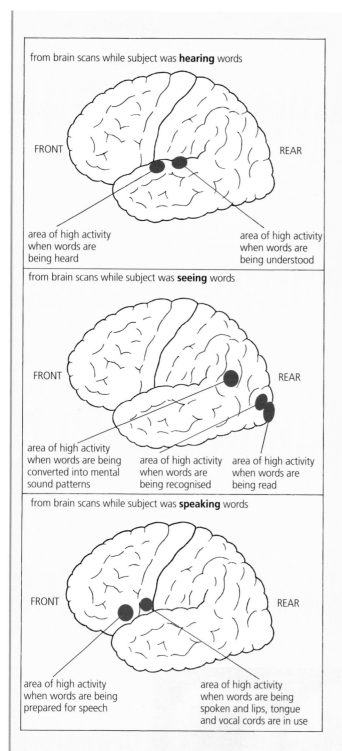

from brain scans while subject was **hearing** words

FRONT   REAR

area of high activity when words are being heard

area of high activity when words are being understood

from brain scans while subject was **seeing** words

FRONT   REAR

area of high activity when words are being converted into mental sound patterns

area of high activity when words are being recognised

area of high activity when words are being read

from brain scans while subject was **speaking** words

FRONT   REAR

area of high activity when words are being prepared for speech

area of high activity when words are being spoken and lips, tongue and vocal cords are in use

Figure 17.22 Language areas from brain scans

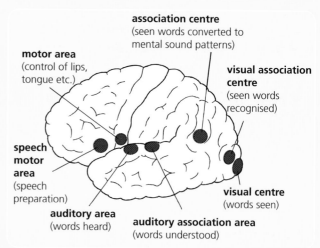

association centre
(seen words converted to mental sound patterns)

motor area
(control of lips, tongue etc.)

visual association centre
(seen words recognised)

speech motor area
(speech preparation)

visual centre
(words seen)

auditory area
(words heard)

auditory association area
(words understood)

Figure 17.23 Map of the brain's language areas

changes in patterns of blood flow to be followed. These indicate activity of brain cells. By this means, experts are able to create **anatomical and functional images** of the brain.

During an **fMRI scan** (see Figure 17.24), a person might, for example, be subjected to a variety of visual, auditory or tactile stimuli. The scan would then reveal those areas of the brain involved in vision, hearing or touch. In addition to research using healthy volunteers, fMRI is used to diagnose disease and disorders such as brain tumours.

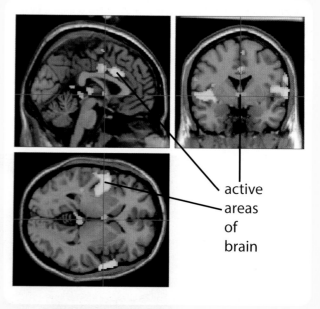

active areas of brain

Figure 17.24 fMRI scan

Testing Your Knowledge 2

1   Decide whether each of the following statements is true or false and then use T or F to indicate your choice. Where a statement is false, give the word(s) that should have been used in place of those in bold print. (6)

   a) The brain's central core contains the medulla and the **cerebrum**.

   b) Balance and muscular coordination are controlled by the **cerebellum**.

   c) Rate of breathing and heartbeat are regulated by the **pituitary gland**.

   d) Emotional and motivational states are influenced by the **limbic system**.

   e) The hypothalamus influences the secretion of hormones by the **cerebral cortex**.

   f) Homeostatic mechanisms involving contraction of smooth muscle are regulated by the **hypothalamus**.

2   a) Name THREE different types of functional area present in the cerebral cortex. (3)

   b) Briefly describe the function carried out by each of these different types of area. (3)

3   a) Which cerebral hemisphere controls
       i) the left, ii) the right side of the body? (1)

   b) i)   What is the *corpus callosum*?
       ii) State its function. (2)

What You Should Know Chapter 17

autonomic	experience	nervous
central	external	parasympathetic
cerebellum	hemisphere	peripheral
cerebral	hypothalamus	pituitary
conscious	integrated	recalled
contractions	involuntary	receptors
corpus	left	somatic
cortex	limbic	sympathetic
effectors	long-term	voluntary
emotional	medulla	water

Table 17.2 Word bank for chapter 17

1   Sensory information from the body's internal and _____ environment is analysed by the _____ system. Appropriate motor responses are then initiated which lead to muscular _____ and glandular secretions.

2   The nervous system can be divided on a structural basis into the _____ nervous system (CNS) and the _____ nervous system (PNS).

3   The nervous system can be divided on a functional basis into the _____ nervous system (SNS) and the _____ nervous system (ANS).

4   The SNS controls _____ movement of the skeletal muscles.

5   The ANS works automatically without involving the person's _____ thought. It regulates the internal environment by exerting _____ control over structures such as the heart and alimentary canal.

6   The ANS is made up of the _____ system which arouses the body in preparation for 'fight or flight' and the _____ system which calms the body and promotes 'rest and digest'.

7   The brain is composed of the central core, the _____ system and the _____ cortex.

8   The central core contains the _____ which regulates heartbeat and breathing, and the _____ which controls balance, movement and posture.

9   The limbic system processes information needed for _____ memories and influences biological motivation and _____ states. It contains the _____ which produces releaser hormones. These make the _____ gland secrete further hormones. The hypothalamus also regulates _____ balance and body temperature.

10 The cerebral _____ is the centre of conscious thought. Its sensory areas receive sensory impulses from _____. Its association areas analyse and interpret information and its motor areas send motor impulses to _____.

11 The right cerebral _____ receives information from the left visual field and controls the _____ side of the body. The reverse is true of the left cerebral hemisphere. Information is transferred from one hemisphere to the other via the _____ callosum, enabling the brain to work as an _____ whole.

12 Nerve cell activity in the cerebral cortex enables decisions to be made, memories to be _____ and behaviour to be altered in the light of _____.

18 Perception and memory

A person's awareness of the components of their external environment is based on information picked up by receptors in sensory organs and passed to the brain. This information is then subjected to analysis and interpretation by the brain and converted into the person's **perception** of their surroundings. Although many perceptual experiences depend on information from sense organs other than the eyes, this chapter concentrates on **visual** perception. This form of perception enables a person:

- to segregate objects from one another and from their background
- to judge how near or far various objects are situated from the person
- to recognise different objects.

Segregation of objects

Perceptual organisation into figure and ground

The first stage in the development of visual perception is the appreciation of certain aspects of an object's shape. In drawing, any line that encloses an area whose shape is recognised as representing an object appears to stand out from the background in an obvious manner. This form of perceptual organisation is called the 'figure-ground' phenomenon. For example, Figure 18.1 shows a black triangular figure that stands out from the white background. The part seen as the figure tends to appear slightly in front of the 'ground', although they are printed on the same two-dimensional surface. Any field of view that contains contrasting components is perceived as 'figural'. One or more parts of it become segregated as figures that stand out from the rest of the field (that forms the ground).

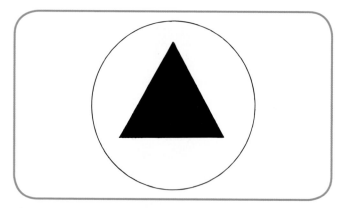

Figure 18.1 Figure-ground phenomenon

Related Activity

Analysing reversible figure-and-ground images

Although shapes that are prominent, interesting, striking or central in the field of view tend, more readily, to become figural, there is no hard-and-fast rule. Nor must the same object or part of the field of view always remain the figure. By fixedly gazing at one part of the field of view or by employing a 'switch of attention' to different parts of the field of view, an alternation often occurs between the figure and the ground, as in the following examples.

Identifiable shapes

Close examination of Figure 18.2 reveals both a black vase figure against a white ground and twin white human profile figures against a black ground.

Figure 18.2 Vase or faces?

Non-identifiable shapes

Geometric shapes and patterns seen against a background appear object-like with contours and boundaries. Very often these figure-ground relationships are reversible. Figure 18.3 shows a flag-like image. Is it a white 'cross' on a black ground or a black 'four-leafed clover' on a white ground? Figure 18.4 shows a pattern of abstract white plant-like shapes against a black ground. Or is it black washing drying on a line against a white ground? The reversible figure-ground effect works best when the two parts of the image are equally meaningful.

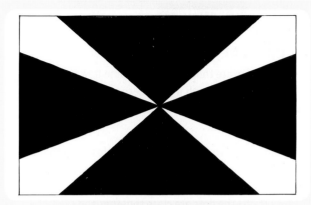

Figure 18.3 Reversible geometric shapes

Figure 18.4 Reversible abstract pattern

Perceptual organisation into coherent patterns

The brain tends to organise visual stimuli that it receives into a coherent pattern rather than into many different parts. Even simple patterns of dots or lines tend to fall into **ordered relationships** when viewed. Figure 18.5 appears to consist of three pairs of lines with an extra unpaired line at the right. Figure 18.6 shows how a modification to the lines tends to reorganise the image as three pairs of lines beginning at the right, with an extra unpaired line at the left.

Figure 18.5 Perception of pattern

Figure 18.6 Reorganisation of perceived pattern following modification

Related Activity

Grouping of visual stimuli

Several factors are thought to determine how the visual system organises stimuli into patterns. They include **proximity**, **similarity**, **closure**, and **orientation**.

Proximity

The mind groups visual stimuli that are close together as part of the same object. Conversely, it groups stimuli that are far apart as separate objects. Consider Figure 18.7. In the left box the dots are close together and form one group (a 'square'). In the box on the right, the spacing of the same set of dots has been changed. Some dots are in closer proximity to one another than to the rest of the dots. The brain now perceives the set as four groups based on the **close proximity** of the members within each group. Similarly, the brain perceives pictures composed entirely of dots as several distinct groups or clusters rather than individual dots. This enables the brain to make sense of a 'dot picture', such as the one in Figure 18.8.

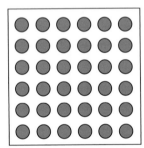
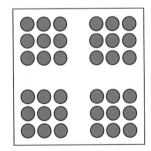

Figure 18.7 Effect of proximity

Figure 18.8 Groups of dots as distinct images

Similarity

The brain organises similar visual stimuli into groups based on factors such as colour, size, shape and tone. In Figure 18.9 some of the dots are blue and some are orange. The brain perceives the dots as three distinct groups based on **similarity of colour** of the members within each group.

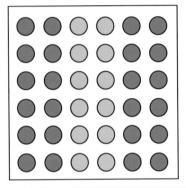

Figure 18.9 Effect of similarity

Closure

When a figure is incomplete or partially hidden by another object, the brain 'fills in' the elements that it does not perceive as visual stimuli. This process is called **closure**

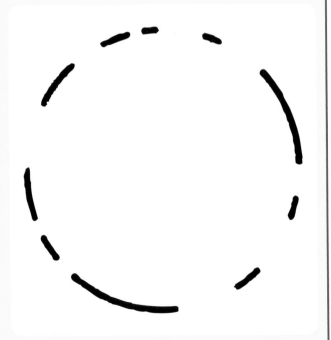

Figure 18.10 Closure

and it is the means by which a whole rather than an incomplete object is perceived. For example, the image in Figure 18.10 is perceived as a circle (rather than just a series of curved, broken lines) as a result of closure.

Orientation

The brain organises visual stimuli from elements that are all seen moving in the same direction at the same rate as one object. For example, the individual birds in a migrating flock are perceived collectively as one object (see Figure 18.11).

Figure 18.11 Orientation

Related Activity

Grouping of stimuli by proximity and similarity

Figure 18.12 is composed of eight parts (**A–H**), each of which is a set of objects. Study each set and decide:

i) the **number** of separate groups of objects present within the set
ii) whether your brain employed **proximity** of objects within a group or **similarity** of objects within a group to arrive at your answer to **i)**.

(Answers below)

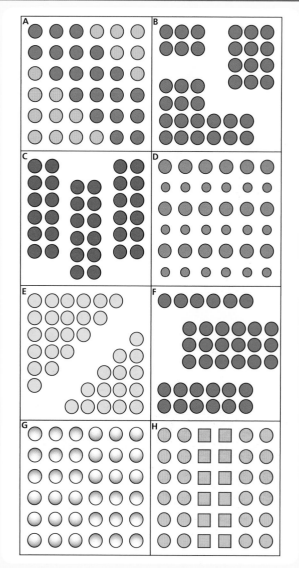

Figure 18.12 Grouping of stimuli by proximity or similarity

Answers:

i) A = 3, B = 3, C = 3, D = 6, E = 3, F = 2, G = 4, H = 3

ii) A = S, B = P, C = P, D = S, E = P, F = P, G = S, H = S

Related Activity

Closure

Study the eight images in Figure 18.13 and work out what the brain perceives when it has filled in the gaps by closure in each case.

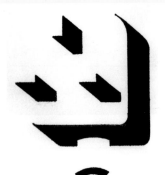

Figure 18.13 Closure activity

Perception of distance

Visual cues

The distance of one or more objects from the eye is indicated by the presence of one or more **visual cues** present in the scene being viewed.

Relative size

The further away an object is situated from the eye, the smaller it is perceived to be. For example, by appearing to decrease in size, the sleepers in the railway line in Figure 18.14 indicate their increasing distance from the eye.

Figure 18.15 Effect of relative height

Relative height in horizontal field

Among a group of objects whose **bases are below the horizon**, those with their base in a relatively higher position appear to be further away and those with their base in a relatively lower position seem to be nearer the eye. For example, in Figure 18.15 the base of fence post X is relatively higher than that of fence post Y. This makes X seem further away than Y. This figure also shows the effect of relative size on the perception of distance.

Texture gradient

A change in the relative size and density of objects is perceived when they are viewed from different distances. The fact that the stones on the left in Figure 18.15 become smaller and less textured leads to the perception that they are more distant.

Figure 18.14 Visual cues

Superimposition

When one object partially blocks the view of another, the blocked object is perceived to be further away. In Figure 18.14, hill A at the front overlaps and cuts off part of the view of hill B behind it. Therefore hill A is perceived to be nearer than hill B.

Linear perspective

Parallel lines such as the railway line in Figure 18.14 and the road in Figure 18.15 appear to converge in the distance.

Investigation

Related Activity

Analysing images of depth perception

Identify and discuss the visual cues that enable the brain to perceive depth in Figure 18.16.

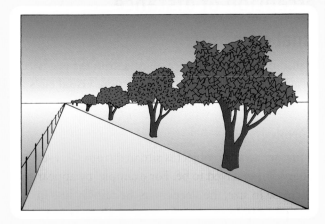

Figure 18.16 Identifying visual cues

Müller-Lyer illusion

For the most part our visual perception serves us very well and seeing really is believing. However, on some occasions such as the viewing of an optical (visual) illusion, perception can be misleading. The **Müller-Lyer illusion** is an optical illusion based on two arrow-like figures. One has two 'tails', the other has two 'heads', as shown in Figure 18.17. Each head or tail is composed of two short lines called **fins**. The length of the arrow with two 'tails' is perceived to be longer than the arrow with two 'heads'.

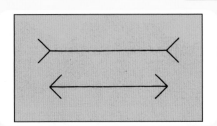

Figure 18.17 Müller-Lyer illusion

Figure 18.18 shows the chart used in this investigation. It has an 'arrow' XY (called the **standard stimulus line**) and a blank line below XY. The figure also shows cards A and B. These begin each trial in boxes A and B. They are picked up simultaneously by the subject and placed onto the blank unlabelled line in an attempt to create 'arrow' AB (the **comparison stimulus line**) which appears equal in length to arrow XY. The length of AB is carefully measured and the reading recorded. The same person repeats the procedure a further four times and an average value is calculated. The difference between the average value for the length of AB and the actual length of XY is calculated. The results are pooled and presented as a graph.

Alternative designs

The investigation can be carried out using arrow heads and tails with fins of different relative length from those in Figure 18.18 or drawn at a different angle, or the arrows drawn using relatively thicker lines. It can also be repeated using the arrows arranged vertically instead of horizontally. Table 18.1 gives the reasons for adopting certain procedures during the investigation.

Procedure	Reason for adopting procedure
The chart and cards A and B are drawn on plain (not squared or lined) paper	To prevent the presence of squares or lines helping the subject to judge distance
Cards A and B begin at the same starting positions each time before being picked up	To prevent each new attempt to create line AB being influenced by the result of the previous attempt
Each person makes five attempts	To increase the reliability of the results
Many students are tested	

Table 18.1 Reasons for adopting procedures during investigation

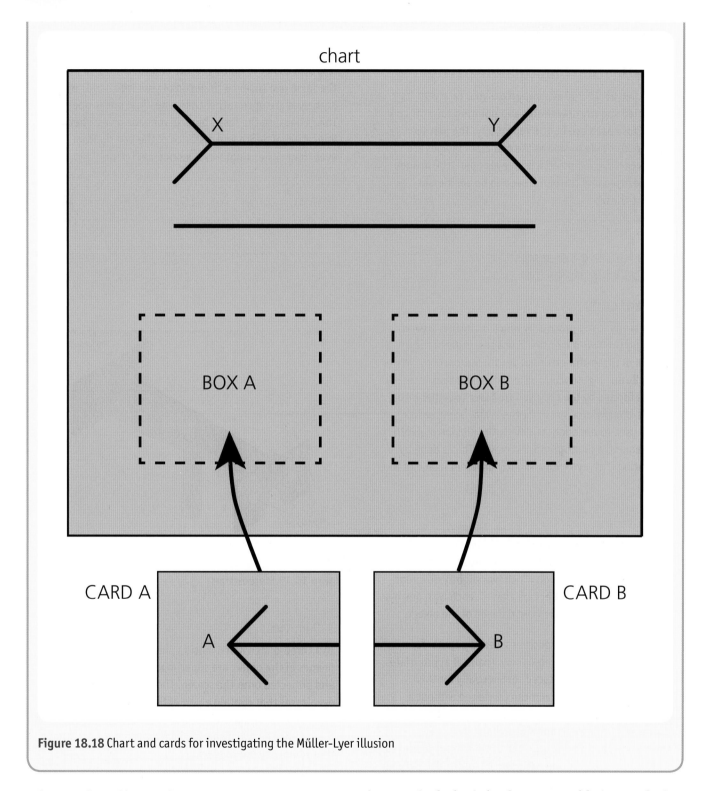

Figure 18.18 Chart and cards for investigating the Müller-Lyer illusion

Binocular disparity

Each eye looks at an object from a slightly different position relative to the other eye. Therefore a slight disparity (difference) occurs between the images of the same object formed by the two eyes. The closer the object is to the viewer, the greater the disparity between the two images. The two images are merged into one in the brain by the process of fusion producing a **binocular** image. This image indicates depth and distance of the object more effectively than either single-eye, **monocular** image. The process may fail to occur if excessive quantities of alcohol have been consumed. Then the person may 'see double'.

Investigating binocular disparity

Hold a pen about 300 mm away from your face with its tip angled slightly downwards and pointing towards your nose. Close each eye in turn repeatedly. The pen seems to jump from side to side as the disparate images are perceived time about. With both eyes open, the images become fused and the pen is seen to be positioned straight ahead.

It is also easier to quickly locate and touch the tip of the same pen (using a fingertip of your free hand) when both eyes are open rather than when just one is open. This is because the fused binocular image gives a better indication of depth and distance of the pen tip from the eyes than either single-eye, monocular image.

Stereoscopy

Stereoscopy is the process by which an image, created by fusing two two-dimensional images, is given the illusion of **three-dimensional depth**. Stereoscopy occurs naturally during binocular vision when the disparate images formed by the two eyes are fused together into a three-dimensional image by the brain. Stereoscopy can be used to enhance artificially the illusion of depth in a diagram, a photograph or a movie. This is done by presenting each eye with a slightly different image of the object.

Stereoscope

A **stereoscope** is a simple optical instrument (see Figure 18.19). It has a binocular eyepiece containing magnifying lenses through which two slightly different pictures of the same object are viewed, each with one eye. This creates a three-dimensional image of the object.

Over 100 years ago, stereoscopes were a popular form of entertainment and stereograms (pairs of appropriate photographs) were produced commercially for this purpose. Some stereograms can be viewed without a stereoscope. Figure 18.20 shows two slightly different images of a molecular model of a channel in a cell membrane. Hold the picture 100–200 mm from your eyes, stare at the space between the two images and allow your eyes to relax. Some people find that the images fuse, producing a third, centrally located, three-dimensional version of the molecular model.

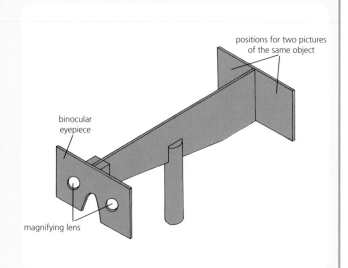

Figure 18.19 Stereoscope

3-D movies

In a three-dimensional motion picture, two moving images of the object are superimposed on one another and projected onto the screen. The viewer wears eyeglasses that contain a pair of filters. Each filter only allows light through from one of the moving images. The brain fuses the two disparate images as before, creating a three-dimensional illusion.

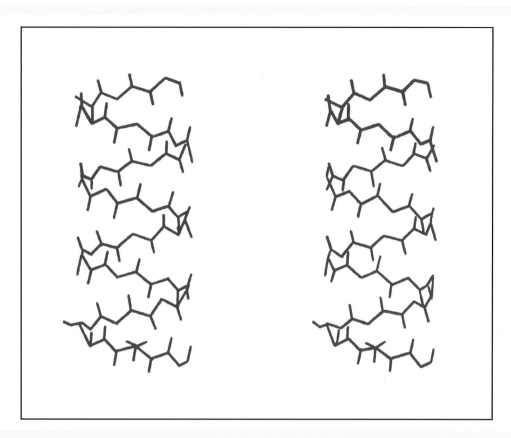

Figure 18.20 Stereogram of molecular model

Perceptual constancy

A person's visual perception of their surroundings remains stable despite the changing images landing on their retinas as they move their eyes or body around. If retinal images were taken at face value, objects would appear to **increase or decrease in size** as they moved towards or away from us. Similarly objects would appear to **change in shape** when viewed from different angles (and colours would seem to alter in response to different levels of illumination).

Size constancy

The perception that an object remains the same size regardless of the size of the image on the retina is called **size constancy**. Imagine you throw a ball for a dog to fetch. As the dog runs away from you, its image on your retina becomes smaller and as it runs towards you, its image on your retina becomes larger. However, you do not believe that the dog has shrunk and then increased in size again.

Size constancy is thought to depend, in part, on **past** experience and **stored knowledge** and, in part, on the cue of **relative size**. Imagine that you move towards a bookcase to fetch a particular book. The retinal image of the book becomes larger as you approach it. However, the retinal image of the bookcase and the other books also becomes larger, so the relative size of the various objects to one another does not change. The book appears constant because it maintains the same size relative to the objects that surround it.

Shape constancy

The perception that an object remains the same **shape** regardless of the changes in the angle at which it is viewed is called **shape constancy**. For example, when a door is viewed as it swings open, it goes through a series of changes with respect to the images of it that land on the retina (see Figure 18.21). However, you do not believe that the door's shape has actually changed from a rectangle to a trapezoid. You perceive an unchanging door. Similarly, the rim of a glass is perceived as round whether it is viewed from above or from the side. Shape constancy is thought to depend largely on past experience and stored knowledge.

to be more important than details such as colour or texture. Figure 18.22 represents eight different types of fruit based on their colour. However, it is not possible to identify the fruits using colour as the sole visual cue. Figure 18.23 represents the same eight fruits using their shape as the visual cue. Now it becomes possible to recognise pear, plum, pomegranate, banana, apple, lemon, orange and grapes.

It is by means of their shape that objects are initially characterised and differentiated from one another during early learning. This information is then stored in the long-term memory (see page 242). The most important feature of an object's shape is its **general outline**. Young children, in particular, gain details about this characteristic by both viewing the object and running their fingers around its edge.

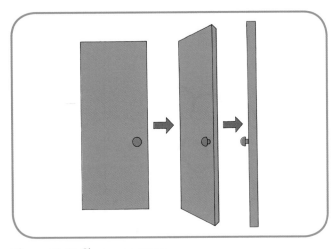

Figure 18.21 Shape constancy

Recognition

Importance of shape

The ability to perceive an object's physical properties such as shape, colour and texture is called **object recognition**. Of these properties, **shape** is considered

Matching perceived shapes

When a person perceives a shape, a subconscious attempt is made to **match** this shape with one of the

Figure 18.22 Colour as a recognition cue

Figure 18.23 Shape as a recognition cue

'**visual descriptions**' already stored in the brain. If the shape is a familiar one, it will be quickly matched and the object recognised. If the shape is unfamiliar it may draw a complete blank.

On the other hand, it may be recognised as being similar but not identical to one or more visual descriptions held in the brain. The brain then infers that the object is related in some way to one or more of these visual images. It may then recognise that the shape represents a familiar object shown from an unusual angle (see Figure 18.24) or it may recognise that the object is a further member of an already familiar group of objects. Imagine, for example, that you have never seen a pink grapefruit (see Figure 18.25). Your brain would not be able to name this type of fruit but it would recognise from its shape and general appearance that it is similar to oranges and yellow grapefruit and therefore it is a type of citrus fruit.

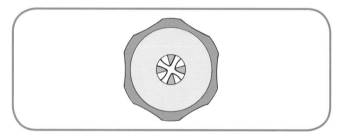

Figure 18.24 Familiar object, unfamiliar angle (answer below)

Figure 18.25 Pink grapefruit

Answer to Figure 15.24: Screwdriver

Perceptual set

Perceptual set is the tendency of a person to perceive certain aspects of available sensory information and ignore others. Perceptual set is affected by **expectation**, **context** and **past experience**. It influences the way in which a stimulus is perceived.

Expectation and context

Figure 18.26 shows a deliberately ambiguous figure in the context of both letters and numbers. When the letters A and C are covered up and the information is read horizontally, the central figure is perceived to be the **number 13** because of the context of the **other numbers**. When the numbers 12 and 14 are covered up and the information is read vertically, the central figure is perceived to be the **letter B** because of the context of the **other letters**. In each case the stimulus is the same but it is perceived differently because the viewer expects to see numbers along with other numbers and letters along with other letters.

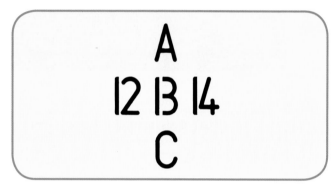

Figure 18.26 Letter or number?

Past experience

Expectations are often, at least in part, the result of **previous experience**. In an investigation, the people in test group A were shown a series of pictures of small mammals (some of them rodents) and those in group B were shown a series of pictures of humans (some bald and some wearing glasses). The members of each group were then shown the ambiguous 'ratman' diagram (see Figure 18.27).

Figure 18.27 'Ratman'

Most people in group A perceived the picture as a mouse or a rat whereas most people in group B perceived it as the face of a bald man wearing glasses. In each case, **perceptual set**, brought about by previous experience, had influenced which sensory data were perceived and which were ignored.

It is interesting to note that when the members of each group were then shown the other set of pictures, it did not change their perception of 'ratman' in most cases.

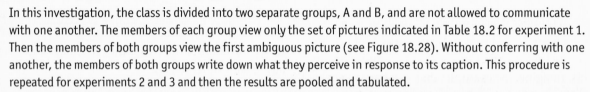

Influence of perceptual set using ambiguous stimuli

In this investigation, the class is divided into two separate groups, A and B, and are not allowed to communicate with one another. The members of each group view only the set of pictures indicated in Table 18.2 for experiment 1. Then the members of both groups view the first ambiguous picture (see Figure 18.28). Without conferring with one another, the members of both groups write down what they perceive in response to its caption. This procedure is repeated for experiments 2 and 3 and then the results are pooled and tabulated.

Experiment	Theme of ten pictures to be viewed during prior treatment		Ambiguous picture to be viewed
	Group A	**Group B**	
1	Human females aged approximately 10–30	Human females aged approximately 60–90	Figure 18.28
2	Birds, some of which are ducks, geese and swans	Small mammals, some of which are hares and rabbits	Figure 18.29
3	Two-dimensional geometric shapes, each containing a smaller, different geometric shape at its centre	Two-dimensional geometric shapes, each containing a smaller, different geometric shape near its edge	Figure 18.30

Table 18.2 Treatments intended to create perceptual set

Figure 18.28 Young woman or old woman?

Figure 18.29 Duck or rabbit?

Table 18.3 shows a typical set of results for this investigation. In each case, **perceptual set** brought about by **previous experience** has influenced the perception of the ambiguous image. For example, in experiment 1, most students in group A exposed in advance to pictures of younger women saw a young woman, whereas those in group B exposed to pictures of older women perceived an older woman. Table 18.4 gives some procedures adopted during this investigation and the reasons for doing so.

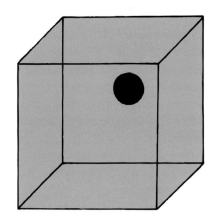

Is the ball in the centre of the back surface or in the corner of the front surface?

Figure 18.30 Where is the ball?

Experiment	Image perceived	Group A		Group B	
		Number of students	% number of students	Number of students	% number of students
1	Young woman	9	90	2	20
	Old woman	1	10	8	80
2	Duck	10	100	3	30
	Rabbit	0	0	7	70
3	Ball in centre	7	70	1	10
	Ball in corner	3	30	9	90

Table 18.3 Results for perceptual set investigation

Procedure	Reason for adopting procedure
No communication allowed between groups A and B	To prevent members of one group viewing the other group's preparatory pictures so that optimal conditions are created to bring about two types of perceptual set
No collaboration allowed among the members of a group once the ambiguous figure has been viewed	To prevent group members influencing one another's perception of the ambiguous image
Large number of students invited to take part	To increase the reliability of the results
Several ambiguous images used	

Table 18.4 Reasons for adopting procedures during investigation

Testing Your Knowledge 1

1 Decide whether each of the following statements is true or false and then use T or F to indicate your choice. When a statement is false, give the word that should have been used in place of the one(s) in bold print. (6)
   a) The process by which the brain analyses and makes sense of incoming sensory information is called **sensitivity**.
   b) A form of perceptual organisation where a shape stands out from its surroundings is called the **figure-ground** phenomenon.
   c) Visual stimuli are organised into coherent **patterns** by the brain.
   d) The perception that an object remains the same shape regardless of changes in the angle at which it is viewed is called shape **segregation**.
   e) The most important visual cue in the recognition of an object is its **texture**.
   f) The tendency to perceive some aspects of sensory information and ignore others is called **perceptual set**.

2 a) Name THREE types of visual cue that enable the brain to perceive distance. (3)
   b) i) What is meant by the term *binocular disparity*?
      ii) By what means is a binocular image formed? (3)

3 Identify TWO aspects of an object that are perceived to remain constant despite the fact that the person viewing it is moving towards it. (2)

4 a) Briefly describe how the brain makes use of stored visual descriptions in order to recognise objects. (2)
   b) In Figure 18.31, the central figure could be seen as the letter I or the number 1. Briefly explain this ambiguity in terms of perceptual set. (3)

Figure 18.31

Memory is one of our major mental faculties. It is the capacity of the brain to **store** information, **retain** it and then **retrieve** it when required. The brain is so **versatile** that it can capture images of sights, sounds, smells, tactile sensations and emotions all experienced at the one time and retain them as memories. For example, a child receiving a present of a new puppy remembers various details of the experience vividly for a long time. The brain is able to store a vast quantity of knowledge, thoughts and detailed information relating to past experiences as **memories**. Memory enables us to deal with future situations in the light of past experience. In the absence of memory we would be helpless, unable to manage even the simplest task without having to first relearn it.

Selective memory

The receptors in the human sense organs are continuously picking up stimuli and transmitting sensory impulses to the brain. However, only a fraction of the sensory images formed become **committed to memory** because the process is highly **selective**. If this were not the case the mind would become cluttered with useless information such as every phone number ever used, every musical note of every tune ever heard whether liked or disliked, and so on.

Encoding, storage and retrieval

To become part of the memory, the selected sensory image must first become **encoded** (converted to a form that the brain can process and store). **Storage** is the retention of information over a period of time. This may last for only a brief spell such as 30 seconds or for a very long period, perhaps a complete lifetime.

Retrieval is the recovery of stored material. This involves the recall of information that has been committed to either the **short-term** or the **long-term** memory (see page 242). Thus when memory is functioning properly, encoding leads to storage of information that can be retrieved later when required.

Different levels of memory

Memory is thought to involve three separate **interacting levels**, as shown in Figure 18.32. All information that

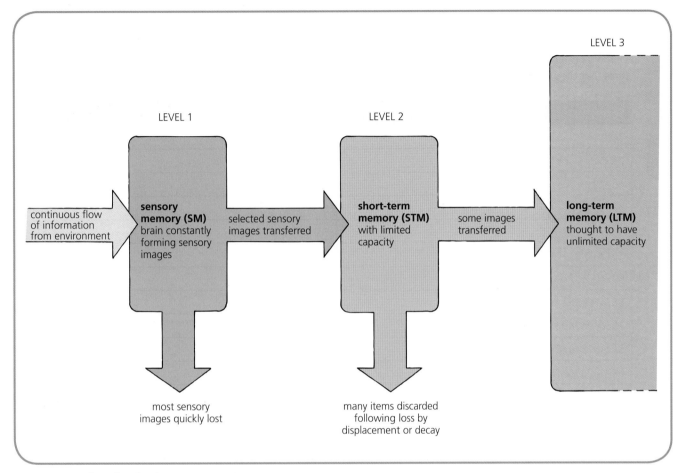

Figure 18.32 Three levels of memory

gains access to the brain must first pass through the **sensory memory (SM)** and then, if selected, enter the **short-term memory (STM)**. From there it may be transferred to the **long-term memory (LTM)** or be discarded.

Level 1 – sensory memory (SM)

Stimuli from the outside world are continuously being perceived as sensory images by the brain. These impressions are very **short-lived** (e.g. 0.5 seconds for visual and 2 seconds for auditory) and only a few are selected and transferred to level 2. The SM provides a detailed representation of the person's entire sensory experience from which relevant pieces of information are sent to the STM.

Level 2 – short-term memory (STM)

Most of the information encoded into this second level of the system consists of visual and auditory images. However, the STM holds only a limited amount of information – about seven items at the one time (see Investigation – Length of memory span). Not only does the STM have a limited capacity but, in addition, the items are held for a short time (approximately 30 seconds). During this time, retrieval of items is very accurate. Thereafter they are either transferred to level 3 or lost by **displacement** (the pushing out of 'old' by new incoming information) or by **decay** (the breakdown of a fragile 'memory trace' formed when a group of neurons briefly became activated).

Length of memory span

A person's **short-term memory span** can be measured by finding out the number of individual 'meaningless' items that they can reproduce correctly, and in order, immediately after seeing or hearing them once.

Series	Number of digits in series
741	3
2835	4
46279	5
584153	6
9082637	7
16136209	8
592403517	9
8076148362	10
78501942493	11
512367509308	12
6821496708754	13

Table 18.5 List of series of digits

In the following investigation there is one tester (e.g. the teacher) and many subjects (the members of the class). The tester reads out the first series of digits (see Table 18.5) clearly and at uniform speed. Immediately after reading out the last digit of the first series, the tester signals that all subjects should lift their pencils, write down the series of digits that they have just heard and then lay their pencils down again.

The tester then reads out the next series, and so on until the end of the list. The responses are checked and each subject's memory span for the first list is established. The procedure is repeated twice using different lists. Each subject's **best overall score** is taken to represent his or her memory span. The class results are pooled and graphed. Figure 18.33 charts a typical set of results for a class of 20 pupils. Table 18.6 outlines the design features of this investigation and the reasons for employing them.

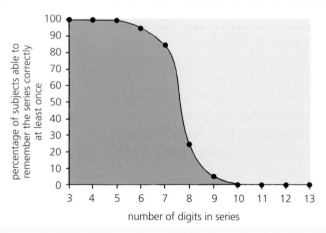

Figure 18.33 Graph of memory span results

From the results in Figure 18.33, it is concluded that for this group of pupils, the poorest (minimum) memory span was 5 digits (one pupil), the best (maximum) memory span was 9 digits (one pupil) and that all of the other subjects had a memory span somewhere in between. On average, the human short-term memory span is found to be 7 ± 2 digits, though some amazing exceptions have been recorded.

Design feature or precaution	Reason
Series of random numbers used	To eliminate easily remembered sequences or groups of numbers
All information read out clearly at uniform speed by the same tester	To ensure that the only variable factor is the number of digits in the series
Pencils laid down between responses	To prevent over-eager subjects starting to write down the series before it has been completely read out
Each subject given three attempts	To obtain a more reliable result for each subject's best score
Many pupils tested and the results pooled as a graph	To further increase the reliability of the results

Table 18.6 Design features for memory span investigation

Chunking

A **chunk** is a meaningful unit of information made up of several smaller units. To most people familiar with the dates of the Second World War, 1945 is one chunk of information not four chunks. However, to most people 4951 is four chunks of information (unless it happens to be something significant such as the PIN number of their bank account). Since short-term memory is only capable of holding about seven new items at one time, **chunking** is a useful method of increasing its memory span. The compilers of all-digit telephone numbers provide users with means to transfer an 11-digit number from directory to telephone by chunking.

Imagine, for example, that a business woman in Aberdeen wishes to phone an unfamiliar Glasgow number (e.g. 01416293801). If she already knows that Glasgow's national code is 0141 then that chunk reduces her task to remembering 8 items. If in addition she has cause to phone Glasgow fairly regularly and recognises 629 as a district code then this becomes a second chunk. The job now demands a memory span of 6 items, which many people can manage comfortably.

Related Activity

Investigating the effect of chunking on memory span

In this investigation, the subjects view list 1 only (in Table 18.7) for 2 minutes and are then allowed 4 minutes to write down as many of the 3-lettered items as they can remember. The procedure is repeated for list 2. The items in list 2 are found to be much more easily remembered than those in list 1. This is because each is an acronym that acts as a meaningful chunk of information.

List 1	List 2
ICL	PIN
TPT	KGB
OML	HIV
MVM	SQA
EZQ	VAT
CPG	FBI
UPR	PTO
DUL	USA
MCA	RAC
SUT	NYC
ATX	UFO
NSE	BBC
RPA	MOT
YAD	RIP
BCU	PLC

Table 18.7 Effect of chunking

Rehearsal

Rehearsal involves repeating to yourself over and over again (silently or out loud) a piece of information that you are trying to memorise. This process helps to extend the time for which the information is maintained in the STM.

Serial position effect

From the results in Related Activity – Investigating the serial position effect, it can be seen that recall is best for the objects shown at the end (**recency effect**), closely followed by those shown at the start (**primacy effect**). Those in the middle of the viewing sequence gain a very poor score. This memory pattern is called the **serial position effect**.

Related Activity

Investigating the serial position effect

In this investigation, the tester informs the subjects that they will be required to memorise 20 fairly similar objects. The tester reveals the first object and allows the subjects to view it for 5 seconds. Object 1 is then removed and object 2 revealed for 5 seconds.

This procedure is repeated for the remaining objects. As soon as the last one has been removed, the subjects are invited to pick up their pencils and write down as many of the objects as they can recall in any order. Figure 18.34 shows a graph of a typical set of results for a group of 100 subjects.

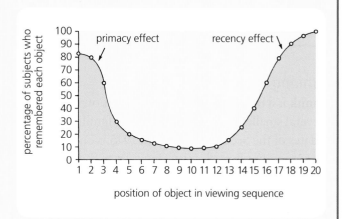

Figure 18.34 Serial position effect

Explanation

Images of the first few objects can be remembered because during the experiment there has been enough time for them to have been well **rehearsed**. In many cases they have therefore become **encoded** and **transferred** to the LTM from where they can be retrieved at the end of the experiment.

The last seven or so objects are remembered because images of them are still present in the **STM** and are quickly 'dumped' onto paper by the subjects as soon as they start writing. (If there is a 1 minute delay and rehearsal is prevented before subjects are allowed to write down the objects that they recall, the recency effect vanishes.) Images of the objects in the middle of the sequence are not well retained by the vast majority of subjects because, by the time these images enter the STM, it is already crowded. Therefore many are forgotten before they can be rehearsed, encoded and stored in the LTM.

Working memory

Working memory is an extension of the short-term memory. It actively processes, manipulates and controls information while it is held in the STM. This enables simple **cognitive tasks** to be carried out. Imagine, for example, that you have been asked to think of all the pieces of furniture containing drawers that are present in your home and then calculate the total number of drawers. To do this, you form a mental image of your home and then go for a visuo-spatial tour, room by room. As you come to each relevant piece of furniture, you employ your working memory to count the number of drawers and add this value to the running total in your STM.

Level 3 – long-term memory (LTM)

This third level in the system (see Figure 18.35) is thought to be able to hold an **unlimited** amount of information. During encoding, the items are organised into **categories** such as personal facts and useful skills.

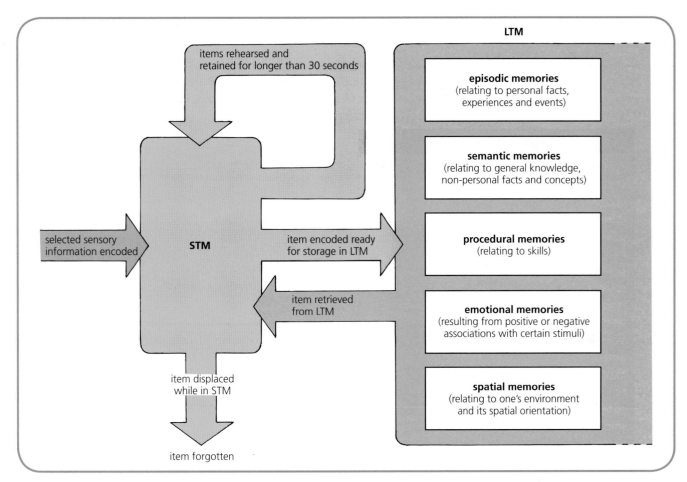

Figure 18.35 Transfer of information

These are then stored for a long time, perhaps even permanently. It must be stressed that this multi-level model of memory and how it works is probably an oversimplification. In addition, the three levels of memory should not be thought of as occupying three distinct regions of the brain.

Transfer of information between STM and LTM

A hypothetical representation of this process is shown in Figure 18.35. Information is constantly being **transferred** between the brain's two storage 'depots' – the **STM** and the **LTM**. If, during its brief stay in the STM, an item is successfully **encoded** (see page 242) then this enables it to be transferred for **storage** in the LTM which has an enormous, perhaps unlimited, capacity. This item may later be **retrieved** from the LTM when required. Successful transfer of information from the STM to the LTM is promoted by the processes of **rehearsal, organisation** and **elaboration of meaning**.

Benefit of rehearsal

In addition to extending the length of time for which a piece of information is held in the STM, **rehearsal facilitates its transfer** from STM to LTM. Research shows that students who regularly stop and rehearse what they are reading (and try to learn) are much more successful at committing the information to memory than students who read continuously and resist taking rehearsal breaks. Reciting in your own words what you have just read forces your attention (probably starting to wander) back to the material. Several short rehearsal breaks during the learning process are found to be more effective than one long rehearsal at the end of a marathon learning session.

Organisation

Information that is **organised** into logical categories is more easily transferred from the STM to the LTM (see Investigation – Effect of organisation on retrieval from LTM).

Effect of organisation on retrieval from LTM

In this investigation, the subjects are divided into two groups, A and B. The members of group A are given 1 minute to memorise the 20 words in list 1 (in Table 18.8). The members of group B are given 1 minute to memorise list 2. All participants then count backwards from 50 to 0 out loud in unison. Finally, all subjects are given 2 minutes to write down as many words as they can remember from their list.

List 1	List 2
apple	apple
skirt	orange
autumn	banana
father	pear
pear	spring
iron	summer
brother	autumn
summer	winter
jacket	mother
lead	father
trousers	sister
winter	brother
orange	copper
tin	lead
sister	iron
shirt	tin
spring	jacket
copper	shirt
banana	trousers
mother	skirt

Table 18.8 Effect of organisation

The members of group B are found to be much more successful. This is because the words in list 2 are much easier to memorise than those in list 1. The reason for this is that the items in list 2 have been organised into **categories**. Grouping items of information in an **organised fashion** increases their chance of being successfully encoded and transferred from STM to LTM. The group headings ('fruit', 'seasons', 'family', etc.) act as **contextual cues** (see page 246), which facilitate the retrieval of the information from LTM to STM at a later stage. Thus **organisation** of material helps to transfer it in both directions.

The experiment is repeated using new lists and giving group A the organised list and group B the disorganised list to memorise. Table 18.9 gives some design features of this investigation and the reasons for employing them.

Design feature or precaution	Reason
Large number of subjects used	To increase the reliability of the results
20 items present on each list	To make the task beyond the scope of the STM
Only 1 minute allowed to memorise list	To reduce the effect of other memory aids such as rehearsal
Subjects count backwards from 50 before writing answers	To prevent the STM contributing answers
Experiment repeated with new lists and groups reversed	To increase the reliability of the results

Table 18.9 Design features for organisation investigation

Elaboration of meaning

Elaboration is a further means of aiding the encoding and transfer of information from STM to LTM. It involves analysing the meaning of the item to be memorised and taking note of its various features and properties. Let us imagine, for example, that you are trying to commit the idea 'limbic system' to your LTM. You could try rehearsing 'limbic system – important part of the brain' a few times and it might become encoded. However, as it stands, this information is sparse and lacking in interest. Therefore it will probably make little impression and is unlikely to be well retained.

Successful **long-term retention** is much more likely if elaboration of meaning is employed, as shown in Figure 18.36. By being **analysed** and **elaborated**, the idea 'limbic system' becomes more interesting and meaningful, enabling it to make a long-lasting impression.

Encoding

Encoding is the conversion of one or more nerve impulses into a form that can be received and held by the brain and retrieved later from the STM or LTM. The quality of the memory is affected by the attention given to the task of encoding the material. Some forms of encoding are shallow; others are deeper. Information encoded by repetition is an example of **shallow encoding**. Information encoded by associating it with other information such as meaning or linking it with previous memories is called **elaborative encoding**. It is regarded as a deeper form of encoding.

It is for this reason that rehearsing material (such as a group of words in a foreign language) simply by repetition (shallow encoding) can be a less effective way of memorising the words than by linking them to their meaning and to other related words already known (elaborative encoding).

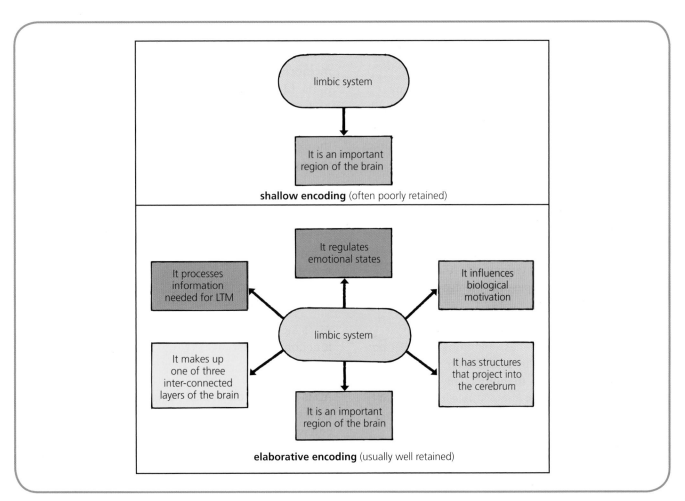

Figure 18.36 Elaboration of meaning

Classification of information in the LTM

The system of storage in the LTM is analogous to a filing cabinet of unlimited capacity, organised into **distinct categories** of information. As items are encoded and transferred to the LTM, they are classified and filed in the appropriate section(s).

Retrieval of items from the LTM

The LTM contains a vast and permanent store of remembered experience which is constantly being revised, reorganised and enlarged as new material flows into it. When a piece of information needs to be called up and **retrieved** from the LTM, a search is mounted (see Figure 18.37). This is aided by **contextual cues.** (A cue is a signal or reminder; contextual means relating to the conditions or circumstances that were present at the time when the information was encoded and committed to the LTM.) It is thought that a contextual cue somehow triggers an impulse through a 'memory circuit'.

If a memory has been stored under several different categories (e.g. dandelion might feature under 'plants', 'flowers', 'leaves', 'clocks', 'weeds', etc.) then it can be retrieved in various ways. This is because many contextual cues for it exist and lead to the different files relating to it. These can then be checked out to see if one contains the information being sought (e.g. names of common weeds with yellow flowers).

Figure 18.37 Attempting retrieval

It is more difficult to retrieve a memory that has been filed under a few categories only, since it will have few contextual cues relating to it. Hence the beneficial effects of organisation and elaboration when trying to memorise information. A memory whose encoding in the LTM is accompanied by unusual, emotional or dramatic events (e.g. the person's wedding day) possesses powerful contextual cues. These enable the experience to be retrieved and recalled clearly throughout life.

| Case Study | Alzheimer's disease |

Alzheimer's disease (AD) is the most common type of dementia (mental deterioration). It is diagnosed most often among people over the age of 65.

Symptoms

The earliest symptoms often take the form of problems with attentiveness, planning and abstract thought, accompanied by partial memory loss. The latter is observed as an **inability** by the person **to acquire new memories** (for example, they cannot recall recently observed events). This state is often wrongly attributed simply to aging.

As the condition continues to develop, some or all of the following symptoms develop:

- confusion
- mood swings
- irritability
- aggression
- loss of LTM
- loss of speech.

AD is incurable and gradually leads to death as bodily functions are lost.

Diagnosis

AD is normally diagnosed using information based on the patient's medical history and on the results of **behavioural assessments** and **neuropsychological screening tests**. Figure 18.38, for example, shows a diagram that a patient would be asked to copy as part of a test. Many AD patients find the task difficult. When AD is suspected, the diagnosis is often followed by a brain scan such as a PET (see chapter 17, page 221).

Cause

In AD, cell-to-cell connections in the brain are progressively lost but the cause of this breakdown is not clearly understood. One hypothesis proposes that AD is caused by reduced synthesis of **acetylcholine**, a neurotransmitter substance. But drugs that are used to treat this deficiency are not very effective. A second hypothesis suggests that AD is caused by the accumulation of proteins called β-amyloids as **plaques**. However, a vaccine that was found to clear the plaques in a clinical trial did not have any effect on dementia. Further studies have shown that deposition of the plaques is not closely correlated to loss of neurons. A third hypothesis proposes that the formation of **tangles of tiny fibres** inside the cell bodies of neurons leads to their degeneration and that the resultant breakdown of synapses (neuron connections) in the cerebrum causes AD.

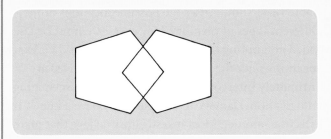

Figure 18.38 Test item

Management

At present no drug that halts or even delays the progression of the disease is available, although some antipsychotic drugs are used to reduce aggression and psychosis in AD patients with behavioural problems. Care for an AD patient is often provided by the person's partner or close relative. However, this places an enormous burden on the carer, especially when the condition progresses and the patient becomes incapable of tending to their own basic needs such as feeding themselves.

Future

Approximately half of new dementia cases each year are AD. Every 5 years after the age of 65, the risk of acquiring AD doubles. As people live longer, the incidence of dementia (including AD) is expected to increase (see Table 18.10).

Year	Estimated percentage of people worldwide with dementia
2005	0.379
2015	0.441
2030	0.556

Table 18.10

Prevention

Some studies suggest that factors such as regular exercise, balanced diet, social interaction and activities that promote mental stimulation may reduce the risk of AD. However, no causal relationship has been established.

Location of memory in the brain

Several **different types of memory** exist within the LTM (see Figure 18.35). These are associated with particular areas of the brain, although evidence suggests that there is a degree of overlap. Close communication certainly exists between these areas of the brain.

Episodic and semantic memories ('remembering that...')

Episodic memory is the recall of **personal** facts, experiences and events. **Semantic** memory is the recall of general knowledge, **non-personal** facts and concepts (e.g. abstract ideas). These two forms of memory are closely associated with specific regions of the **cerebral** cortex such as the temporal lobes at the sides of the cerebrum (see Figure 18.39).

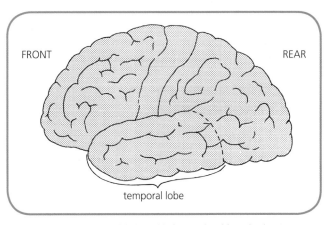

Figure 18.39 Temporal lobe of left cerebral hemisphere

Electrical stimulation

Electrical stimulation of the temporal lobes of patients undergoing brain surgery (to relieve epilepsy) resulted in them recalling, in minute detail, personal facts and events (and even songs) from their distant past. This suggests that the **temporal lobes** were the site of at least some of their **episodic memory**.

Left or right temporal lobe

Patients who have had their left temporal lobe removed are found to have great difficulty remembering unfamiliar words and associations between pairs of words. Patients who have undergone a right temporal lobectomy have great difficulty remembering unfamiliar geometrical figures, new faces and musical sequences. These findings suggest that memories of different categories of general facts are stored in **different sides** of the brain.

Many experts now believe that stores of episodic and semantic memories may be **widely distributed** across many areas of the cerebral cortex, with each memory being stored in the cortical region that first received the information.

Procedural memories ('remembering how to...')

Procedural memories contain the information needed to perform **motor skills** (e.g. remembering how to swim) and **mental skills** (e.g. remembering how to read). Procedural memories are retrieved and put to use without the need for conscious control. Many of these skills (especially those involved in muscular coordination) are closely linked to the **cerebellum** and to long-term modifications of the **motor area of the cerebral cortex**.

Emotional memory

Emotional memories are formed as a result of positive or negative associations with certain stimuli. Often these are not recalled consciously but may be invoked subconsciously during 'automatic' attraction towards, or avoidance of, a previously met stimulus. Emotional memories involve the **cerebral cortex** and the **limbic system** (see chapter 17) and close links between the two regions.

Spatial memory

Spatial memory is responsible for holding a record of information that can be recalled about a person's environment and its **spatial orientation**. Think of the kitchen in your home. Where is the sink in relation to the cooker and the fridge? You can immediately bring to mind their spatial relationship. In doing so you are retrieving information stored in your spatial memory. This type of memory is located in the limbic system. Amnesiac patients suffering damage to part of the limbic system called the hippocampus (see chapter 17, page 215) cannot remember spatial layouts and are often severely impaired with respect to their spatial navigation.

Table 18.11 summarises the possible locations of the different types of memory. However, care must be taken when attempting to draw conclusions about their exact locations since the many brain circuits involved are **intimately interconnected** and constantly exchanging information. Damage to one part may have a knock-on effect on another. Most investigators believe that many different regions of the brain are involved in memory with some regions playing larger roles than others.

Type of memory	Possible location in the brain
episodic and semantic	many regions of cerebral cortex
procedural	motor region of cerebral cortex
emotional	limbic system and cerebral cortex
spatial	limbic system

Table 18.11 Possible locations of memory types

Related Topic

Smart drugs

Smart drugs improve mental functions such as **memory**, **cognition** (the mental process by which knowledge is acquired) and **concentration**. They are used to treat people with Alzheimer's disease, Parkinson's disease and attention-deficit hyperactivity disorder (ADHD). For example, methylphenidate (Ritalin – see Figure 18.40) acts by causing an increase in the level of **dopamine** at nerve cell junctions. This enhances the flow of nerve impulses between certain neurons in the brain and improves attention levels in ADHD patients.

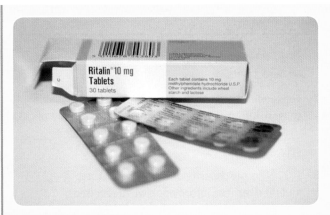

Figure 18.40 Ritalin

Some other memory-enhancing drugs called **cholinergics** work by increasing the level of acetylcholine present at the junctions between nerve cells in the brain. Acetylcholine is a neurotransmitter substance (see page 256). Its presence at the junction of two neurons enables a nerve impulse to pass from one neuron to another. It is therefore described as a 'memory facilitator' since memories can only be formed, stored and retrieved

if nerve impulses can pass from one part of the brain to another. Cholinesterase is an enzyme that breaks acetylcholine down to non-active products under natural conditions in the cell. Under these circumstances a nerve impulse is unable to pass across the junction in the absence of acetylcholine. Cholinergic drugs act by **inhibiting cholinesterase**, thereby sustaining the effect of acetylcholine.

Smart drugs such as Modafinil (known to boost memory and brain power) are used by some soldiers and pilots to help them stay awake on long missions. Sleep-deprived surgeons have been found to think more clearly and be better at solving problems and carrying out simulated operations when given Modafinil. However, this smart drug has not yet been subjected to long-term safety tests. Some students have also begun taking smart drugs in an attempt to improve their memory and level of concentration. But this is a risky business because the efficacy of these drugs and the extent of their side effects on healthy people have not yet been established conclusively.

Testing Your Knowledge 2

1   a) Approximately how many items can be held in the STM at any one time? (1)
    b) Is the memory capacity of the STM limited or unlimited? (1)
    c) For approximately how long are items held in the STM? (1)
    d) Identify TWO ways in which items may be lost from the STM. (2)
    e) Explain what is meant by the term *chunking* and include an example in your answer. (2)

2   a) The three stages involved in memorising facts are *storage*, *retrieval* and *encoding*. Arrange these into the order in which they occur. (1)
    b) Explain the reasons for the *primacy* and *recency* effects found to occur during an investigation into the serial position effect. (2)

3   a) What is meant by the term *rehearsal* in relation to pieces of information to be memorised? (1)
    b) Suggest why rehearsal aids the transfer of information from the STM to the LTM. (1)

4   a) Explain the meaning of the terms *organisation* and *elaboration* in relation to the transfer of information from the STM to the LTM. (4)
    b) Why is it easier to retrieve information from the LTM if its components were organised and elaborated prior to their transfer to the LTM? (1)

5   a) Distinguish between the terms *episodic, semantic* and *procedural* memory. (3)
    b) Where in the brain is the *spatial* memory thought to be located? (1)

What You Should Know Chapter 18

binocular	height	retain
brain	long-term	retrieval
chunking	memory	segregate
contextual	organisation	semantic
displacement	past	sensory
distance	patterns	shape
elaboration	perception	short-term
emotional	perceptual	span
episodic	procedural	spatial
expectation	rehearsal	visual
ground	relative	working

Table 18.12 Word bank for chapter 18

1 The process by which the _____ analyses and interprets incoming sensory information is called _____.

2 _____ perception enables a person to _____ objects from their background, judge how far away they are and recognise them.

3 Images may be perceived as figure and _____, and as coherent _____.

4 The distance of an object from the eye is indicated by visual cues such as its _____ size and its relative _____ in the horizontal field.

5 The two different images of the same object formed by the eyes are merged by the brain into one _____ image that indicates depth and _____.

6 The most important characteristic in the recognition of an object is its _____.

7 The tendency to perceive certain aspects of available sensory information and ignore others as a result of _____ experience, context or _____ is called _____ set.

8 _____ is the capacity of the brain to store, _____ and then retrieve information when required.

9 Information entering the _____ memory lasts for a few seconds on its way to the _____ memory (STM).

10 The STM has a memory _____ of about seven items which it holds for about 30 seconds. This time can be increased by _____ and the number of items remembered can be increased by _____.

11 If information is not passed to the _____ memory it is lost by _____ or decay.

12 An extension of the STM used to perform cognitive tasks is called _____ memory.

13 Transfer of information from STM to LTM and its _____ from LTM at a later stage are aided by rehearsal, _____ of meaning and _____ during encoding.

14 _____ cues aid the retrieval of information from LTM.

15 Different types of memory exist. _____ memory (personal facts) and _____ memory (general knowledge and concepts) are stored in the cerebral cortex. _____ memories (motor skills) are stored in the motor area of the cerebral cortex. _____ memories involve both the limbic system and the cerebral cortex. _____ memory is located in the limbic system.

The cells of the nervous system and neurotransmitters at synapses

Cells of the nervous system

The nervous system consists of a complex network of nerve cells called **neurons** which receive and transmit electrical signals (nerve impulses), and **glial cells** which support and maintain the neurons.

Neurons

The efficient working of the human body and all of its parts depends on the coordinated activity of billions of neurons. These cells provide the body with rapid means of communication and control. They are structurally adapted to suit their function of conducting nerve impulses from one part of the body to another. There are three types of neuron – **sensory, inter** and **motor** – as shown in Figure 19.1.

Although these appear to be very different, they all share the same basic structures. Each consists of a **cell body** and associated processes: one **axon** and

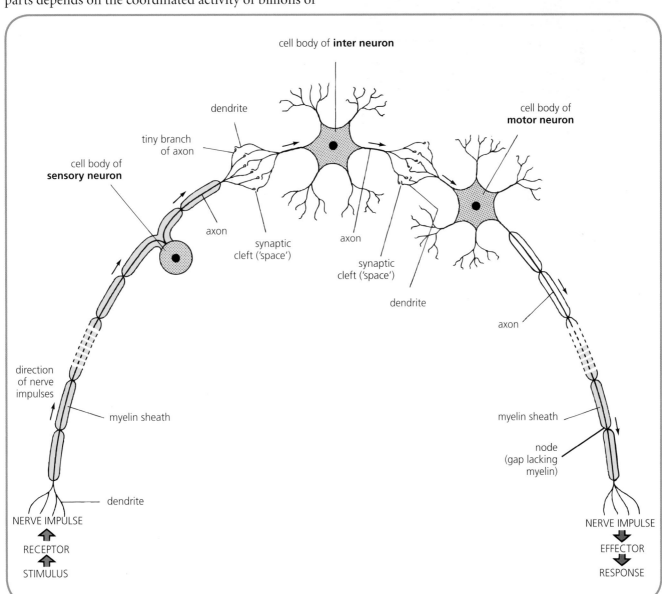

Figure 19.1 Three types of neuron

several **dendrites** (see Figure 19.2). These thread-like extensions of the cytoplasm are often referred to as **nerve fibres**.

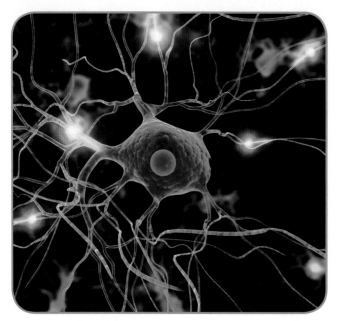

Figure 19.2 Neuron

Dendrites

Dendrites are nerve fibres that receive nerve impulses and pass them **towards** a cell body. A sensory neuron's dendrites gather into one elongated fibre, which transmits information from **receptors** (in contact with the environment) and sends it to the cell body. Inter and motor neurons have several short dendrites that collect messages from other neurons and send them to their respective cell bodies.

Cell body

The cell body of a neuron contains the **nucleus** and most of the **cytoplasm**. It is the **control centre** of the cell's metabolism and contains clusters of ribosomes. These are required to make various proteins including the enzymes needed for the synthesis of neurotransmitters (see page 256). The cell bodies of inter neurons are situated in the central nervous system.

Axon

An axon is a single nerve fibre that carries nerve impulses **away from** a cell body and, in the case of sensory and inter neurons, on to the next neuron in the sequence. The axons of motor neurons are extremely long. For example, those that connect with distant parts of the body (e.g. toes) can be more than a metre in length! Each axon from a motor neuron carries a message from the cell body to an effector.

The direction in which a nerve impulse travels is always:

dendrites → cell body → axon

At the two points in Figure 19.1 where information passes from the axon of one neuron to the dendrites of the next, there is great potential for successful transmission because in reality one neuron ends in many tiny axon 'branches' and the next neuron normally begins as many tiny dendrite 'branches'.

Myelin sheath

The **myelin sheath** is the layer of fatty material that insulates an axon. The small gaps in the myelin sheath along an axon are called nodes. A nerve fibre lacking myelin is described as **unmyelinated**.

Speed of transmission of impulse

The presence of the myelin sheath greatly **increases the speed** at which impulses can be transmitted from node to node along the axon of a neuron. In unmyelinated fibres, the axon is exposed to the surrounding medium and the velocity at which impulses are conducted is greatly reduced.

Myelination

Myelination, the development of myelin round axon fibres of individual neurons (see Figure 19.3), takes time and is not complete at birth but continues during postnatal development until adolescence.

The **hypothalamus** is not fully myelinated until about 6 months. For this reason a very young baby does not have a fully effective 'thermostat' able to bring about finely tuned control of body temperature. Similarly, an infant is unable to control fully the lower body because the neurons in the spinal cord that transmit impulses from the brain to the lower body are not fully myelinated until the child is about 2 years old.

Diseases

In some diseases the myelin sheath around axons becomes damaged or destroyed. This leads to problems such as a loss of muscular coordination. (See Related Topic – Analysis of three diseases.)

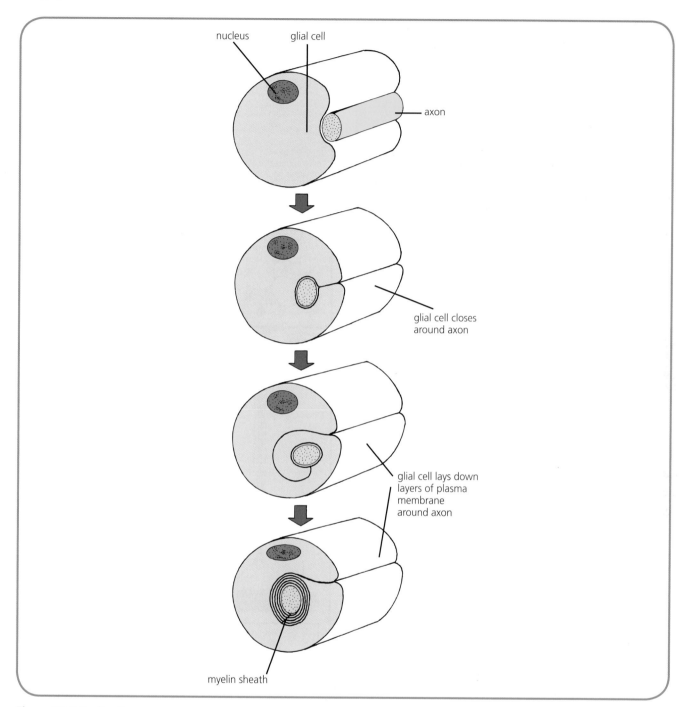

Figure 19.3 Myelination

Glial cells

There are several types of **glial cell**. They do not transmit nerve impulses but are essential to provide neurons with physical support. Some are responsible for the process of **myelination**. This production of the myelin sheath occurs when a type of glial cell lays down successive, tightly packed layers of plasma membrane around an axon (see Figures 19.3 and 19.4).

Other glial cells provide neurons with some of the chemicals that they need to function. They also help to **control the chemical composition** of the fluid surrounding the neurons. By this means these glial cells maintain a **homeostatic environment** around the neurons.

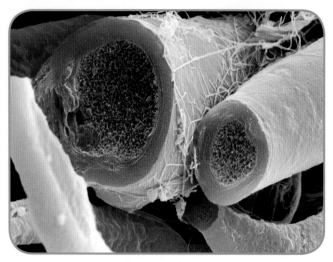

Figure 19.4 Myelin sheath

The intimate association between projections from certain glial cells (see Figure 19.5) and the cells forming the walls of capillary blood vessels is thought to contribute to the make-up of the **blood-brain barrier** (**BBB**). This is a layer that lines blood capillaries and is composed of cells that fit together very closely. It keeps blood that is circulating in capillaries separate from the extracellular fluid in the brain and prevents movement of microorganisms and large molecules into the fluid from the bloodstream. Some glial cells are phagocytic. They remove foreign and degenerate material from the CNS by phagocytosis.

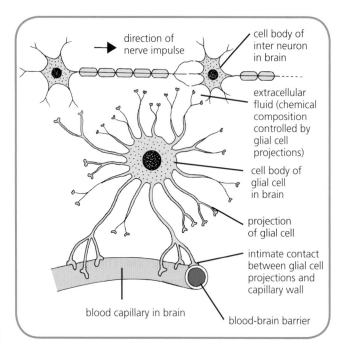

Figure 19.5 Role of glial cell

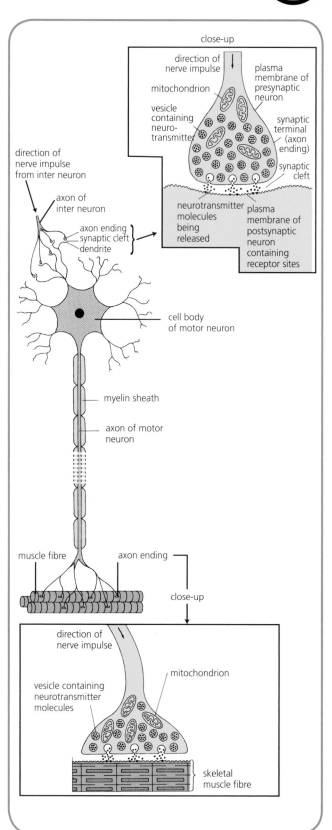

Figure 19.6 Synaptic cleft

Analysis of three diseases

Table 19.1 gives an analysis of three diseases that result from problems affecting neurons.

	Poliomyelitis (polio)	Multiple sclerosis (MS)	Tay-Sachs disease (TSD)
Cause	Caused by a virus that is transmitted via oral-to-oral and faecal-to-oral routes. In a minority of cases the virus spreads along nerve pathways destroying motor neurons in the spinal cord and brain.	The cause is unknown. It is thought that a complex interaction between the person's genotype and some unidentified environmental factor or agent results in the person's own immune system damaging the myelin sheaths around axons. This produces demyelinated nerve fibres unable to transmit nerve impulses efficiently (especially between the brain and spinal cord).	This is a genetic disorder involving a single mutated gene. The sufferer inherits a copy of the defective recessive allele from both parents. The mutation leads to insufficient activity of an enzyme that normally breaks down certain fatty acid derivatives. In a TSD sufferer, harmful quantities of these molecules gather in neurons of the CNS, making the neurons become distended and incapable of performing their function.
Symptoms	These take the form of fever and pains in the neck and back muscles. In a minority of cases, asymmetrical weakness of certain muscles occurs, followed by paralysis of one or more body parts, commonly a limb.	The patient suffers numbness, walking difficulties, impaired vision and progressive loss of coordination as the ability to control muscles is lost. Some forms of MS involve episodic attacks interspersed with spells when symptoms decrease and may disappear temporarily. In other forms the symptoms persist and slowly accumulate.	In TSD patients, a region of the retina shows up as a red spot. This is the centre of the fovea, the only part of the retina whose cells are not distended by fatty acid derivatives. The most common form of the disease (infantile TSD) results in the continuous deterioration of physical and mental abilities from the age of around 6 months. The child becomes deaf and blind. Muscles fail to work and paralysis follows. The child normally dies at about age 4 years.
Treatment	There is no cure. Treatment focuses on providing relief of symptoms by using painkillers and promoting moderate exercise and consumption of a healthy diet. In most cases, paralysis is temporary, though complete recovery may take years. In a few cases paralysis is permanent.	There is no cure for MS. Treatment involves trying to prevent the patient from suffering new attacks and helping them to regain function when they have suffered an attack. Several types of medication are available that help to decrease the number of attacks but they are often accompanied by adverse side effects. Some patients resort to alternative treatments unsupported by clinical evidence, such as medicinal cannabis.	There is no cure. Treatment takes the form of palliative care given to patients to ease their symptoms.

Table 19.1 Analysis of three diseases

Synaptic clefts and neurotransmitters

The tiny region of functional contact between an axon ending of one neuron and a dendrite (or sometimes cell body) of the next neuron in a pathway is called a **synapse**. The plasma membranes of the two neurons at a synapse are very close to one another and separated only by a narrow space called a **synaptic cleft** (see Figure 19.6).

The nerve cell before the synaptic cleft is called the **presynaptic** neuron; the one after the synaptic cleft the **postsynaptic** neuron. Neurons also connect with muscle fibres and endocrine gland cells via spaces similar to synaptic clefts. Messages are relayed across synaptic clefts from neuron to neuron both inside and outside the brain by chemicals called **neurotransmitters**. Two examples are acetylcholine and norepinephrine (noradrenaline).

Action of neurotransmitter

Each synaptic terminal of an axon holds a rich supply of **vesicles** containing a store of one type of neurotransmitter (see Figures 19.6 and 19.7). When a nerve impulse passes through the presynaptic neuron and reaches the synaptic terminal, it stimulates several vesicles. These simultaneously move to the terminal's surface, fuse with its membrane, form openings and discharge their contents (about 10 000 molecules of neurotransmitter per vesicle) into the synaptic cleft.

Once in the cleft, the neurotransmitter molecules briefly combine with **receptor molecules** at sites on the membrane of the postsynaptic dendrite. This process alters the membrane and its electrical state. For example, the binding of acetylcholine to receptor sites makes 'gates' in the postsynaptic membrane open (see Figure 19.8). This allows increased flow of ions to occur through the membrane, resulting in a nerve impulse being initiated in the postsynaptic membrane.

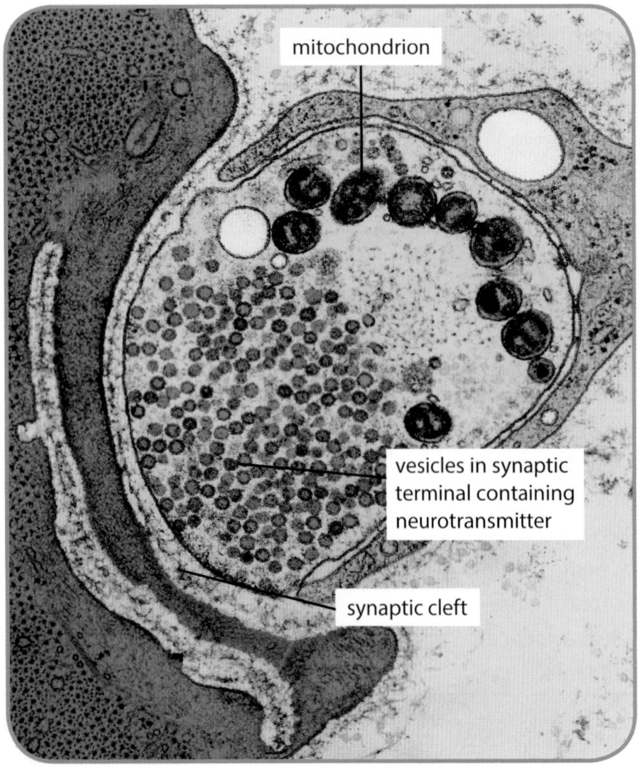

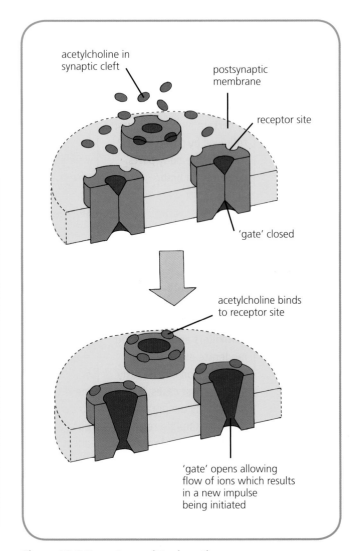

Figure 19.7 Electron micrograph of a synaptic cleft

Labels: mitochondrion; vesicles in synaptic terminal containing neurotransmitter; synaptic cleft

Figure 19.8 Neurotransmitter in action

Labels: acetylcholine in synaptic cleft; postsynaptic membrane; receptor site; 'gate' closed; acetylcholine binds to receptor site; 'gate' opens allowing flow of ions which results in a new impulse being initiated

Excitatory and inhibiting signals

The type of alteration to a postsynaptic membrane that occurs following the binding of a neurotransmitter to its receptors depends on the **type of receptor** present. The signal generated is determined by the receptor and may be **excitatory** or **inhibitory**. The same neurotransmitter may even have an excitatory effect in one situation and an inhibitory effect in another, as in the following example.

Acetylcholine released into the cleft between a motor neuron and a skeletal muscle fibre binds to receptors that have an excitatory effect on the muscle fibre and make it contract. Acetylcholine released into the cleft between a motor neuron and a heart muscle fibre binds with receptors that have an inhibitory effect. This reduces the rate and strength of contraction of cardiac muscle (and heart beat).

Direction of impulses

Vesicles containing neurotransmitter occur on one side only of a synapse. This ensures that nerve impulses are transmitted in **one direction** only.

Need for removal of neurotransmitter after transmission of impulse

To ensure precise control of the system and allow for the successful transmission of each short-lived impulse, the postsynaptic membrane must remain excited for only the brief moment required to pass on that impulse. This is achieved by the neurotransmitter being **rapidly removed** as soon as the impulse has been transmitted, thereby **preventing continuous stimulation of postsynaptic neurons**.

There are two types of mechanism for the removal of neurotransmitters. These are **enzyme degradation** and **re-uptake**.

Acetylcholine, for example, is broken down into non-active products by an enzyme present on the postsynaptic membrane, as in the following equation:

$$\text{acetylcholine} \xrightarrow{\text{cholinesterase}} \text{non-active products}$$

The non-active products are **reabsorbed** by the presynaptic neuron and **resynthesised** into active neurotransmitter, which is stored in vesicles ready for reuse. The energy required is supplied by the **mitochondria** present in the synaptic terminal. Norepinephrine (noradrenaline), on the other hand, undergoes re-uptake by being **reabsorbed** directly by the presynaptic membrane that secreted it and is stored in the vesicle ready for reuse.

Filtering out weak stimuli

A nerve impulse is only transmitted across a synapse and on through the postsynaptic neuron if it first brings about the release of a certain **minimum number** of neurotransmitter molecules. This critical number

---

Related Information

Mode of action of neurotransmitters

Table 19.2 gives some examples of neurotransmitters and their modes of action.

Neurotransmitter	Mode of action
Acetylcholine	Excitatory at vertebrate skeletal muscle sites; excitatory or inhibitory at other sites depending on type of receptor present
Norepinephrine (noradrenaline)	Excitatory or inhibitory depending on type of receptor present
Serotonin	Excitatory or inhibitory depending on type of receptor present
GABA (gamma aminobutyric acid)	Inhibitory
Dopamine	Excitatory or inhibitory depending on type of receptor present
Glycine	Inhibitory

Table 19.2 Neurotransmitters and their mode of action

is needed to affect a sufficient number of receptor sites on the membrane of the postsynaptic neuron. Achievement of this is called reaching the membrane's **threshold**.

Weak stimuli that fail to do so are called **subthreshold** stimuli. They are **filtered out** by the synapse acting as an unbridgeable gap. The continuous low-level drone of machinery, for example, fails to evoke a response because the weak 'background' stimuli do not bring about the release of enough neurotransmitter to create an impulse in the postsynaptic membrane. However, a sudden change in the stimulus (such as an increase in volume) brings about the normal response and makes the person aware of the machinery.

Summation

The electrical change in a postsynaptic membrane that results from the binding of a neurotransmitter to its receptors is called an **EPSP** (**excitatory postsynaptic potential**). An EPSP at a synapse can be too weak to enable threshold to be reached. However, a postsynaptic cell may receive information via synapses from several neighbouring neurons. If many synaptic terminals of many presynaptic neurons discharge their neurotransmitter simultaneously or in rapid succession, then enough of the chemical is released to fire an impulse. This **cumulative effect** of a series of weak stimuli that together bring about an impulse is called **summation** (see Figure 19.9).

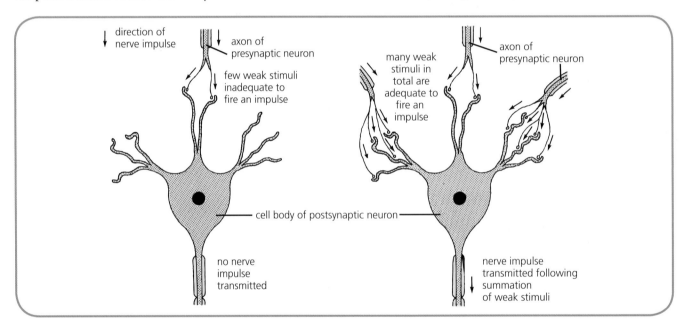

Figure 19.9 Summation

Related Topic

Action of curare and strychnine

Curare

Curare is the general name given to several poisons originally found in South America. These are produced by certain plants and have been used traditionally as arrow and dart poisons by South American native forest dwellers to paralyse prey (see Figure 19.10).

Acetylcholine is a neurotransmitter that relays messages needed to bring about muscular contraction in vertebrates. When a muscle is wounded by a poison dart or injected with curare, the poison binds to acetylcholine receptors on the postsynaptic membranes of muscle fibres. This results in the neural pathway becoming blocked and no nerve impulses reaching the muscles. Therefore these fail to contract and remain relaxed and in a state of paralysis. Curare is therefore described as a **muscle relaxant**.

Curare-poisoned prey normally dies as a result of asphyxiation since its respiratory muscles no longer work. Curare-poisoned prey is safe to eat because curare molecules are unable to pass through the lining of the consumer's gut and into their bloodstream.

Figure 19.10 Curare is used on blowpipe darts

Strychnine

Strychnine is a highly poisonous chemical made by many members of the plant genus *Strychnos*. It is used to control pests such as rats, which suffer muscular convulsions and eventually die of asphyxiation.

Glycine is a neurotransmitter that occurs naturally in the bodies of many types of animal including mammals. When it binds to its receptors on the postsynaptic membranes of motor neurons in the brain and spinal cord, it has an **inhibitory** effect, which prevents excessive contraction of skeletal muscles. Strychnine acts as an **antagonist** (see page 268) and competes with glycine for the same receptor sites. The higher the concentration of strychnine absorbed into the body, the fewer the number of glycine molecules that manage to combine with their receptors and bring about the normal, natural inhibitory effect.

In the absence of the inhibitory effect, an **excitatory** state results and motor neurons are able to transmit an unchecked flow of nerve impulses to skeletal muscles. Therefore the victim suffers continuous spasms of muscular contraction that can affect the whole body. In addition, skeletal muscles may become fully contracted. In very low doses strychnine acts as a stimulant and some Olympic athletes in the past have risked using it in an attempt to improve their performance.

Testing Your Knowledge 1

1 a) Draw a simple diagram of a motor neuron and label the parts: *cell body, dendrite* and *axon*. (3)
   b) State the function of each of the labelled parts in your diagram. (3)
2 a) What effect does the presence of a myelin sheath around a nerve fibre have on the speed at which the fibre can transmit a nerve impulse? (1)
   b) Why are children unable to exert full control of their lower body before the age of 2 years? (1)
3 Figure 19.11 shows a synapse.
   a) Match numbered parts 1–4 with the following terms: *synaptic cleft, axon, synaptic terminal, membrane of dendrite.* (4)
   b) i)   Identify structure P.
      ii) Give the name of a neurotransmitter that could be released at Q.
      iii) To what structures would these neurotransmitter molecules briefly combine?
      iv) In which direction would the nerve impulse pass in this diagram?
      v)   State the fate of the neurotransmitter that you gave as your answer to part ii) once the nerve impulse has been transmitted.
      vi) Identify structure R and state its function. (7)

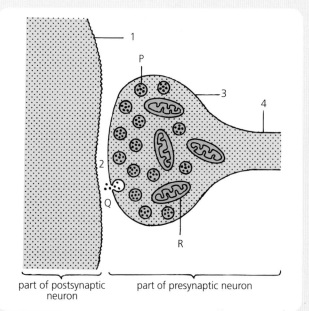

part of postsynaptic neuron   part of presynaptic neuron

Figure 19.11

4 What is meant by the term *summation* with reference to nerve impulses? (2)

Complex neural pathways

Neurons are found to be connected to one another in many different ways in the CNS. The various combinations allow many types of complicated interaction to occur between neurons. This enables the nervous system to carry out its many complex functions. Examples of neural pathways are as outlined below.

Converging neural pathway

To **converge** means to come together and meet at a common point. In a convergent neural pathway, impulses from several sources are channelled towards, and meet at, a common destination, as shown in Figure 19.12. This brings about a concentration of excitatory or inhibitory signals at a common neuron. (See Related Topic – Convergence of neurons from rods.)

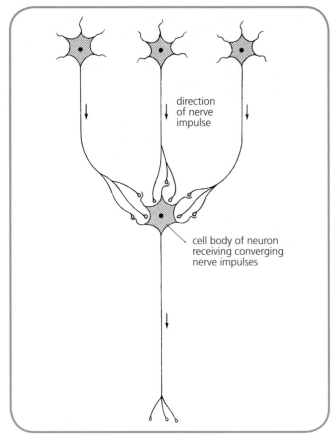

direction of nerve impulse

cell body of neuron receiving converging nerve impulses

Figure 19.12 Converging neural pathway

Convergence of neurons from rods

Rods and cones

Rods and **cones** are visual receptors present in the retina of the eye. They contain pigments that break down in the presence of light. In each case, this breakdown forms a chemical that triggers nerve impulses along a pathway of neurons. The pigment present in cones is not very sensitive to light. Bright light (e.g. daylight) is needed to break it down and trigger the transmission of nerve impulses. The pigment in rods, on the other hand, is so sensitive to light that it even reacts in very **dim light** and fires off impulses. It is quickly rendered temporarily inactive in bright light.

Convergence of signals from rods

As the intensity of light entering the eye decreases, cones cease to respond and rods take over. Unlike cones, several rods form synapses with the next neuron in the pathway, as shown in Figure 19.13. The nerve impulse transmitted by one rod in dim light is weak. On its own it would be unable to bring about the release at the synapse of enough neurotransmitter to raise the postsynaptic membrane to threshold. However, the **convergent arrangement** of several rods allows several impulses to be transmitted simultaneously and these have the combined effect of releasing enough neurotransmitter. The postsynaptic membrane now reaches **threshold** and transmits the nerve impulse on through the neural pathway of the optic nerve to the brain.

This process increases the human eye's **sensitivity** to low levels of illumination and allows vision in conditions of almost total darkness. Furthermore, we gain a reasonably comprehensive view of the surroundings because the rods are thoroughly distributed throughout the retina (except for the fovea).

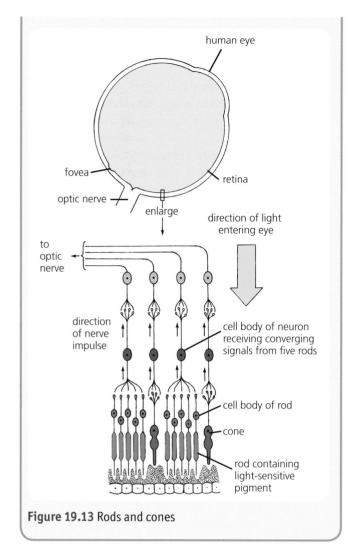

Figure 19.13 Rods and cones

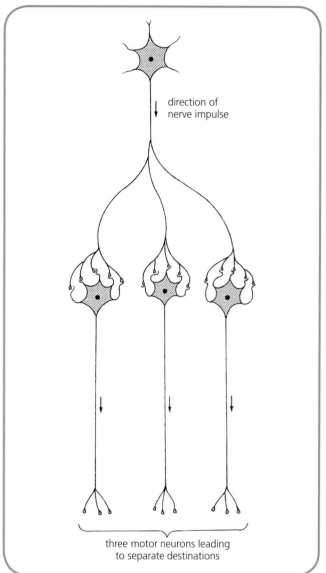

Figure 19.14 Diverging neural pathway

Diverging neural pathway

To **diverge** means to branch out from a common point. In a diverging neural pathway, the route along which an impulse is travelling divides. This allows information from the original single source to be transmitted to several destinations. Figure 19.14 shows a simplified version of this principle.

Examples of diverging neural pathways

Fine motor control

Movement of those parts of the body operated by skeletal muscles is controlled by the motor area of the cerebrum (see Figure 17.15, page 217). The cerebrum communicates with the muscles by sending impulses via motor neurons in neural pathways. Divergence of these pathways from a common starting point allows impulses to be simultaneously transmitted to different muscles of the hand, for example. This brings about fine motor control of the fingers and thumb by allowing them to operate in unison when required to do so.

Temperature control

Similarly, a neural pathway that begins in the **hypothalamus** is found to diverge into branches that lead to sweat glands, skin arterioles and skeletal muscles. This enables the hypothalamus to exert **coordinated** control over the structures involved in temperature regulation. For example, vasoconstriction, shivering and decreased rate of sweating can all be initiated simultaneously if the body temperature begins to drop.

Reverberating pathway

A **reverberation** means a sound that occurs repeatedly, as in an echo or a vibrating tuning fork. In a reverberating neural pathway, neurons later in the pathway possess axon branches that form synapses with neurons **earlier** in the pathway (see Figure 19.15). This arrangement enables nerve impulses to be recycled and to repeatedly stimulate the presynaptic neurons. Once the circuit is activated and reverberating as a result of this feedback of impulses, it continues to give out signals until the process is brought to a halt when no longer required.

Examples of reverberating pathways

Complex reverberating circuits in the brain are involved in the control of rhythmic activities such as breathing. They are also thought to be involved in short-term memory but not long-term memory. Electroconvulsive shock, which brings electrical activity in the nervous system to a temporary halt, affects breathing and STM but not LTM. A pathway can reverberate and transmit impulses for seconds, for hours or, in the case of breathing, for a lifetime.

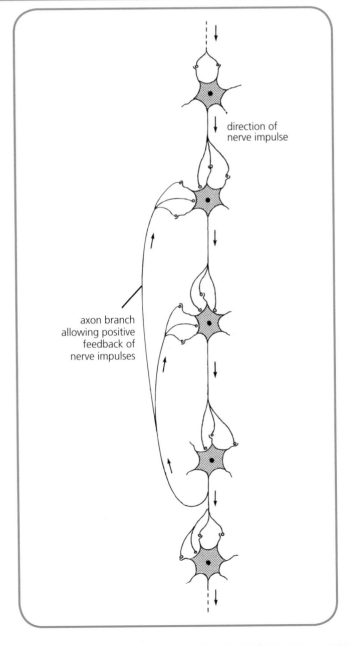

direction of nerve impulse

axon branch allowing positive feedback of nerve impulses

Figure 19.15 Reverberating pathway

Development of new neural pathways

The brain is a dynamic structure containing a profusion of neural connections at synapses. It is not 'hard-wired' with fixed, immutable neural pathways. In fact, neurons in the brain **undergo change** in their synaptic network during a person's lifetime. Depending on sensory input, some synapses lose connections while others retain them or even develop additional ones. This remarkable ability of brain cells to become altered as a result of new environmental experiences is called **plasticity of response**. (Plastic in this biological sense means able to be influenced or become changed.) Plasticity of response enables new neural pathways to form, especially during:

- early development of the brain
- the learning of new skills
- response to brain injury.

Related Topic

Effect of type of environment on early development of brain

As a young mammal, such as a rat, develops and interacts with its environment, neurons in its brain are activated. If the environment is enriched and **stimulating** (e.g. contains ladders to climb, wheels to spin, other rats to play with, etc.) then certain synaptic connections between brain neurons are made frequently. As a result, the synaptic connections become reinforced and **increase in number** as the neurons develop extra dendrites. By this means **new neural pathways** develop, some of which endow the animal with **increased cognitive skills**.

When synaptic connections between brain cells are not used, they become redundant and are removed by **'pruning'**. If the young mammal's environment is one of sensory and/or social deprivation (e.g. small, bare cage, no toys, solitary confinement, etc.) then few new synaptic connections are made between brain neurons and many of the original dendrite connections are pruned. The animal's brain does not develop as fully and the animal is found to be **less intelligent** than the animal in the stimulating environment. It is thought that, in a similar way, the brains of human babies reared in seriously deprived environments fail to develop to their full potential. The process of neural plasticity in response to new environmental stimuli is never completely lost. It continues throughout life but becomes more limited later in adulthood.

Related Activity

Analysing data on neural development in rat brains

The graph in Figure 19.16 shows the results from an investigation into the effect of two different environments on brain development in rats. Those reared in a stimulating, enriched environment were found to have a larger number of synapses (up to 25% more) than rats reared in a deprived environment. In addition, the former were found to possess a cerebral cortex that was 3.3–7.0% thicker than that of the latter.

In a further investigation, one group of rats was reared in an enriched environment and another group in an unenriched environment. The rats were in turn released individually into a maze (see Figure 19.17) several times a day for 20 days.

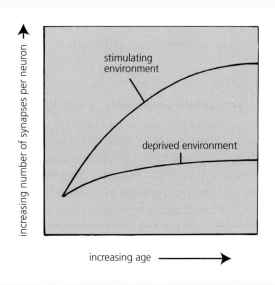

Figure 19.16 Stimulating versus deprived environment

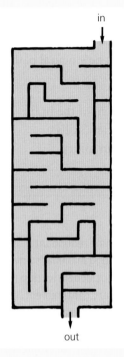

Figure 19.17 Typical rat maze

The bar graph in Figure 19.18 shows the results for the final maze run. From the results it is concluded that members of the group from the enriched environment were much more successful than those from the

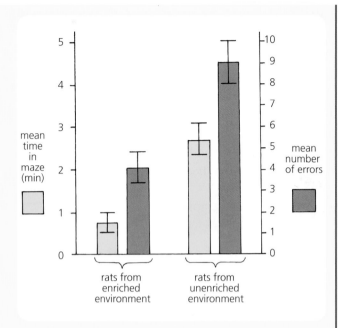

Figure 19.18 Results from maze experiment

unenriched environment at solving maze problems. They were also found to have a significantly **thicker cerebral cortex**, containing enlarged neurons (with a greater spread of dendrites) and a higher number of glial cells than the deprived rats.

Related Topic

Brain development and sensory deprivation

Failure of environmental stimulation during critical periods of brain development can lead to irreversible deficiencies in certain functions of the brain.

'Blind' kitten

In one of several related experiments, scientists temporarily blocked the visual input to one eye of a kitten during a period of brain development when neural connections were being formed in response to sensory experiences. They found that this led to **irreversible** structural and functional **changes in the visual cortex** of the animal's brain. It caused permanent impairment of vision in the affected eye because in the absence of visual stimuli, the requisite neurons had been pruned of their unused synaptic connections and were unable to develop these connections again (when visual input returned). When the experiment was repeated using adult cats, it had no effect on their vision.

Feral children

These are children who have lived in the wild without any human contact during their formative years. Often this survival has depended upon the support of wild animals that have taken them into their group. The normal development by young children of human communication skills (e.g. spoken language, meaningful eye contact, physical gestures, etc.) is dependent upon continuous:

● observation of the activities of other human models
● hearing of sounds made by other humans
● mimicking of human sounds and gestures that evoke responses (e.g. praise) from other humans.

By growing up in conditions of **sensory deprivation** and **lack of human contact**, feral children fail to develop these communication skills. Even if the children are successfully retrieved from the wild, they may not be able to develop the skills fully because the requisite brain neurons have passed their critical period of development and are no longer sufficiently plastic to form all the necessary synaptic connections.

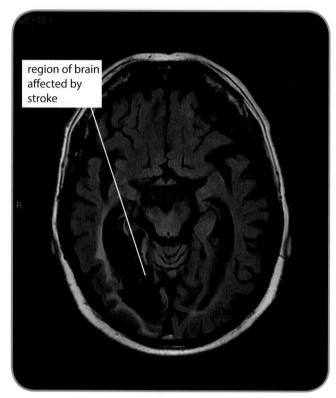

Figure 19.19 Brain scan of stroke victim

region of brain affected by stroke

Plasticity of response following brain injury

Figure 19.19 shows a brain scan of a person who has suffered a **stroke**. The black area indicated by the arrow shows the region of the brain affected in this case. People who have suffered serious brain damage of this type are found to be severely affected. Depending on the specific area affected, they may, for example, be unable to speak or be unable to move one or more of their limbs.

However, during the first few months following the injury, some sufferers are found to make a significant recovery and regain speech or use of the limb, at least in part. Neurons in the damaged region of the brain have not regained their functional state. What has happened is that neurons in some other region of the brain that has escaped damage have formed **new neural pathways** enabling them to take on these jobs. People who show this type of improvement are demonstrating **major plasticity** of response where new neural pathways enable an area of brain damage to be bypassed. (See Case History – Brain injury and sensory substitution.)

Case History — Brain injury and sensory substitution

After surgery, a patient was given an antibiotic to aid her recovery from an infection. A few days later she found that she had completely lost her sense of balance. The antibiotic had damaged the region of her brain responsible for visual and gravitational stability.

Four years later, the woman volunteered for an experimental treatment devised by a leading exponent of **neuroplasticity**. He used an apparatus called a brainport. This consisted of a strip of tape containing microelectrodes wired to a 'spirit-level' structure attached to a hard hat. When the strip was attached to the woman's tongue and the hard hat put in place, the level on the hat was able to determine her spatial coordinates and send information to her tongue. This in turn sent impulses to her brain. Eventually the woman regained her balance and no longer needed to use the brainport.

Her recovery is attributed to neuroplasticity of the brain in response to **sensory substitution**. The region of the brain normally responsible for processing sensory information from chemoreceptors on the tongue was able to respond and **adapt** to the arrival of the completely new set of impulses. By doing so, that region of the brain became able to process information and send impulses to the effectors responsible for balance. Remarkably, that region of the brain also continued to perform its normal function and the woman did not lose her sense of taste.

Minor plasticity

The ability of the brain to suppress reflexes (such as blinking) or responses to sensory impulses (such as visual distractions) is called **minor plasticity** of response. It is investigated in the Investigations that follow.

Investigation

The brain's capacity to suppress the blinking reflex

A **reflex action** is a rapid, automatic, involuntary response to a stimulus. Blinking the eye by contraction of the eyelid muscle in response to a real or imaginary danger is a reflex action. In the experiment shown in Figure 19.20, ten attempts are made to make the volunteer blink using their right eyelid. An interval of 10 seconds is allowed between each trial to allow the volunteer to compose themselves and summon maximum willpower. The eye is held at the same distance from the pipette at each trial.

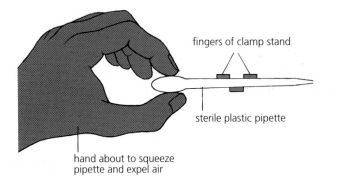

fingers of clamp stand

sterile plastic pipette

hand about to squeeze pipette and expel air

eye of volunteer who is to try to resist blinking

Figure 19.20 Resisting blinking

It is found that some people are very good at suppressing this reflex action while others cannot resist blinking, however hard they try. This form of **minor plasticity** is thought to occur in the following way. Since the nervous system consists of many interconnecting nerve cells, two conflicting types of message (one 'saying' *Blink!*; the other 'saying' *Resist blinking!*) meet in a converging pathway. If the overall effect at the synapses affected is **excitatory** then an impulse is fired and the blinking response occurs. If the overall effect is **inhibitory** then no impulse is fired and blinking fails to occur. Thus some people cannot resist blinking while others can successfully 'stare out' the stimulus.

Investigation

The ability of the brain to suppress sensory impulses

The person being tested is given 2 minutes to attempt a task that requires a reasonable amount of concentration (e.g. to correctly solve as many simple arithmetical problems as possible from a long list). The first trial is conducted in optimum conditions (e.g. silence, good lighting, etc.). The second trial, using a fresh list of problems of equal difficulty to the first, is carried out under conditions of **auditory distraction** (e.g. sound of a car alarm). The final trial is performed under conditions of **visual distraction** (e.g. flashing light).

Some people are found to be very good at suppressing the sensory impulses from the distractions and perform very well each time. Other people find it very difficult to concentrate when distracted and fail to block out the sensory impulses.

Discussion

From the above two investigations, it can be concluded that responses shown by a normal, healthy person's nervous system are not necessarily fixed and unchangeable. Sometimes the brain can be persuaded to temporarily suppress a reflex action or block out certain sensory impulses. This demonstrates **minor plasticity of response** of the nervous system.

Testing Your Knowledge 2

1 Briefly explain what is meant by the terms *converging* neural pathway, *diverging* neural pathway and *reverberating* neural pathway. (6)

2 a) Give an example of major plasticity of response of the brain. (1)

   b) Give an example of minor plasticity of response of the brain. (1)

Endorphins

When the body is injured or affected by illness, nerve impulses pass to the brain, resulting in the sensation of pain. **Endorphins** are chemicals that function like neurotransmitters and act as natural painkillers by combining with receptors at synapses and blocking the transmission of pain signals. Endorphins are produced in the hypothalamus and their level of production increases in response to:

- physical and emotional stress
- severe injury
- lengthy periods of vigorous exercise
- certain foodstuffs such as chocolate.

Depending upon circumstances, increased levels of endorphins may bring about:

- regulation of appetite
- release of sex hormones
- feelings of euphoria.

Dopamine and the reward pathway

Dopamine is a neurotransmitter produced in several regions of the brain. Two of these centres, 'V' and 'N', are shown in a simplified way in Figure 19.22. They are located in the limbic system and are connected by nerve fibres which form a neural circuit. When a survival-related urge such as hunger, thirst or sexual need is being satisfied by current behaviour, neurons in centre 'V' release dopamine which is carried to centre 'N'. Neurons in centre 'N' also release dopamine and induce a pleasurable feeling. It is for this reason that V and N are often referred to as **pleasure centres** and the route from V to N is called the brain's **reward pathway**. It is thought to have evolved because it reinforces forms of beneficial behaviour (such as eating when hungry) that are of survival value. Several other circuits that release dopamine are also present in the brain, although only one of these is shown in Figure 19.22. It leads to the frontal area of the cerebral cortex which is responsible for cognitive appreciation of pleasure.

Endorphins and pain threshold

In an investigation, the endorphin level in the blood plasma of each of the participants was measured. Then each person's **pain threshold** was tested. This was done using a blood-pressure cuff (see Figure 19.21) pumped up until the volunteer indicated that they could not take any more pressure. The participants were split into groups A and B. The members of group A exercised vigorously for 10 minutes whereas those in group B remained at rest.

After exercise, the **endorphin level** of the members of group A was found to have risen in every case and, when the blood-pressure cuff test was repeated, their pain threshold was found to have increased. They were, on average, able to stand the cuff at a higher pressure for a longer time. The endorphin level and pain threshold of group B was found to have remained unchanged. The result of this and many similar experiments support the theory that endorphins act as **natural painkillers**.

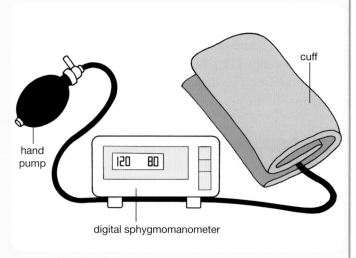

Figure 19.21 Digital sphygmomanometer with manually inflated cuff

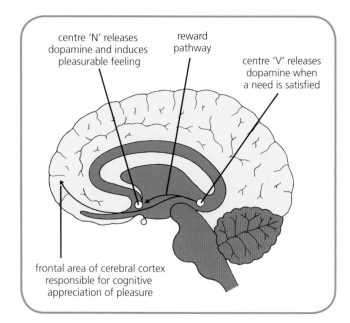

Figure 19.22 Reward pathway

Addiction

When rats were allowed to press a lever that stimulated area N in their brain, they became addicted to this form of stimulation of their pleasure centre. Even when kept on a starvation diet, they chose the brain-stimulation lever over food until completely exhausted. Female rats, addicted in this way, even abandoned their young in order to gain access to the lever.

Treatment of neurotransmitter-related disorders

Agonists and antagonists

An **agonist** is a chemical that binds to and stimulates specific receptors on the membrane of postsynaptic neurons in a neural pathway. Since the agonist **mimics the action** of a naturally occurring neurotransmitter, it triggers the normal cellular response. Therefore, nerve impulses are transmitted, sometimes at an enhanced level (see Figure 19.23).

An **antagonist** is a chemical that binds to specific receptors on the membrane of postsynaptic neurons in a neural pathway. By **blocking the receptor sites**, an antagonist prevents the normal neurotransmitter from acting on them. Therefore normal transmission of nerve impulses in that neural pathway is greatly reduced or brought to a halt.

Many drugs used to treat neurotransmitter-related disorders are very similar in chemical structure to

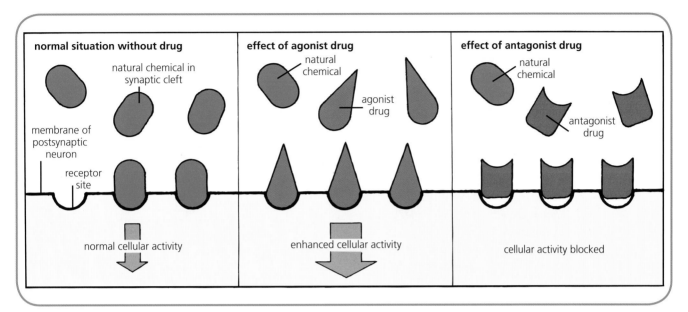

Figure 19.23 Action of agonist and antagonist

neurotransmitters. Therefore some are employed to act as agonists where appropriate and others are used as antagonists.

Inhibitors

Other drugs act in a different way by preventing the removal of the neurotransmitter from synaptic clefts. Some of these work by inhibiting the enzyme (e.g. cholinesterase) that would normally degrade the neurotransmitter (for example, acetylcholine). Others act by inhibiting the normal reabsorption of the neurotransmitter (for example, norepinephrine) by presynaptic neurons.

Related Information

Examples of neurotransmitter-related disorders and their treatment

Alzheimer's disease

Alzheimer's disease (AD) is a form of dementia that is incurable and eventually terminal (see also page 246). The sufferer's brain gradually degenerates (see Figure 19.24). One theory proposes that AD is caused by the loss, in several parts of the brain, of neurons that make **acetylcholine.** Under normal circumstances this neurotransmitter crosses synaptic clefts, binds to receptors and allows nerve impulses to be transmitted through neural pathways. After transmission of the nerve impulse, acetylcholine in a synaptic cleft is broken down by the enzyme cholinesterase (see page 257).

Several drugs have been developed to treat AD by acting as **cholinesterase inhibitors.** The intention is that the use of one of these drugs will bring about an increase

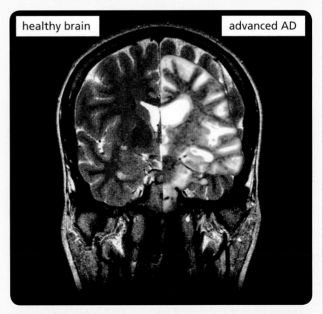

Figure 19.24 Degeneration of the brain

in concentration of acetylcholine in the sufferer's synaptic clefts and lead to improved communication between those neurons that use acetylcholine as a neurotransmitter. These treatments do improve symptoms in some patients and may temporarily slow down the progression of the condition, but for the most part they have not been found to be very effective.

Parkinson's disease

Parkinson's disease (PD) is a major cause of neurological disability in older people. It is caused by the loss of neurons in the midbrain that synthesise **dopamine**. In addition to activating the brain's reward pathway, this neurotransmitter has other functions. For example, it plays a key role in the control and coordination of movement. Therefore, severe loss of dopamine-producing cells can lead to:

- muscle tremors (shaking)
- rigidity (stiffening of muscles)
- difficulty in the initiation of movement and speech
- slowness of movement
- reduced control over fine motor movement.

These are all symptoms typical of PD.

Production of dopamine

Under normal circumstances, dopamine is synthesised from tyrosine (an amino acid), as shown in Figure 19.25. The dopamine is stored in vesicles in neurons in preparation for release into a synaptic cleft when required to combine with receptors and bring about transmission of a nerve impulse. Following transmission, dopamine is broken down by an enzyme called monoamine oxidase (MAO).

Treatment

L-dopa

Patients suffering PD are not given dopamine to treat their condition because dopamine cannot cross the blood–brain barrier (see page 254). However, **L-dopa** can be used as a treatment because it is able to pass through the barrier and is then converted to dopamine. In a healthy brain, dopamine is released only in very specific areas. Unfortunately, the use of L-dopa is accompanied by unpleasant side effects such as involuntary movements and nausea because it penetrates the whole brain.

Inhibition of MAO

Drugs that act by inhibiting the action of MAO (see Figure 19.25) are also used in the treatment of PD. By preventing the action of MAO, an **MAO inhibitor** makes any dopamine that is present (for example, from

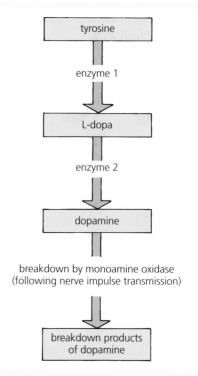

Figure 19.25 Synthesis and breakdown of dopamine

administered L-dopa) have a longer-lasting effect.

Dopamine agonists

Agonists that mimic the action of dopamine are also used to treat PD but they can lead to unpleasant side effects.

Potential use of adult stem cells

Researchers have taken samples of skin cells from many patients at an early stage of PD. These skin cells are being used to produce induced pluripotent stem cells (see chapter 1 page 12) which are then induced *in vitro* to develop into cultures of brain neurons affected by PD. This work enables scientists to **compare normal with diseased nerve cells** and gain a better understanding of why dopamine-synthesising neurons die and PD develops. In the future, stem cells themselves may be used to treat PD which at present remains incurable.

Schizophrenia

Schizophrenia is a complex psychotic disorder in which the sufferer's contact with reality may become highly distorted. The condition is characterised by:

- emotional instability
- delusions
- hallucinations
- social apathy
- progressive deterioration of personality.

It is thought to be caused by a combination of inherited factors and environmental factors (such as the use of certain recreational drugs) that lead to a biochemical imbalance in the brain.

Dopamine antagonists

An overactive dopamine system contributes at least in part to some forms of schizophrenia when specific pathways in the brain (responsible for memory, self-awareness and social behaviour) become blocked. Drugs that act as **dopamine antagonists** bring about relief of this condition.

Generalised anxiety disorders

Chronic anxiety is the central component of this group of conditions which is expressed by symptoms of impaired concentration, sleep disturbance, irritability, restlessness and panic attacks.

Serotonin and GABA agonists

Serotonin is a neurotransmitter with various functions including regulation of mood, appetite, sleep and intestinal movements. It is known to contribute to feelings of well-being when it is at the optimum level. When an imbalance occurs, this may trigger a generalised anxiety disorder.

GABA is the body's chief inhibitory neurotransmitter. When it combines with its receptors, it causes an inhibitory effect on neurotransmission. **GABA agonists** (such as valium) are drugs that mimic the effect of GABA. By suppressing neural firing, they inhibit other neurotransmitters when an imbalance exists. This results in a **reduction of anxiety** in the patient.

Norepinephrine and beta blockers

Norepinephrine is a neurotransmitter that promotes the transmission of nerve impulses, some of which have an **excitatory** effect on certain parts of the body. For example, it brings about the increase in force and rate of heartbeat that occurs during the 'fight or flight' response (see chapter 17). An imbalance in norepinephrine is also implicated in generalised anxiety disorders. This condition can be treated using **beta blockers** which bind with receptor sites and prevent them being stimulated by norepinephrine. This treatment results in a significant reduction in anxiety and in some cases it may even enhance performance such as that of musicians, actors or dancers suffering from performance anxiety (stage fright).

Depression

MAO and re-uptake inhibitors

Very low levels of serotonin and/or norepinephrine often cause **depression**. Many prescribed antidepressant drugs alter serotonin levels. Like dopamine and norepinephrine, serotonin is broken down by monoamine oxidase (MAO). Therefore **MAO inhibitors** are used to prevent the breakdown of serotonin and increase the length of time that it remains in the synaptic clefts of the depressed person's neural circuits.

Some drugs called selective serotonin **re-uptake inhibitors** (such as Prozac) decrease the re-uptake of serotonin, making it stay longer in the synapses. Other antidepressant drugs inhibit re-uptake of both serotonin and norepinephrine enabling them to sustain their effects for a longer period.

Mode of action of recreational drugs

Many people choose to alter their state of consciousness by using **recreational drugs** (some legal, some illegal) that affect the transmission of nerve impulses in the reward circuit of the brain. The subsequent alteration in the person's neurochemistry may lead to changes in:

- **mood** (the person feels happier, more confident or more aggressive)
- **cognitive thinking** (the person becomes poorer at carrying out complex mental tasks such as problem solving and decision making)
- **perception** (the person misinterprets environmental stimuli – sounds, colours, and sense of time seem altered)
- **behaviour** (the person is able to stay awake for longer and talk about him/herself endlessly).

Recreational drugs mimic or interact with neurotransmitters in different ways. For example, they can:

- stimulate the release of a natural neurotransmitter
- act as an agonist by initiating the action of a neurotransmitter
- act as an antagonist by binding with receptors and blocking the action of a neurotransmitter
- inhibit the re-uptake of a neurotransmitter
- inhibit the breakdown of a neurotransmitter by an enzyme.

271

Related Information

Examples of mode of action of recreational drugs

Cocaine

This drug, extracted from the coca plant, acts as a **psychostimulant**. It produces feelings of well-being in the user and gives the person the impression that they have untapped reserves of energy available to tackle any task with confidence (usually *over*confidence – see Figure 19.26). **Cocaine** can induce hallucinations, such as the sensation that there is sand under the skin surface or that thousands of bugs are crawling over it. Continued use of cocaine leads to social withdrawal, depression and dependence on ever higher dosages. People intoxicated with cocaine, especially 'crack' (the form that is smoked), readily become aggressive and violent. They also risk taking a life-threatening overdose.

Cocaine works by **inhibiting the dopamine re-uptake channels** (see Figure 19.27). This creates an over-abundance of dopamine in neural pathways such as the reward circuit and causes them to become **overstimulated**. Drugs such as cocaine that can increase normal dopamine levels by more than 10 times can cause severe mental disorders such as **paranoia**.

Cannabis

This drug, obtained from the Indian hemp plant (see Figure 19.28), acts first as a pleasurable **stimulant** and then as a **sedative**.

The user may feel excited, restless and uninhibited at first but later feels drowsy and normally falls into a deep sleep. The nature and intensity of the effects of **cannabis** depend on:

- the dose
- the strain of the source plant
- the method of consumption
- the physical and mental state of the user.

Like cigarette smoke, cannabis smoke contains chemicals that cause lung diseases. Heavy use of cannabis may trigger schizophrenia in some susceptible individuals.

Cannabis contains chemicals called **cannabinoids**, which work by binding to **cannabinoid receptors**. Under normal conditions, these receptors become occupied by a natural neurotransmitter during the transmission of nerve impulses that bring about control of muscles and regulation of pain sensitivity. Since cannabis mimics these effects, its use in a **medical context** (for example,

Figure 19.26 Cocaine-fuelled delusions of self-importance

in the treatment of multiple sclerosis and arthritis) is approved of in many countries.

MDMA

MDMA ('ecstasy') is a synthetic drug sold illegally in tablet form (see Figure 19.29). It makes the users feel more alert, energetic and in tune with their surroundings and the people around them. Dancing for long periods while under the influence of MDMA increases the chance of **overheating** and serious **dehydration**. Other adverse effects include anxiety, panic attacks and feelings of paranoia and depression during the days that follow use of the drug.

Serotonin is a natural neurotransmitter whose effects include a contribution towards feelings of well-being (see also page 270). Under normal circumstances re-uptake of serotonin occurs in synapses following the transmission of nerve impulses. MDMA works by **inhibiting serotonin's re-uptake**, thereby causing an increase in its level in synaptic clefts and promoting temporarily heightened sensations of well-being.

Nicotine

This drug is the active ingredient of tobacco plant products. Cigarette smokers report that nicotine has a

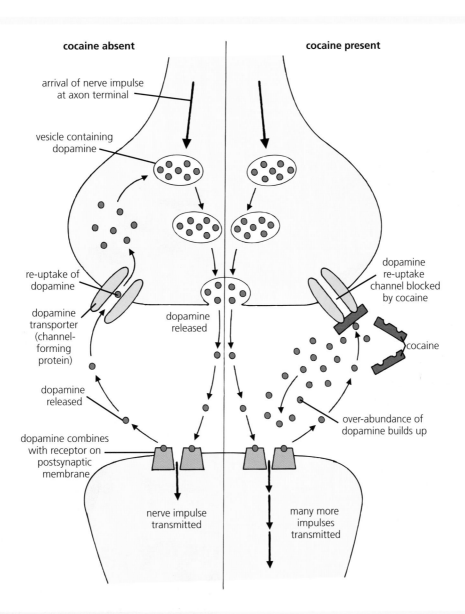

Figure 19.27 Effect of cocaine on dopamine re-uptake

soothing effect and helps them to concentrate. However, **nicotine** is highly addictive and cigarette smoke (see Figure 19.30) is responsible for many diseases, including **lung cancer.**

Under normal circumstances nicotinic acetylcholine receptors on brain neurons become occupied by the neurotransmitter acetylcholine causing transmission of nerve impulses. Such transmission leads to an increase in the level of dopamine, serotonin and norepinephrine. Since nicotine mimics this effect and **increases the activity of these nicotinic acetylcholine receptors,** one of its effects is to increase the level of dopamine in

the reward circuit of the brain, resulting in euphoria, relaxation and eventual addiction.

Alcohol

Alcohol, which is a **depressant,** is one of the most widely used recreational drugs in the world. Although its use is almost universally legal, it is not safer than many other drugs that remain illegal. The effect of alcohol on the brain varies from feelings of relaxation and good humour after a drink or two, to complete loss of consciousness following excessive consumption (see Figure 19.31). Short-term effects of drinking alcohol include decreased

273

reduces feelings of anxiety. Alcohol can also lead to **activation of dopamine-synthesising neurons** thereby elevating dopamine levels and making the person feel good temporarily while the reward system in the brain continues to be over-stimulated.

Figure 19.29 MDMA tablets

Figure 19.28 Indian hemp plant

inhibitions, motor impairment, confusion and drowsiness. Long-term effects include **liver and brain damage**.

GABA receptors normally have an inhibitory effect on neural pathways when the neurotransmitter GABA binds to them. Alcohol **mimics the effect of GABA** and

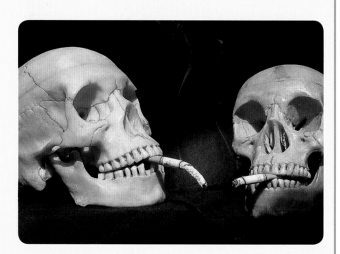

Figure 19.30 Not long to go!

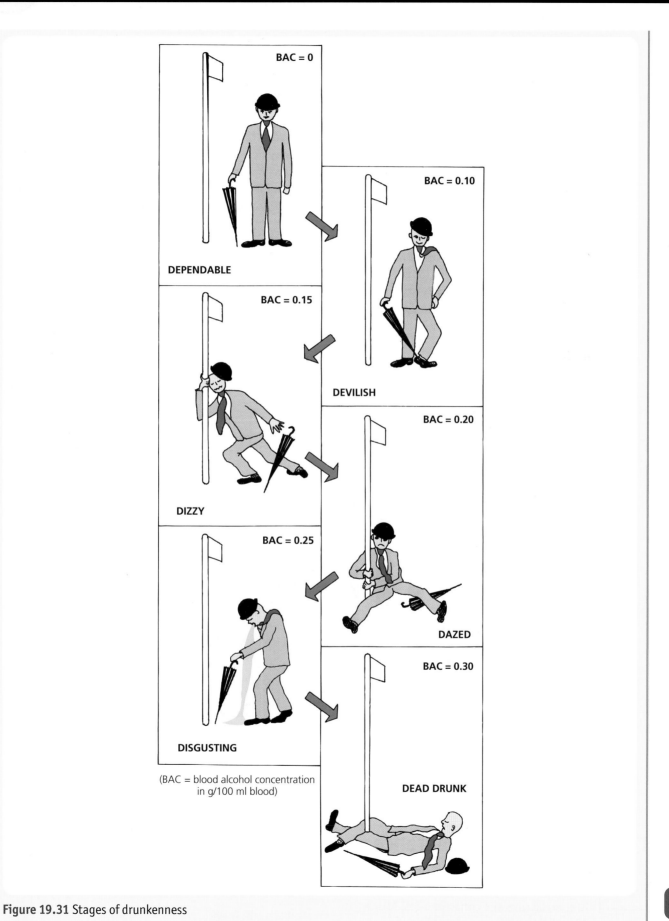

Figure 19.31 Stages of drunkenness

Alcohol and reaction time

Alcohol is a water-soluble drug that is rapidly absorbed through the mouth and stomach linings into the blood. It affects brain functions within minutes of consumption.

Reaction time is the time between the occurrence of an event and a person's reaction to it. Figure 19.32 shows the results of an investigation into the effect of alcohol on the reaction time to complete a simulated emergency stop of a vehicle. 10% of subjects were found to show impaired reaction time with a **BAC (blood alcohol concentration)** as low as 0.02 g/100 ml. From a BAC of 0.06 g/100 ml, the number of subjects showing impaired reaction time outnumbered those showing no impairment.

Driving

Imagine a person driving down a street when a car suddenly pulls out of a driveway in front of them. The driver is faced with a choice of braking or swerving round the vehicle. However, there could be oncoming traffic and the safer option might be to brake, provided that there are no vehicles behind the driver. The average driver's reaction time in such a situation doubles from 1.5 to 3.0 seconds when their BAC is 0.08 g/100 ml. Experts estimate that this makes the person about four times more likely to crash than if their BAC had been zero. In UK the **maximum legal BAC threshold** for driving is 0.08 g/100 ml (i.e. 80 mg/100 ml blood). Some people would like to see this limit reduced to a lower level or even to zero.

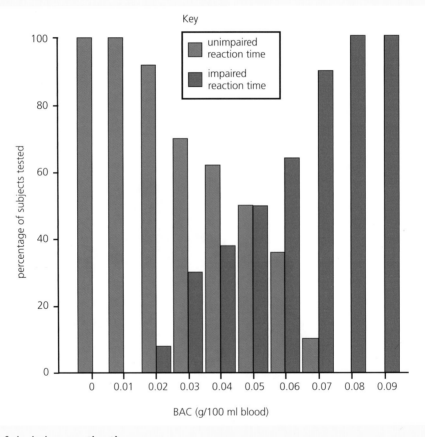

Figure 19.32 Effect of alcohol on reaction time

Drug addiction and tolerance

Drug addiction can be defined as a chronic disease that causes the sufferer to compulsively seek out and use the drug regardless of the consequences. Although the initial decision to take the drug is voluntary, subsequent changes that take place in the person's brain soon override their self-control. This makes the person incapable of resisting the overpowering urge to take more of the drug.

A drug user is said to have built up a **drug tolerance** when their reaction to an addictive drug is found to have decreased in intensity compared with previous times, even though the concentration of the drug has remained unaltered. A larger dose is now required to bring about the original effect.

Drug desensitisation

Repeated use of a drug that acts as an **agonist** (see page 268) results in certain neuroreceptors (e.g. those that promote the release of dopamine) being repeatedly stimulated, causing increased feelings of well-being or euphoria. The nervous system compensates for the overstimulation of these receptors by reducing their number. In addition, the remaining receptors become less sensitive to the agonist drug. This leads to drug tolerance because a larger dose of the drug is now required to stimulate the reduced number of less sensitive receptors. This process involving a decrease in number and sensitivity of receptors as a result of repeated exposure to a drug acting as an agonist is called **desensitisation** (and is also referred to as drug tolerance).

Drug sensitisation

Repeated use of a drug that acts as an **antagonist** (see page 268) by blocking certain neuroreceptors prevents the normal neurotransmitter from acting on them. The nervous system compensates for the reduced stimulation of the receptors by increasing their number. In addition, the receptors themselves become sensitive to the antagonist drug.

This process involving an increase in number and sensitivity of receptors as a result of repeated exposure to a drug acting as an antagonist is called **sensitisation**. It appears that sensitisation results in other psychological changes, which transform ordinary sensations of 'wanting' into excessive drug-craving and addiction.

Related Topic

Genetic components of addiction

Alcoholism

The development of **alcoholism** among individuals with a family history of this problem is found to be four to eight times more common than among individuals with no such family history. So is alcoholism the result of **inherited** factors or **environmental** factors (such as imitation of parental behaviour) or a combination of both?

Twin studies

Many studies involving identical and non-identical **twin pairs** have been carried out. The results indicated that when one identical twin was addicted to alcohol, there was a high probability (about 76%) of the other twin also being an alcoholic. When one non-identical twin was an alcoholic, on average, the other twin ran a much lower risk (about 26%) of being an alcoholic. These studies suggest that there is a **large genetic component** to addiction to alcohol. Clearly **environmental factors** also play a large part in the process. This relationship is thought to be true for other addictive drugs.

Cocaine addiction

Dopamine transporter (DAT) is the channel-forming protein that controls the re-uptake of dopamine through its channels (see Figure 19.27) following the transmission of nerve impulses at synapses. Cocaine blocks the re-uptake channels and this leads to an over-abundance of dopamine in the synaptic clefts and over-stimulation of the brain's reward pathway.

A **mutant allele** for the gene that codes for DAT has now been identified. People who are homozygous for this mutant allele are found to make less DAT and to be even more susceptible to the effects of cocaine. Evidence suggests that they are 50% more likely to become addicted to the drug (if they choose to consume it).

Work with mice

Several **genes involved in drug addiction** have been discovered in mice. Some examples are given in Table 19.3. Researchers are attempting to identify comparable genes in humans. This work may eventually lead to the development of treatments for drug addiction. It must be kept in mind that inheriting some genes for

→

susceptibility to addiction does *not* mean that addiction is inevitable. Environmental factors also contribute largely to the risk of addiction.

Type of mutant mouse	Result
Bred with defective *Per2* gene	Animals found to consume three times more alcohol than normal mice
Bred to lack serotonin receptor gene *Htr1b*	Animals display a greater urge to consume alcohol and cocaine than normal mice
Bred to lack cannabinoid receptor gene *Cnr1*	Animals display a reduced reward response to morphine compared with normal mice

Table 19.3 Genetic factors involved in drug addiction in mice

Related Topic

Drug rehabilitation programmes

Drug rehabilitation programmes ('rehab') are forms of treatment for people who have become addicted to psychoactive drugs. The aim of the programme is to enable the addict to recover and to cease substance abuse. Some programmes specialise in combating **physical tolerance**, others in dealing with **psychological dependency**.

Physical tolerance

This problem is tackled using **medications**. For example, methadone and buprenorphine are used to treat people addicted to opiates such as heroin and morphine. The medications are intended to stabilise the patient, prevent withdrawal symptoms and bring use of the illegal drug to a halt. The patient, sustained on doses of medication, is able to lead a 'normal' life, free from the pressure of financing the next 'fix'.

Psychological dependency

This problem is addressed by encouraging addicts to carry out some or all of the following:

- Stop using the drug.
- Disassociate themselves from friends who are still using the drug.
- Adopt the twelve-step programme.
- Receive counselling to identify behaviours and problems related to their addiction.
- Examine their habits and lifestyle and make changes in relation to their addiction.
- Accept that complete abstention from the drug (not moderate use of it) is necessary for recovery.
- Adopt a cognitive-behavioural approach to prevent relapse in the future.

Success of the **cognitive-behavioural approach** depends on the patient's ability to deal effectively with high-risk, relapse-provoking situations. The former addict is instructed on how to develop and employ problem-solving and decision-making skills in order to resist temptation and prevent a relapse.

Testing Your Knowledge 3

1 a) Identify the chemical substances made in the hypothalamus that function like neurotransmitters and act as natural painkillers. (1)
   b) Give TWO examples of situations to which the body would respond by increasing the level of production of these chemicals. (2)

2 Choose the correct answer from the underlined choice given in each of the following statements. (7)
   a) Acetylcholine/dopamine is a neurotransmitter that activates the reward pathway and induces feelings of pleasure.
   b) Endorphin production is increased/decreased in response to physical and emotional stress.
   c) A drug that mimics a neuroreceptor by binding to and stimulating certain receptors is called an agonist/antagonist.
   d) A chemical that binds to specific receptors and blocks the action of the neurotransmitter is called an agonist/antagonist.
   e) Some drugs work by promoting/inhibiting the action of enzymes that degrade neurotransmitters.
   f) The euphoric effect of recreational drugs is the result of altered neurotransmission in the brain's cognitive/reward pathway.
   g) An increase in the number and sensitivity of neurotransmitter receptors as a result of exposure to a drug acting as an agonist is called sensitisation/desensitisation.

What You Should Know Chapter 19

addiction	endorphins	presynaptic
agonist	enzyme	receptor
altered	excitatory	recreational
antagonist	glial	removed
back	inhibiting	re-uptake
behaviour	mimics	reverberating
body	mood	reward
cleft	myelin	sensitisation
converging	myelination	sheath
dendrites	neurons	speed
desensitisation	neurotransmitter	summation
diseases	painkillers	support
diverging	plasticity	tolerance
dopamine	postsynaptic	weak

Table 19.4 Word bank for chapter 19

1 The nervous system is composed of sensory, inter and motor _____ which transmit electrical signals, and _____ cells.

2 Each neuron consists of a cell _____ and associated nerve fibres: one axon and several _____.

3 An axon is surrounded by a _____ sheath of insulating material whose presence greatly increases the _____ at which nerve impulses can be transmitted through the fibre.

4 Glial cells _____ neurons, maintain a stable environment around neurons and produce the myelin _____.

5 _____ continues until adolescence. The myelin sheath is destroyed by certain _____ causing loss of coordination.

6 A synaptic _____ is a tiny space between two neurons. Information is transmitted at a synapse by a chemical called a _____ being released from vesicles in the _____ neuron. The neurotransmitter combines with _____ sites on the postsynaptic membrane.

7 The receptor determines whether the signal generated is _____ or inhibitory.

8 To prevent continuous stimulation of _____ neurons, neurotransmitters are _____ from the synaptic cleft by _____ action or re-uptake.

9 The cumulative effect of a series of _____ stimuli that together bring about an impulse is called _____.

10 In a _____ neural pathway, nerve impulses from several sources meet at a common destination.

11 In a _____ neural pathway, the route along which a nerve impulse travels divides, allowing information to pass to several destinations.

12 In a _____ pathway, later neurons form synapses with earlier ones, allowing the nerve impulse to be sent _____ through the circuit.

13 The ability of brain cells to become _____ as a result of new environmental experiences is called _____ of response.

14 _____ are chemicals that function like neurotransmitters and act as natural _____.

15 _____ is a neurotransmitter secreted by neurons in the brain's _____ pathway which is activated by certain types of beneficial behaviour.

16 Chemicals that act like neurotransmitters are used in the treatment of some disorders. An _____ is a chemical that stimulates specific receptors on postsynaptic neurons and _____ the action of the naturally occurring neurotransmitter. An _____ blocks receptors and prevents the neurotransmitter from acting on them.

17 Some drugs act by _____ the enzyme that degrades the natural neurotransmitter or by inhibiting its _____.

18 Many _____ drugs bring about their effect by affecting the brain's reward circuit thereby altering the person's _____, perception and _____. The drugs may act as agonists, antagonists or inhibitors.

19 An increase in the number and sensitivity of neurotransmitter receptors following repeated exposure to a drug that is an antagonist is called _____ and leads to _____.

20 A decrease in the number and sensitivity of neurotransmitter receptors following repeated exposure to a drug that is an agonist is called _____ and leads to drug _____.

20 Communication and social behaviour

Humans are **social** animals. The vast majority prefer to live in communities rather than lead a solitary existence. To operate successfully the members of a group must be able to **communicate** with one another. Such **social behaviour** involves transmitting and receiving information using **signs** and **signals** (such as verbal, written and body language). Communication between humans begins at birth and continues throughout life.

Infant attachment

In humans the period of dependency of the infant upon the adult members of the species is a very lengthy one. Under normal circumstances nature provides the newly born infant with a mother (or other primary carer) who is able to satisfy the baby's needs such as **food** and **contact comfort**. The newborn baby's activities, such as suckling, clinging and crying, help to trigger in the mother a desire to protect and care for the child. As she does, a strong emotional tie develops between the baby and the mother. The tie that binds the baby to the carer is called **infant attachment**.

At first, attachment is **indiscriminate** on the baby's

part but as the months go by the baby narrows down its interest to selected people. **Specific** attachment to the mother (and a few other carers) becomes evident between 6 and 9 months. As specific attachment develops, indiscriminate attachment weakens. This is shown in Figure 20.1 where attachment is measured as the amount of protest shown by the baby on being separated from the carer.

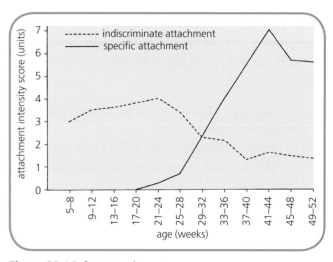

Figure 20.1 Infant attachment

Related Topic

Contact comfort

For many years it was thought that babies became attached to their parents principally because the parents provide food. However, in recent years, the additional importance of **contact comfort** has become appreciated. This was first demonstrated experimentally using infant monkeys exposed to two types of substitute mother. The first 'mother' was constructed of bare wire; the second 'mother' was made of the same wire covered in thick, soft towelling (see Figure 20.2).

In one experiment, only the bare wire mother supplied food. However, once the infant monkeys had finished feeding, they spent much more of their time clinging to the cloth mother than the bare one. They always ran to the cloth mother when frightened. Cuddling the cloth

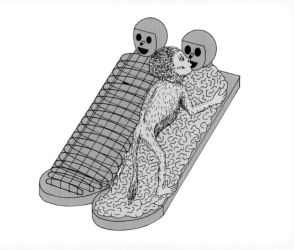

Figure 20.2 Contact comfort with cloth 'mother'

mother's soft towelling material calmed the infants down. These experiments demonstrate a high-level need for **close bodily contact** and the sensation of physical well-being and safety. Contact comfort similarly plays a basic role in establishing attachment between human infants and their primary carers.

Effects of deprivation

In a variation of the experiment illustrated in Figure 20.2, some infant monkeys were denied access to the cloth mother. Such **deprivation** of contact comfort led to the development of **disturbed adults**. Some were over-aggressive, others withdrawn and uncommunicative and all became **inadequate parents**. Human infants who receive plenty of food and warmth but are denied contact comfort may also exhibit maladjusted behaviour.

The 'strange situation'

The '**strange situation**' is a research tool devised to investigate infant attachment. It allows experts to assess whether a child is **securely** or **insecurely** attached to their primary carer. It involves the following sequence of events (each of which lasts 3 minutes, except the first one):

- A mother brings her baby into an unfamiliar room supplied with toys.
- The mother and child are left alone in the room and the child may explore and investigate the toys if they wish to do so.
- A stranger enters the room, talks with the mother and after a few moments tries to play with the infant.
- The mother leaves the baby with the stranger who tries to interact with the child. (This stage is called the **first separation** episode.)
- The mother returns and plays with the child while the stranger leaves. (This stage is called the **first reunion** episode.)
- The mother leaves the baby alone. (This stage is called the **second separation** episode.)
- The stranger returns and tries to engage with the child.
- The mother returns as the stranger leaves. (This stage is called the **second reunion** episode.)

This series of episodes allows hidden observers to study the behaviour of the baby:

- with the mother
- with the stranger
- alone.

Table 20.1 shows different categories of children identified by the experts.

Importance of secure attachment

The long period of dependency on the parent provides the infant with a secure base from which to operate. The more secure the attachment of infant to their mother, the more likely they are to explore their immediate environment, all the time safe in the knowledge that the parent is present as a haven of safety. During these explorations the infant will come across many opportunities for **learning** and for the development of **cognitive abilities**. The number and variety of these opportunities, and the level of stimulation that they provide, will depend on what the infant's environment has to offer (see Figure 20.3). A securely attached infant is more likely to benefit from these opportunities than an insecurely attached one.

Infant attachment is of fundamental importance because it lays down the foundations for the formation of **stable, trusting relationships** in the future. In addition it provides the basis for social, linguistic and intellectual skills to develop and be built upon for the rest of life. Happy, securely attached infants

Figure 20.3 Early development of cognitive abilities

		Type of attachment		
		Secure	**Insecure**	
			Detached (avoidant)	**Resistant (ambivalent)**
Examples of baby's behaviour during the 'strange situation'	**Response to toys and chance to explore new environment**	Child explores freely and plays with toys if mother is present	Child hardly explores or plays with toys regardless of who is present	Child does not explore freely or play with toys even when mother is present
	Response to departure of mother	Displays major distress	Displays indifference or mild distress	Displays major distress
	Response to presence of stranger in absence of mother	Resists offers of comfort from stranger	Accepts comfort from stranger if required	Resists offers of comfort from stranger
	Response to mother's return	Goes to mother immediately for comfort and then calms down and returns to play	Ignores mother or approaches her while looking away	Displays inconsistent behaviour by seeking and resisting comfort at the same time (e.g. approaches mother to be picked up but then struggles to be released again); may show signs of resentment or anger and try to hit mother
Psychologists' interpretation from the study of many case histories		Baby is more attached to mother than to stranger. This situation is thought to arise because the mother is capable, demonstrative of her love and sensitive to the baby's needs (she can often interpret these from the baby's various forms of crying).	Baby treats mother and stranger equally in a detached manner. This situation is thought to arise because the mother is inept and insensitive to the baby's needs, though not rejecting him/her. (Often the mother lacks perception and does not know how to relieve the baby's distress.)	Baby is more attached to mother but in an erratic way. This situation is thought to arise because the mother tends to be irritated by the baby. On occasions she is insensitive to the baby's needs and, by expressing controlled anger towards the baby, is rejecting him/her.

Table 20.1 Results of the 'strange situation'

begin life with an excellent start. Insecurely attached infants deprived of normal social contact, affection and cuddling often suffer long-lasting ill effects.

Socialisation and social competence

Socialisation is the gradual modification of a developing individual's behaviour in order to accommodate the demands of an active social life led within the community. Compared with other mammals, young humans are dependent on adults for a very long period of development during childhood and adolescence. This is advantageous in that it provides time for socialisation and learning to occur and for **social competence** to develop. Social competence is the foundation upon which successful interaction with others depends. A socially competent person possesses a combination of attributes, such as those shown in Table 20.2.

Attribute	Details
Behavioural skills	Ability to react appropriately in social situations, for example by being sympathetic, assertive, submissive, forceful, affectionate, demanding, modest, approving or supportive, depending on circumstances
Cognitive skills	Ability to gain the knowledge and develop the problem-solving skills necessary to function effectively, for example to gain employment within society
Emotional skills	Ability to form stable relationships and demonstrate feelings towards others

Table 20.2 Attributes of social competence

Methods of control

The quality of a developing child's social competence is affected by the **method of control** adopted by their parents (and other influential adults in their life). Three different methods of control are shown in Table 20.3. Research indicates that children raised by parents who exert authoritative control, on average, show higher levels of social competence than those exposed to authoritarian or permissive control. Children with authoritative parents are more likely to develop into self-reliant, academically successful and socially accepted adults.

Method of control	Behaviour adopted by parent
Authoritarian ('unreasonably strict')	• exerts an extremely high level of control • never explains the reasons for the rules to the child • expects the child to obey orders without question • does not engage in verbal give-and-take • demonstrates little or no warmth towards the child • uses shaming or withdrawal of love as a means of discipline
Authoritative ('demanding but responsive')	• is warm, nurturing and emotionally supportive towards the child • sets limits, rules and high standards and explains the reasons for them • gives direction and expects responsible behaviour and cooperation in return • explains the consequences of unacceptable behaviour • reasons with the child and encourages verbal give-and-take • demonstrates respect for the child as an independent individual
Permissive ('excessively lenient')	• is warm and nurturing towards the child • is responsive to the child's needs and wishes • does not set limits, lay down rules or assign responsibilities • adopts a 'no discipline' approach and does not try to keep the child under control • does not encourage the child to aim for high standards of behaviour • allows the child to regulate their own behaviour

Table 19.3 Methods of control

Effect of communication

Communication is the exchange of information, facts, feelings, ideas and opinions between people. Most people spend a large part of their time each day communicating (at various levels) with other people. To do so they make use of **verbal** and **non-verbal** means of contact.

Non-verbal communication

Non-verbal communication (sending and receiving wordless messages) plays an important part in the establishment of relationships between individuals. On some occasions, it reinforces verbal messages, on others, it adds to the information being transmitted (see Figure 20.4). This form of communication may also indicate an individual's **attitude** (for example,

their degree of like or dislike for something). It may display **emotion** (for example, as a facial expression). Non-verbal communication comes in many forms.

Smiling in infants

A baby is capable of taking part in a non-verbal dialogue based on sounds and visual signals such as **smiling**. At about 6 months, smiling becomes a selective social act normally reserved for the mother (and other close members of the family) whom the baby recognises (see Figure 20.5). Smiling is important because it makes the baby especially appealing and lovable, thereby strengthening the bond between the carers and the baby. This is of **survival value** because it helps to ensure that the baby (who is almost totally helpless at this stage) will receive the food, care and attention that they need.

www.CartoonStock.com

Figure 20.4 Non-verbal communication

Figure 20.5 Smiling as a social act

Facial expressions

A **facial expression** is the result of the activity of many muscles. It conveys the emotional state of the individual to observers. **Smiling**, for example, indicates pleasure shared by the people exchanging smiles. Many other facial expressions are also used to act as indicators of **emotions**. The six main types are shown in Figure 20.6. It is interesting to note in particular the varied use to which the eyes and mouth are put to convey these non-verbal messages. Cartoonists are especially skilful at employing facial expressions to communicate attitudes and emotions essential to the understanding of a joke (see Figure 20.7). Surveys show that on average women are better than men at correctly recognising the emotion represented by a facial expression.

Eye contact

The eyes are used to communicate information by the length of time that they are allowed to continue

Figure 20.7 Cartoonists use facial expressions to good effect

making contact with those of another person. This maintenance of gaze between two people is called **eye contact** (see Figure 19.8). People who are in a close relationship exchange glances and meet one another's direct gaze much more often than people who are strangers. Extended eye contact is one method by which people communicate sexual interest in one another. Therefore strangers (who wish to remain strangers) feel embarrassed if eye contact extends beyond a mere exchange of glances. They tend to play safe by avoiding unnecessary eye contact.

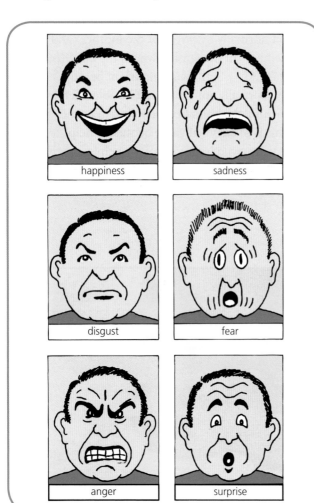

Figure 20.6 Facial expressions indicating emotion

Figure 20.8 Eye contact can be used to communicate sexual interest

Looking as signalling

In addition to various eye movements such as eyes popping, eyes narrowing, etc., the eyes are used to convey **signals** by simply looking at another person in a meaningful way. For example, **winking** (see Figure 20.9) at someone only slightly known to you might indicate friendliness (or overfriendliness!) and the wish to become better acquainted. On the other hand, winking discreetly at a good friend in the company of others might indicate the mutual enjoyment of a private joke.

Figure 20.9 Winking provides a quick signal

When the process of looking normally at another person continues for a period of time beyond that required for routine information gathering, the signals become loaded with further meaning. Consider an individual (e.g. male person A in Figure 20.10) who wants to make contact with another individual (e.g. female person B).

One way of indicating this is for A to catch B's eye. If, in return, B wishes to signal that she is not interested in making contact, she will avoid his gaze. Then A will either give up or persist. If he persists, B may then decide to adopt an angry defiant stare to signal rejection. On the other hand, B may wish to signal that she does want to make contact with A and therefore will allow him to catch her eye. This may lead to a conversation (verbal communication).

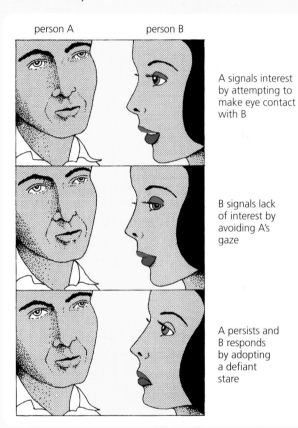

person A   person B

A signals interest by attempting to make eye contact with B

B signals lack of interest by avoiding A's gaze

A persists and B responds by adopting a defiant stare

Figure 20.10 Looking as signalling

Body language

Often people are unaware of the extent to which they use their bodies to communicate with one another non-verbally. Such **body language** is expressed by posture, gestures and certain other activities. Several examples of **posture** and the possible attitude of the person adopting the bodily position are given in Table 20.4.

Posture	Suggested attitude
Sprawled back in easy chair with legs spread	Relaxed
Sitting up stiffly with legs together	Tense
Standing with feet apart and arms akimbo	Confrontational
Slumped forward in chair with arms folded	Bored or sad
Lying curled in 'fetal' position	Frightened

Table 20.4 Communication by posture

Some **gestures** have definite meaning. Beckoning someone forward, nodding the head and pointing (see Figure 20.11) all give clear signals. Other gestures indicate emotional states of mind. For example, continuous drumming of the fingers or fidgeting suggests tension or boredom; wringing of the hands indicates anguish; clenched fists signal pent-up anger. Some people engage in certain activities that give an indication of their state of mind. Nail-biting and hair-chewing are outward signs of nervous tension or stress.

Figure 20.11 Pointing is an effective gesture

Physical proximity

Within each culture there is a generally accepted **distance** that two people keep between them while conducting a normal conversation. Increase in this distance may suggest dislike or even repugnance between the two participants; decrease in the distance may indicate sexual attraction or aggression (in other words an invasion of someone's '**personal space**' – the invisible 'bubble' that surrounds each human body).

Effect of physical proximity on eye contact

In an experiment, a group of female volunteers each agreed to take part in three brief interviews with a female stranger who would be sitting at a different distance from the volunteer each time. The interviewer had been trained to stare continuously at the volunteer during each interview. Each volunteer's percentage eye contact was measured for each interview. The results are shown as a graph in Figure 20.12.

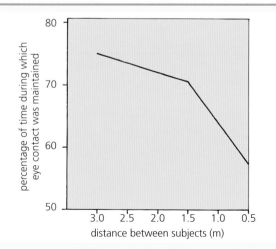

Figure 20.12 Effect of physical proximity on eye contact

As **distance** between the two people **decreased**, the amount of **eye contact** also **decreased**. At a distance of 0.5 m the two people were physically and psychologically much **closer** than the generally accepted level for strangers. They were within the region associated with aggression or sexual attraction. In each case the specially trained interviewer continued to stare while talking; the volunteer appeared uncomfortable and embarrassed, and tried to avoid making eye contact.

Touching

Touch is a powerful form of non-verbal communication. An impersonal situation, such as an interview for a job, may begin or end with a **handshake** between employer and potential employee. If the person gets the job and a satisfactory working relationship develops over time, the dominant employer may feel free to touch, as a supportive gesture, the arm, for example, or shoulder of the subordinate employee. However, the employee knows that they are not free to touch their employer.

An increased amount of touching occurs between people who have close personal relationships, such as the members of a family.

Figure 20.13 Touching as a form of non-verbal communication

The process of touching may communicate **different meanings** including support (see Figure 20.13), playfulness, appreciation, affection and sexual interest.

Mode of delivery of speech as a form of non-verbal communication

Tone of voice, accent, emphasis, speed of delivery and timing of speech are **auditory signals** that depend on spoken language for their existence. They often indicate the person's frame of mind. A monotonous voice suggests fatigue and boredom, loudness may indicate anger and high speed often signals excitement or nervousness.

Verbal communication

Language

A **language** is a system that combines basic sounds (in themselves meaningless) into **spoken words** usually also represented by **written symbols** (see Figure 20.15). These sounds and symbols represent information that can then be arranged into simple categories (words) and more complex hierarchies (phrases, sentences and paragraphs). Since the sounds and symbols have meaning to the members of the society in question, they enable its members to express thoughts and feelings and convey information to one another. The ability to make sophisticated use of language is one of the distinguishing features that make much of our behaviour unique and set us apart from other animals.

Related Topic

Go and no-go touch areas

The results of detailed research indicate that people regard different areas of their body in different ways with respect to accessibility to others to touch. In general, arms and shoulders are regarded as 'non-vulnerable', **touchable** areas. Other parts, especially those close to sex organs, are regarded as 'vulnerable' and **untouchable** by most people. Figure 20.14 shows the results of a survey carried out among a large group of college students. Some parts of the body were touchable to one category of person but not to another.

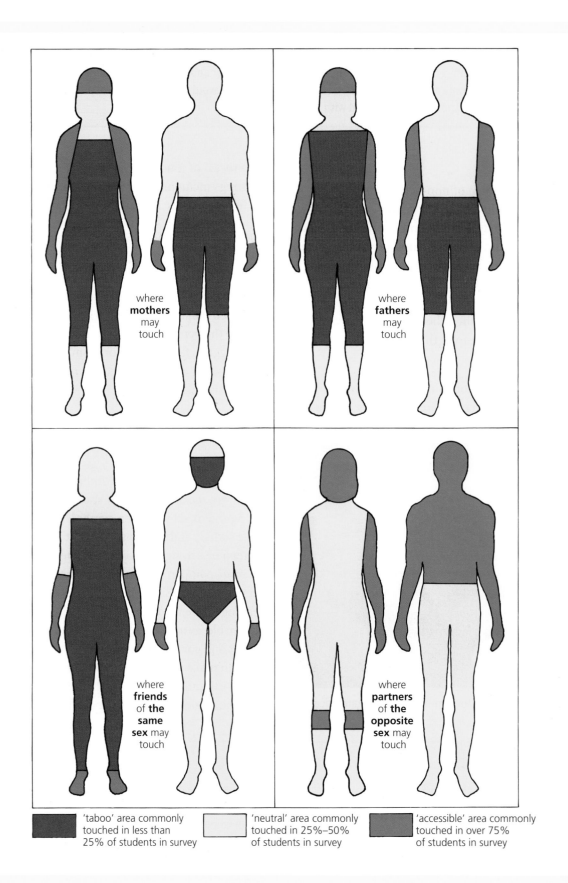

where **mothers** may touch

where **fathers** may touch

where **friends** of **the same sex** may touch

where **partners** of **the opposite sex** may touch

'taboo' area commonly touched in less than 25% of students in survey

'neutral' area commonly touched in 25%–50% of students in survey

'accessible' area commonly touched in over 75% of students in survey

Figure 20.14 Go and no-go touch areas

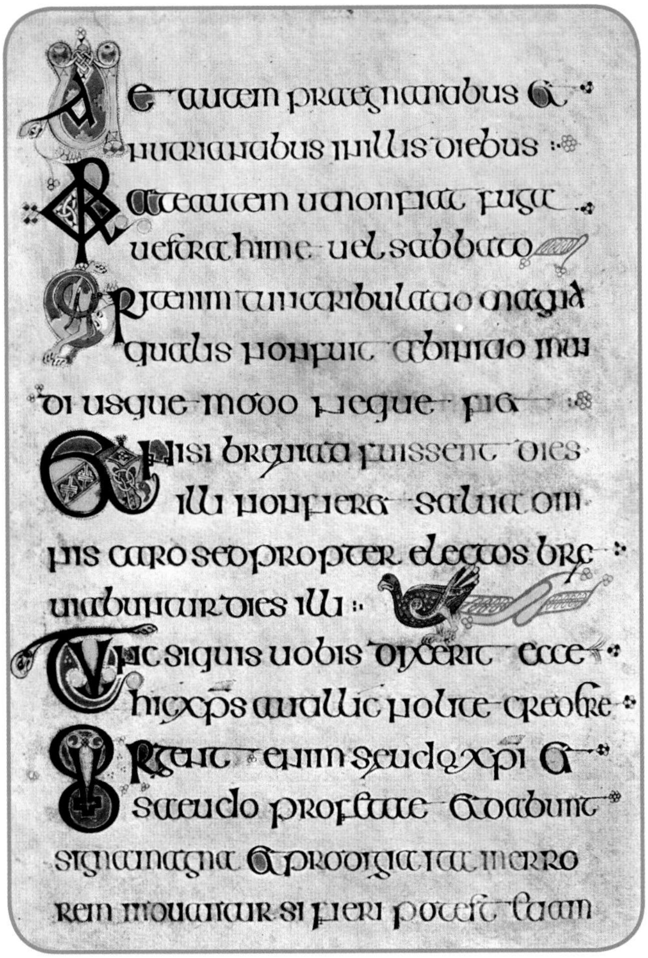

Figure 20.15 Language as written symbols

Transfer of information

On a **short-term** basis, language allows humans to convey to one another the information necessary for successful day-to-day living. We are able, for example, to request our basic needs, express feelings about others and handle a variety of interpersonal situations. We soon learn how to employ certain **linguistic strategies** to our advantage. For example, adopting a friendly tone of voice, speaking clearly and not too quickly, varying the tone of voice and listening to other people's point of view are all techniques employed by good communicators.

On a **long-term** basis, language allows the transfer and receipt of information from generation to generation. For most people in Britain this occurs during 11 or more years of schooling, often followed by several years of further education.

Knowledge possessed and discoveries made by one generation are passed on by **spoken** and **written** word to the next generation without each new generation having to make each discovery for itself. This saves time and enables successive generations to go on and make new discoveries which then become added to the existing body of human knowledge.

Thus language promotes the acceleration of **learning** and the development of **intellect**. It allows us to make detailed plans that, in turn, affect and benefit future generations by promoting their continued **cultural** and **scientific progress** and **social evolution**.

Testing Your Knowledge 2

1 a) In general, what is meant by the term *communication* in relation to human social behaviour? (1)
   b) Identify the TWO types of communication. (2)
   c) Why is smiling by an infant of survival value? (2)

2 a) Give TWO examples of emotions that are often indicated by facial expressions. (2)
   b) For each of these expressions, describe the state of the person's mouth when the facial expression is adopted. (2)

3 a) Some people find it more difficult to communicate important information to a stranger by telephone than by talking face to face. Suggest why, by referring to TWO examples of ways in which information can be communicated by non-verbal means. (2)
   b) Other people are described as having an 'excellent telephone manner'. Suggest TWO possible features of such a person's verbal skills that this might include. (2)

4 a) What is *language*? (2)
   b) State TWO methods by which language can be transferred from person to person. (2)
   c) Why is transfer of language from generation to generation important? (2)

Learning

The term **learning** is often used to mean knowledge gained by study or the act of gaining knowledge by studying. However, psychologists define learning as **any relatively permanent change in behaviour that occurs as a direct result of experience.**

Effect of practice on motor skills

Once a **motor skill** (e.g. riding a bicycle) has been mastered, repeated use of it promotes the establishment of a **motor pathway** in the nervous system. Repetition of the skill is thought to result in an increased number of synaptic connections being formed between the neurons in the pathway. This leads to the formation of a '**motor memory**' for the skill. Practice improves performance; lack of practice results in the skill becoming 'rusty' (but not being completely lost).

Investigation

Learning using a finger maze

The apparatus shown in Figure 20.16 is used by a learner who is blindfolded throughout the experiment. The learner's task is to proceed through the maze from entrance to exit using the tip of their forefinger. The observer's job is to measure the time taken for each trial by the learner. The procedure is repeated to give a total of 10 trials. Table 20.5 lists some design features (and precautions) and the reasons for adopting them in this investigation.

After several trials, the time required to pass through the maze is found to decrease until eventually a minimum is reached. When graphed, the results give a **learning curve** (see Figure 20.17). From this investigation it is concluded that practice improves performance of a motor skill. In the case of the finger maze, a best time is eventually reached and this cannot be improved upon. By trial-and-error learning, the person has formed a picture (cognitive map) of the route through the maze in their 'mind's eye' and a certain minimum amount of time is required to physically run the finger tip through it.

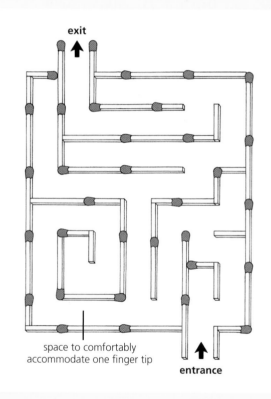

Figure 20.16 Finger maze

space to comfortably accommodate one finger tip

exit

entrance

Types of learning curve

The curve shown in Figure 20.17 from the results of a finger maze investigation charts the decrease in time taken as an indication of the effect of practice on learning. Counting the decreasing number of errors made by the learner in successive trials is an alternative method of carrying out this investigation and gives a learning curve of similar shape.

Design feature or precaution	Reason
Same learner used for each group of 10 trials; same finger used each time; same design of maze used each time	To ensure that only one variable factor is included in the investigation
10 trials per learner	To give learner opportunity to reach best score
Learner blindfolded throughout all 10 trials	To prevent learner improving their performance artificially
Path between matchsticks just wide enough to accommodate one finger tip comfortably	To prevent two fingers being used to explore simultaneously two routes at a junction and establish the correct one more quickly
Experiment repeated with many learners and learning curves compared	To obtain a more reliable set of results

Table 20.5 Design features for investigation

Learning can also be measured as an increase in the number of correct responses achieved per unit time (e.g. the number of three-lettered words that a person unfamiliar with a keyboard can key in in 1 minute from a long list). When graphed, the results give a second type of learning curve (see Figure 20.18). Again practice is found to improve performance. Eventually, after much practice, a maximum level of performance is reached that cannot be improved upon. However skilled the person, there is a physical limit to the number of words that he or she can key in in 1 minute.

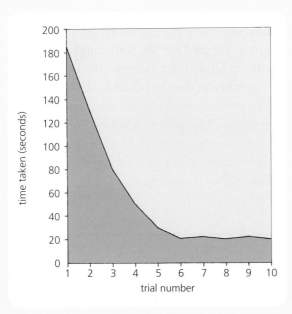

Figure 20.17 Learning curve of typical set of results

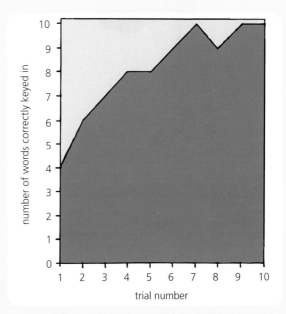

Figure 20.18 Second type of learning curve

Imitation

Children learn by observing and **imitating** (copying) adults and other older children. This often occurs during play. A child dressing up in adult clothes (see Figure 20.19) is imitating the behaviour of adult relatives and friends. Learning by imitation is not restricted to children. Throughout life, many aspects of human behaviour are learned by observing and imitating the behaviour of others. When faced with a new task (e.g. learning how to operate a smart phone), it is much easier and takes less time to learn by watching and then imitating an expert than by reading the manual.

Figure 20.19 Learning by imitation

Imitation is an especially effective method of learning if the expert breaks up the demonstration into several small parts and allows the learner to try to **repeat** what they have seen after each part. Learning by imitation is further promoted if the expert is perceived by the learner as an attractive role model whose status is enhanced by the possession of the skill being demonstrated.

For most people, copying a **demonstration** is the preferred method of learning a new skill. To try to compensate for the lack of a live expert to imitate, instruction manuals often include many diagrams of human models which aid comprehension.

Imitation of social skills

Behaviour acquired by imitation is not restricted to the learning of physical tasks and skills; it is also involved in the learning of **social skills** and **attitudes**. Parents, other adults and perhaps older brothers and sisters provide children with a variety of possible **models** to imitate.

Children tend to imitate many aspects of cultural and social behaviour. Once learned, many of these **values** and **traditions** (such as being kind and generous to others, belonging to a certain religion and so on) might be adopted for life. Some might be accepted during childhood (for example, the belief that smoking damages health) but be rejected during adolescence (when smoking might be made to seem attractive by a peer group). However, in adulthood the person might resume the original belief. Most people eventually embrace many of the cultural and social traditions and values held by their parents.

Investigation

Speed of performance of a new task

The task in this investigation is learning to find the way through a trial-and-error cavity maze. The class, working in pairs, is divided into two groups, A and B. Each pair in group A is given a copy of Figure 20.20 and timed as they carry out the task by **following written instructions**. Each pair in group B is timed as they first watch the teacher demonstrate the task and then carry it out by **imitation**.

On average it is found that the members of group B are faster at performing the task than those in group A. It is therefore concluded that imitation is more effective than following a set of written instructions. It is interesting to note that no significant difference is found between the groups in terms of the number of errors made during the first attempt or the reduction in number of mistakes made during the second attempt. This suggests that the two methods of instruction are equally effective.

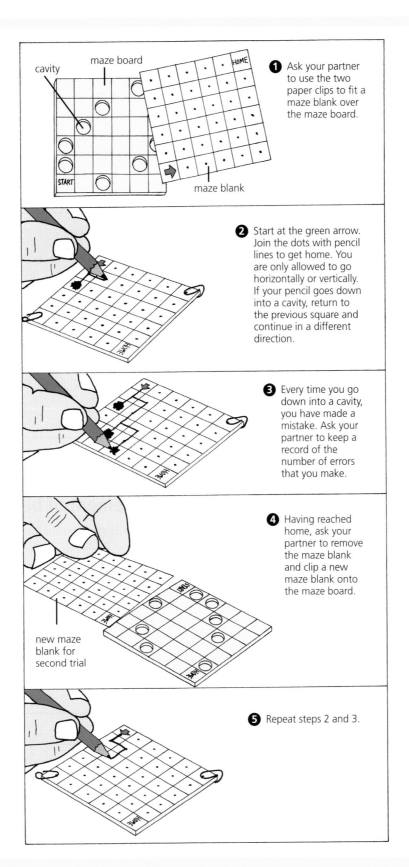

Figure 20.20 Investigation procedure

Social techniques

Among adults, imitation is an especially effective method of learning certain **social techniques**. For example, tone of voice, sympathetic manner and oral delivery carrying authority cannot be described easily. They need to be experienced to be learned.

Trial-and-error learning

If a rat is placed in a specially designed box (see Figure 20.21) it responds in various ways such as exploring the box, touching the floor and leaning against the sides. Sooner or later the animal pushes the lever and food immediately appears in the food tray. If the rat is only rewarded with food when it presses the lever, it soon learns to associate its own behaviour with the delivery of food.

Figure 20.21 Trial-and-error learning in rats

Motivation

Motivation is the 'inner drive' that makes an animal want to participate in the learning process. Animals are motivated by many factors such as hunger, thirst, sexual drive and curiosity. The effect of motivation on an animal's ability to learn can be investigated by comparing hungry and well-fed rats, which must **negotiate a maze** before receiving a food reward. From the graph in Figure 20.22, it can be seen that the number of errors made per trial by the hungry rats quickly decreased since they were motivated to learn. The performance of the well-fed rats failed to improve because they lacked motivation.

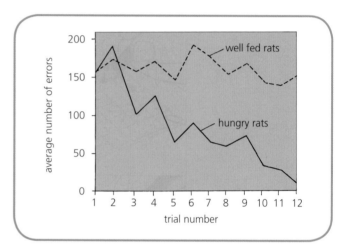

Figure 20.22 Effect of motivation on learning

Reinforcement of behaviour

In the above examples, the behaviour pattern (trial-and-error learning) has positive consequences for the animal (it gets fed). The behaviour is therefore repeated and, as a result, becomes reinforced. **Reinforcement** (see Figure 20.23) is the process that makes an organism tend to repeat a certain piece of behaviour. During reinforcement, the **reinforcer** (such as a food reward) increases the probability of the response being repeated.

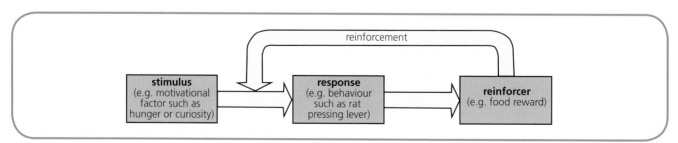

Figure 20.23 Reinforcement

Related Information

Positive and negative reinforcement

Positive reinforcement occurs when something pleasant or **positive** is received by the organism after a particular response has been made, thereby increasing the chance of the response being repeated. For example, a hungry rat rewarded with food for pressing the lever repeats the operation.

Negative reinforcement occurs when something unpleasant or **negative** is brought to a halt when the organism makes a particular response, thereby increasing the chance of the response being repeated. For example, a human discovering that a painkiller cures a headache is likely to use this remedy when pain strikes in the future. (Negative reinforcement should not be confused with punishment, which is the application of a penalty or sanction following the wrong response by the learner.)

Related Information

Continuous and intermittent reinforcement

Behaviour is said to be **continuously** reinforced when the response is always reinforced (for example, the hungry rat receives food every time it presses the lever).

Behaviour is said to be **intermittently** reinforced if the response is reinforced on only some of the occasions that it occurs. For example, a child's mother is not always present to reward him or her with praise for following the Green Cross Code. However, the influence of the reinforcements that do occur is (hopefully) strong enough to persist and make the child cross the road safely when on their own.

Superstition

People develop a **superstition** as a result of a favourable event (which has occurred merely by chance) apparently occurring in response to some piece of behaviour by the person. The person repeats the behaviour and occasionally the favourable event also occurs thereby reinforcing the probability of the person repeating the behaviour. A gambler who blows on the dice muttering 'Luck, be a lady tonight', might be rewarded with a win sufficiently often to reinforce the belief that their behaviour is affecting the response.

Shaping of behaviour

Shaping is the process by which a desired pattern of behaviour is eventually obtained from the learner by the trainer **reinforcing successive approximations** of the desired response. Normally the desired response would have a low probability of occurring spontaneously.

Case Study Shaping in learning

The proper use of a knife and fork by a child has almost no probability of occurring of its own accord. By using **shaping**, the parents direct the child's behaviour along the desired route by praising (and therefore reinforcing) those responses that are approximations of the required response.

The flow chart in Figure 20.24 shows how **reinforcement** of responses that are successively more and more similar to the final desired response results in the child learning the new skill. Shaping also features in toilet-training, learning to dress oneself, tying shoe laces and many other skills.

Animals
Animal trainers use shaping to teach dolphins to balance balloons on their snouts, pigeons to dance in patterns, and parrots to ride bicycles. An especially useful application of shaping is the training of dogs to act as the eyes of the blind (see Figure 20.25).

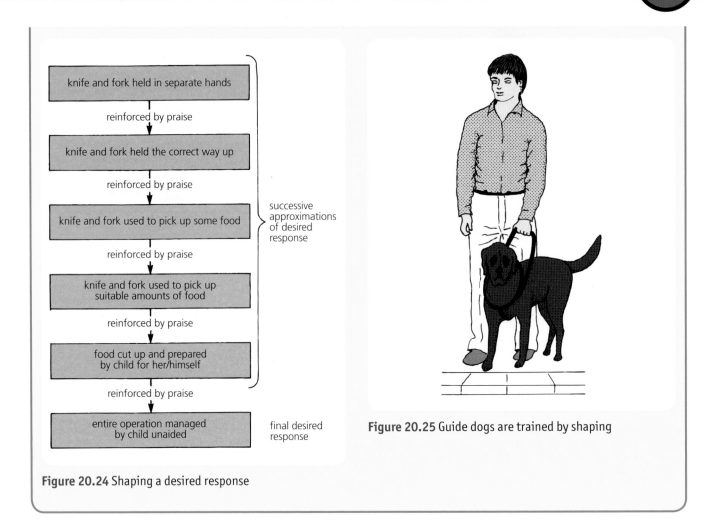

Figure 20.24 Shaping a desired response

Figure 20.25 Guide dogs are trained by shaping

Extinction of behaviour

Extinction is the name given to the eventual **disappearance** of a behaviour pattern when it is no longer reinforced by rewards. Consider the rat in Figure 20.21, which has learned that pressing the lever gives food **every time**. If this rat is now put into a situation where pressing the lever fails to give the food reward, the rat is found to press it less and less frequently. Eventually the rat does not press it at all and the earlier behaviour pattern (which had been reinforced continuously and learned quickly) is said to have become **extinct**.

Related Information

Effect of intermittent reinforcement on extinction

Rats that have learned that pressing the lever **sometimes** gives food (in other words, have been subjected to intermittent reinforcement) take longer to learn the behaviour pattern but also take much longer to give it up completely. They tend to persist in giving the lever an occasional 'hopeful' press. They had learned not to expect food every single time and are not so easily put off by the lack of reward. Thus, intermittent reinforcement is **more resistant** to extinction than continuous reinforcement.

Related Topic

Rewarded and unrewarded behaviour

Responsible parents try to teach their children the difference between acceptable and unacceptable behaviour. They tend to encourage the development of 'good' behaviour by **reinforcing** it. For example, parents might be of the opinion that being truthful, showing consideration for others and trying hard at school are all forms of desirable behaviour. They would therefore reinforce these behaviour patterns with **rewards** such as attention, praise, outings, birthday presents and pocket money.

Parents also have to deal with their children when their behaviour is unacceptable. In theory, behaviour that goes **unrewarded** should become **extinct**. In some cases unacceptable behaviour (such as a child nagging a parent for sweets every time they arrive at the supermarket checkout) does disappear provided that the parent can summon the patience to ignore it and ensure that the behaviour always goes unrewarded.

However, most parents find that many forms of unacceptable behaviour shown by their children are impossible to ignore. Responsible parents normally attempt to modify such behaviour on the part of their children by resorting to punishment. This might involve a long talk followed by removal of the rewards stated above, accompanied perhaps by further sanctions such as being 'grounded' (see Figure 20.26).

Figure 20.26 Grounded!

Generalisation

Generalisation is the ability to respond in the same way to many different but related stimuli. In an experiment, an 11-month-old boy who liked furry animals was put in a room with a tame white rat. He reached out for the rat showing no fear. Each time just as his hand went to touch the animal, a very loud noise was deliberately made using a steel bar. As a result, the child developed an aversion for white rats. It was also discovered that the child had developed a fear for many other furry objects that he had not seen before (see Figure 20.27). The spread of the response (in this case fear) to different but related stimuli is an example of **generalisation**.

Discrimination

Discrimination is the ability to distinguish between different but related stimuli and give different responses. Discrimination is taught by **reinforcing the desired response** (for example, a mother giving her baby hugs and kisses when she is addressed as 'Mama') and by not reinforcing the wrong response (for example, the father not responding when he is addressed by the baby as 'Mama').

The baby is soon able to tell the difference between similar stimuli (such as adult family members) and determine whether or not the correct stimulus (the mother) is present before saying 'Mama'. The baby has learned to **discriminate**. Learning to discriminate is an essential part of a child's preparation for coping with everyday life. It is important for the child to appreciate, for example, that some loud noises such as thunder do not indicate danger, that some green fruit such as Granny Smith apples are ripe and ready for eating and that some dogs are unfriendly and might bite.

stimulus	response by child
	reaches out to rat showing no fear
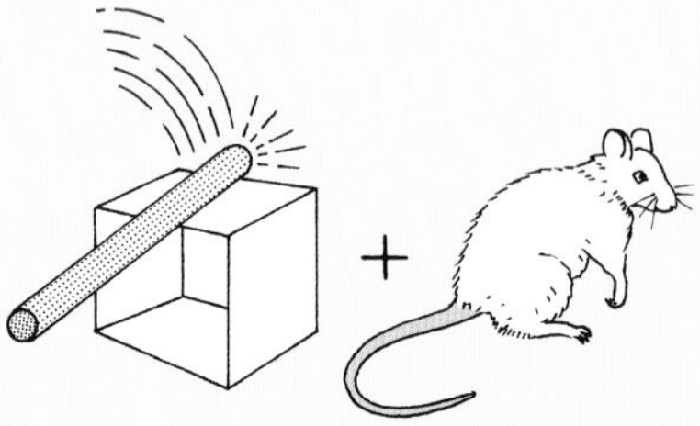	refuses to reach out and shows fear of white rats
	refuses to reach out and touch any furry object

Figure 20.27 Generalisation

Social influence

Social groups

Human beings are social animals. A large part of the average person's life is spent **interacting** with other people who act as stimuli and offer responses to the person's behaviour. Almost without exception, people belong to one or more **social groups** of different types and sizes. These could include, for example, the family, the teenage gang, the sports team, the trade union, the army regiment, the cub pack, the football supporters' club, the political party, the religious sect, the school orchestra, and so on.

Social groupings provide people with a feeling of belonging and of being accepted. Many groups are held together by **rules**, written and unwritten, and **symbolism** such as a uniform that sets the group apart from the non-members. Groups provide support for their members especially in times of need. However, the group also affects the behaviour of the individual members by setting standards and even deciding what should be done in certain situations. An individual who accepts these conditions must **behave in the same way** as the other members of the group.

Social facilitation

One of the factors that motivates many people is the need for **status**. They want to impress and be admired

---

Testing Your Knowledge 3

1 a) Give an example of a motor skill. (1)
   b) Why does practice improve the performance of motor skills? (1)
2 a) With reference to the learning of a new skill, what is meant by the term *imitation*? (1)
   b) Give an example of a cultural tradition that a person may learn by imitation from their parents and adopt for life. (1)
3 a) What is meant by the term *reinforcement*? (1)
   b) What name is given to the rewarding of behaviour that approximates to the desired behaviour? (1)
   c) What is meant by the *extinction* of behaviour? (1)
4 a) Define *generalisation* and give an example to illustrate your answer. (3)
   b) Define *discrimination* and give an example to illustrate your answer. (3)

5 Choose the correct answer from the underlined choice given in each of the following statements. (6)
   a) Knowledge/learning is a change in behaviour as a result of experience.
   b) Repeated use of a motor skill establishes a motor/sensory pathway in the nervous system.
   c) Learning can be measured as an increase/a decrease in the number of errors made per unit time.
   d) For most people, the preferred method of learning a new skill is to copy a demonstration/manual.
   e) If a behaviour pattern is not rewarded, it is likely to become extinct/reinforced.
   f) When a child who has been bitten by a dog fears all dogs, this is called discrimination/generalisation.

by other members of a social group to which they belong. It is interesting therefore to consider whether or not the presence of other people affects an individual's **performance**. Research shows that in **competitive** situations, subjects do tend to work faster and achieve a higher level of productivity and energy output than they do when working alone.

Figure 20.28 shows the results from a survey done by an athletics club on its members (in the absence of spectators). The presence of other competitors seems to spur the individual on to heights not achieved on their own. This increased performance in the presence of others (especially when the situation is competitive) is called **social facilitation**. Even in non-competitive situations, individuals are found to achieve more when working in a group than when working in isolation.

Familiarity with the task

The presence of others is especially effective at improving an individual's performance if they are already very **familiar** with the task or the task is very **simple**. However, the presence of others tends to interfere with progress when the person is trying to learn something **new** or perform a more **complex** task. Then they imagine that the others are monitoring their progress, feel stressed and tend to make more mistakes. Both of these aspects of social facilitation are summarised in Figure 20.29.

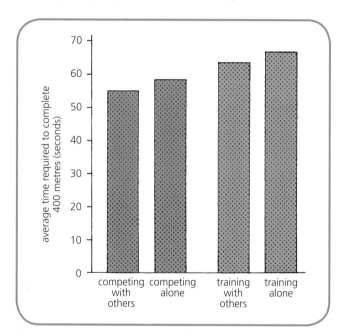

Figure 20.28 Effects of social facilitation

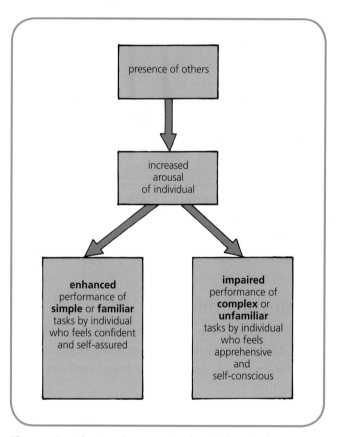

Figure 20.29 Contrasting aspects of social facilitation

Related Topic

Group pressure

The following experiment was set up to investigate the effect on an individual of **majority opinion** even when it was clearly **contrary** to the fact. A group of eight people together in the same room were invited in turn to solve the problem of visual judgement shown in Figure 20.30. Unknown to the eighth member of the group, the other seven were part of the experiment and had agreed in

advance to give wrong answer C. When it came to the turn of number eight (the experimental subject), they were placed in the position of having to disagree with the majority or doubt their own judgement sufficiently to conform to group opinion. The experiment was carried out on many subjects and repeated many times with each one. The results are shown in Figure 20.31.

During the discussion sessions that followed the investigation, some interesting points emerged. Many of the subjects who had agreed with the majority had not done so simply to bring the confusing episode to an end. At the time they had expressed genuine respect for the (wrong) judgement of the majority! They had come to the conclusion that there had been something wrong with their own eyesight or that they had been experiencing an optical illusion.

The process of exerting such a strong influence on an individual that they abandon their own views or ideas in favour of those held by the social group is called **group pressure**. If susceptible individuals can be pressurised into agreeing to the wrong answer in a straightforward situation like the one above, the possibility of influencing a person's judgement is increased dramatically when the controversial subject is an **opinion** or an **attitude**.

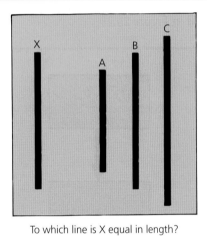

To which line is X equal in length?

Figure 20.30 Visual judgement problem

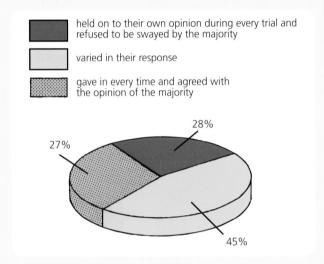

held on to their own opinion during every trial and refused to be swayed by the majority

varied in their response

gave in every time and agreed with the opinion of the majority

28%

27%

45%

Figure 20.31 Results of group pressure experiment

Deindividuation

In real life, group pressure is a powerful force. People find it difficult to resist going along with the decisions made by the group. Members of a group conform because they:

- **identify** with other group members and want to be like them
- desire the **personal gain** that membership often brings
- want to be **liked** and not to be thought of as unpopular.

Once under the influence of group pressure, individuals think and act differently from the way that they would if they were on their own. Decisions and behaviour now depend less on the members' individual personalities and more on the **collective influence** of the group. This loss by an individual of personal identity when in a group is called **deindividuation** and it can lead to **diminished restraints** on behaviour.

"Which embassy's this?"

Figure 20.32 Deindividuation

Deindividuated people feel indistinguishable from others in the group and are more likely to act **mindlessly** and do things that they would never consider doing on their own. This often takes the form of anti-social, aggressive behaviour by a 'faceless' mob whose members have temporarily lost awareness of their own individuality and responsibility (see Figure 20.32).

Risk-taking

The members of a group will take bigger risks when in the group than when alone. A gang of teenagers will taunt and dare one another into pursuing **extreme activities** (such as playing 'chicken' on a railway line, experimenting with drugs, fighting a rival gang and so on) which they would be unlikely to attempt on their own.

Anonymity

Being an anonymous member of a crowd (for example, by being hooded or masked) offers the individuals

possible protection from punishment. If the members of a mob do not have to worry about being caught breaking the law, then they are more likely to pursue their irresponsible activities.

Influences that change beliefs

Internalisation

Internalisation is the process by which an individual incorporates within him or herself an enduring changed set of beliefs, values or attitudes. Radio, television and newspapers subject listeners, viewers and readers to a mass of information that may or may not become **internalised**. By attempting to persuade people to change their current beliefs and **adopt a different set of beliefs**, politicians, government departments and big businesses try to effect internalisation.

Related Information

Examples of internalisation

Televised party political broadcast

Politicians regularly use this method to try to persuade viewers to vote for their party. Research shows that viewers are most likely to pay attention if the presenter:

- has a pleasant appearance
- has a good vocal delivery, including warmth and humour

- presents a two-sided argument but favours one side
- gives convincing reasons for supporting their side
- appears to be an expert on the subject being discussed.

However, it is debatable how many people actually alter their beliefs and vote for the speaker's political party, having been fully convinced that unemployment will

Figure 20.33 Anti-drugs message

Figure 20.34 Anti-smoking message

plummet, the economy will boom, the national debt will shrink and so on.

Health warning

Drug education programmes and anti-smoking campaigns often use posters (see Figures 20.33 and 20.34) in an attempt to persuade people to alter their behaviour.

Internalisation and advertising

One method of advertising attempts to create a feeling of dissatisfaction with our current situation while at the same time presenting us with the allegedly better alternative. By this means big businesses try to bring about **internalisation** and persuade us that a particular brand is faster, smoother, sexier, cheaper, etc. than the 'inferior' one that we are currently using. Table 20.6 lists a few products and suggests the altered belief that the advertiser is hoping to create.

Product	Altered belief intended by advertiser
Foodstuff	It will nourish the family and make the person who serves it a 'good mum or dad'
Soap powder	It will get clothes cleaner and softer than ever before
Item of clothing	It will make the wearer fashionably dressed and the envy of their peer group
Toothpaste	It will clean teeth and reduce fillings because it has been proved scientifically
Double-glazing	It will improve the quality of life and make the neighbours envious
Aftershave	It will make a man smell so sexy that women will find him irresistible
Cosmetic	It will make a woman look so beautiful that men will find her irresistible
Motor car	It will offer excitement, freedom and escapism previously only dreamed of

Table 20.6 The power of advertising

Identification

Identification is the process by which person A deliberately changes their beliefs in an attempt to be like person B. Person B exerts a strong influence over person A, because A admires B enormously and makes B the object of hero-worship. In extreme cases, A attempts to enhance their self-esteem by behaving, in fantasy and/or in real life, as if they actually were person B. To a greater or lesser extent most people tend to identify with one or more of the personalities who dominate the worlds of entertainment, big business, politics and sport.

Related Information

Identification and advertising

One method of advertising sets out to exploit the process of **identification** by employing a personality with whom an enormous number of potential customers (with disposable income) identify. This 'superstar' is made the focus of a massive, nationwide (or even international) advertising campaign covering all sections of the mass media.

If the superstar is a reasonable actor, he or she will be able to **endorse** the product with apparent sincerity, perhaps combined with humour and other memorable

Figure 20.35 Celebrity endorsement

gimmickry. Such endorsement greatly increases the product's desirability (see Figure 20.35). It is therefore eagerly purchased by the legions of fans who identify with Ms or Mr 'Wonderful' and their achievements, beliefs and lifestyle.

Testing Your Knowledge 4

1 a) Give TWO examples of a social group. (2)
   b) Give ONE reason why people like being members of a social group. (1)
2 a) Explain what is meant by the term *facilitation*. (2)
   b) Under what circumstances does competition not result in increased performance? (1)

3 a) Give TWO reasons why the members of a social group agree to go along with the decisions and rules made by the group. (2)
   b) What term is used to refer to the loss by an individual of personal identity when subjected to group pressure? (1)

What You Should Know Chapter 20

attachment	emotions	linguistic
authoritative	explore	motivated
beliefs	extinction	motor
cognitive	facilitation	non-verbal
communicate	foundations	pathway
competence	groups	reinforcement
competitive	identification	response
contact	identity	shaping
deindividuation	imitating	signals
dependency	internalisation	stable
deprived	learning	verbally

Table 20.7 Word bank for chapter 20

1 Humans are social animals and _____ with one another throughout life by means of signs and signals.

2 The strong emotional tie that binds a baby to the mother is called _____.

3 Early infant attachment allows _____ to be laid for the development of _____ relationships later in life.

4 Securely attached infants are likely to _____ their environment which helps them to develop their _____ skills; insecurely attached infants, _____ of normal social contact, tend to suffer long-lasting ill effects.

5 The period of _____ of children on adults is lengthy and provides the basis for the development of social, _____ and intellectual skills.

6 Compared with authoritarian and permissive control of children, _____ control normally produces adults with greater social _____.

7 People communicate by verbal and _____ means.

8 Adults communicate non-verbally by employing facial expressions, eye _____, postures and touching to express attitudes and _____.

9 Adults communicate _____ by using language that allows information to be transferred from generation to generation, thereby accelerating _____ and the development of culture and social evolution.

10 Once a _____ skill has been mastered, its repeated use results in the establishment of a motor _____.

11 Most people learn a new task more quickly by _____ an expert than by following written instructions.

12 During trial-and-error learning, animals are _____ to learn by factors such as hunger and thirst.

13 _____ is the process that makes an animal tend to repeat a certain piece of behaviour.

14 _____ is the process by which a desired pattern of behaviour is eventually obtained from the learner by the trainer reinforcing successive approximations of a desired _____.

15 _____ is the eventual disappearance of a behaviour pattern when it is no longer reinforced.

16 Most people belong to one or more social _____ of different types and sizes.

17 In general, individuals are found to perform familiar tasks better in _____ situations than on their own. This process is called social _____.

18 _____ occurs when an individual undergoes loss of personal _____ and is unable to resist going along with the behaviour of the group regardless of whether it is acceptable or not.

19 _____ is the process by which an individual incorporates within her/himself a set of _____.

20 During _____, individuals deliberately change their beliefs to try to be like some other person whom they strongly admire.

Applying Your Knowledge and Skills

Chapters 17–20

1 Imagine that you are at home when the doorbell rings. You respond by heading towards the door. Identify FOUR areas of the cerebral cortex shown in Figure 17.15 on page 217 that are in use during this entire operation and, for each, briefly describe the role that it plays in the process. (8)

2 Describe the antagonistic nature of the autonomic nervous system and compare the effect of its two branches on heart beat and blood distribution. (9)

3 Design an investigation into the effect of perceptual set on the perception of the ambiguous picture in Figure 20.36. State your reasons for carrying out the various procedures that would need to be adopted during this investigation. (10)

4 In an investigation into memory span, 40 students were asked to listen to and then attempt to write down each of several series of letters, the first series containing three letters. The results are shown in Figure 20.37.

a) What relationship exists between the number of letters in a series and the percentage of students able to remember the series? (1)

b) i) What was the best memory span recorded in this experiment?

   ii) How many students possessed this memory span? (2)

c) i) What was the poorest memory span recorded in this experiment?

   ii) What percentage of students possessed this memory span? (2)

Figure 20.36

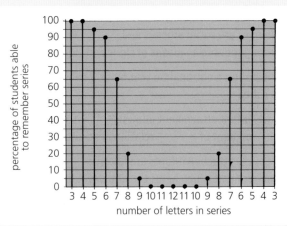

Figure 20.37

d) i) Does presenting the series of letters in descending order of length produce a different set of results from presenting them in ascending order of length?

ii) Explain how you arrived at your answer to i). (2)

e) Give a reason for the adoption of each of the following design features in this investigation:

i) nonsense groups of letters used to make up the series rather than proper words

ii) each series of letters read out by the same tester at a uniform speed

iii) 40 students invited to take part rather than just a few. (3)

5 In an experiment to investigate minor plasticity of response of the brain, six students were each given 2 minutes to correctly solve as many mathematical problems as possible under three different sets of conditions. The results are shown in Table 20.8.

a) Identify the i) auditory, ii) visual distraction. (2)

b) i) Which student(s) demonstrated plasticity of response of the brain?

ii) Explain how you arrived at your answer. (2)

c) Briefly explain how such plasticity is thought to occur. (4)

d) i) Why are the two types of distraction investigated separately rather than simultaneously?

ii) Why is a fresh set of problems needed for each trial?

iii) Identify a possible source of error in this experiment.

iv) Suggest TWO ways in which the reliability of the experiment could be improved. (5)

6 Table 20.9 shows the results from four surveys investigating the incidence of schizophrenia among twins. The members of each pair of twins were raised together by their natural parents.

a) What is schizophrenia? (1)

b) Calculate the average percentage (accurate to two decimal places) of twin pairs affected with schizophrenia when the twins are i) identical, ii) non-identical. (2)

c) What conclusion about the part played by inherited factors can be drawn from these findings? (1)

d) Which survey's data do you consider to be:

i) the most reliable

ii) the least reliable?

iii) Explain your choices for i) and ii). (4)

Student	Number of problems solved in 2 minutes		
	Silence + uninterrupted lighting	Car alarm + uninterrupted lighting	Silence + flashing light
A	20	18	21
B	12	12	11
C	14	3	5
D	18	5	7
E	22	23	21
F	25	8	6

Table 20.8

Survey	Identical twins		Non-identical twins	
	Number of twin pairs	Percentage of affected twin pairs	Number of twin pairs	Percentage of affected twin pairs
1	21	66.69	60	4.67
2	41	68.32	101	14.91
3	41	76.07	115	14.06
4	268	85.16	685	14.52

Table 20.9

7 Figure 20.38 shows the results from a survey in which a large number of mothers of newborn babies were asked to indicate the side on which they normally held their infant.

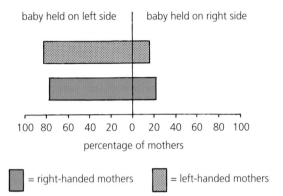

Figure 20.38

a) i) On which side did the majority of right-handed mothers hold their babies?
   ii) On which side did the majority of left-handed mothers hold their babies? (2)

Based on these results, scientists constructed the hypothesis that the mother's heartbeat offers a familiar rhythm and comforts the infant. This hypothesis was put to the test in a large maternity hospital over a period of several days. Babies in group A were exposed night and day to the sound of normal heartbeat; those in group B were exposed to irregular sounds. The bar graph in Figure 20.39 shows the results.

b) Draw TWO conclusions from the graph. (2)
c) Was the hypothesis supported? (1)
d) Identify THREE factors that should be kept constant during an investigation of this type. (3)

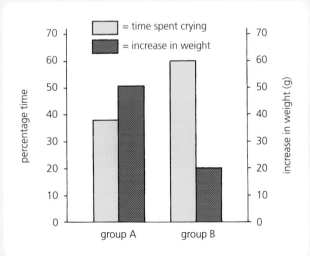

Figure 20.39

8 Two experiments investigating the effect of physical proximity on eye contact were carried out using male interviewers (trained to stare continuously) first with male volunteers and then, later, with female volunteers. The results are summarised in the graph in Figure 20.40.

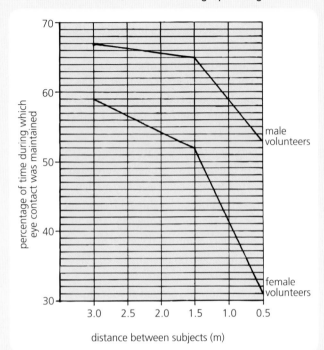

Figure 20.40

a) i) State the reduction in the percentage of eye contact time that occurred when distance decreased from 3 to 0.5 metres for male volunteers.
   ii) By how many more times did the percentage eye contact time decrease for female volunteers over the same distance? (2)

b) i) Give a possible reason why male volunteers show a decrease in percentage eye contact time as distance decreases.
   ii) Give an additional possible reason why female volunteers show an even greater decrease in percentage eye contact time as distance decreases. (2)

9 Each of the following paragraphs describes a situation involving some aspect of behaviour. Give a brief explanation for each using one or more of the terms that you have met in chapter 20.
   a) Certain animals can be trained to do 'clever' tricks. A group of rats, for example, were taught to cross a drawbridge, climb a ladder, crawl through a tunnel, slide down a chute and finally press a lever for a food pellet. (2)

b) Many years ago, cigarette smoking was regarded by most people as a harmless habit. A survey taken at the time showed that there was a much higher incidence of smoking among people who had grown up in a family of smokers than those raised by non-smoking parents. (1)

c) A girl suffering from anorexia nervosa was admitted to hospital and put in a private room on her own without magazines, mobile phone or TV. She was only allowed a visitor, a phone call, a magazine or a TV programme if she agreed to eat something. She gained 3 kg in 10 days. (2)

d) A small girl was bitten by a Scots terrier (a small black dog). She developed a fear of all small black dogs. Her younger brother was bitten by a West Highland terrier (a small white dog). He developed a fear of all dogs. (2)

10 The posters shown in Figures 20.33 and 20.34 on page 303 are designed to attempt to change people's beliefs.

a) For each poster, state a possible version of the altered belief. (2)

b) What name is given to this type of social influence? (1)

Note: Since this group of questions does not include examples of every type of question found in SQA exams, it is recommended that students also make use of past exam papers to aid learning and revision.

Unit 4

Immunology and Public Health

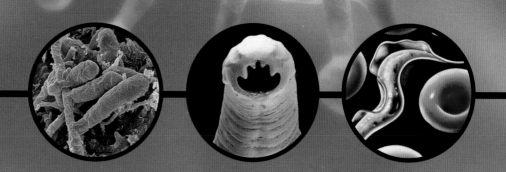

21 Non-specific defences

The body defends itself against disease-causing organisms (pathogens), some toxins (poisons produced by living things) and cancer cells by means of its immune system. **Immunity** is the ability of the body to resist infection by a pathogen or to destroy the organism if it succeeds in invading and infecting the body.

Line of defence	Specific or non-specific	Mechanisms employed	Location in Figure 21.1
First	Non-specific	Use of skin as a physical barrier to keep pathogens out	1
		Secretion of acid by internal lining of stomach to kill microbes	2
		Secretion of mucus by epithelial lining of trachea to trap microbes	3
Second	Non-specific	Inflammatory response	4
		Cellular response – phagocytosis	4
		Cellular response – action of natural killer cells	4
Third	Specific	Response by T lymphocytes from thymus gland (see page 321)	5
		Production of antibodies by B lymphocytes from bone marrow (see page 325)	6

Table 21.1 Three lines of defence

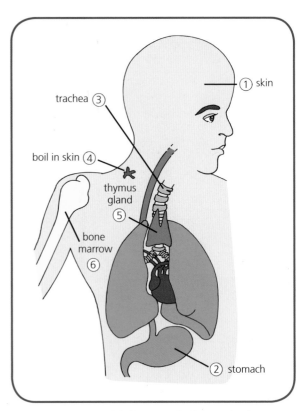

Figure 21.1 Lines of defence in the human body

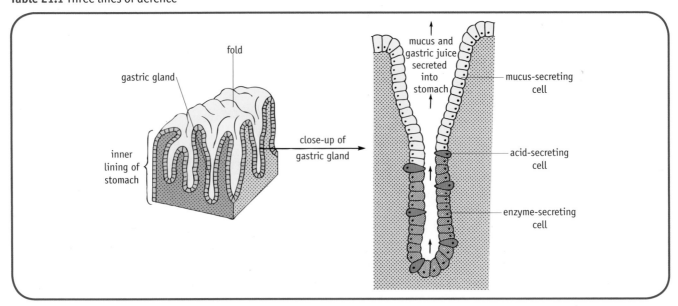

Figure 21.2 Epithelial lining of the stomach

Three lines of defence employed by the body are shown in Table 21.1 and Figure 21.1. The first two are **non-specific**. This means that they work against any type of disease-causing agent. The third line of defence is **specific**, meaning that its components each work against a particular pathogen.

Non-specific defences

Physical and chemical defences by skin and mucous membranes

The surface of the **skin** is composed of layers of closely packed **epithelial cells**, which offer physical protection against bacteria and viruses provided that the skin remains intact. In addition, **mucous membranes** that line the body's digestive and respiratory tracts are composed of epithelial cells that form a protective physical barrier.

The skin and mucous membranes also provide chemical defences against potential pathogenic microorganisms. Secretions from the skin's sweat glands and sebaceous glands keep the skin at a pH that is too low for most microbes to thrive. Secretions such as tears and saliva contain the enzyme **lysozyme**, which digests the cell walls of bacteria and destroys them.

Cells in the mucous membranes secrete sticky **mucus**, which traps microorganisms. The epithelial cells lining the trachea are ciliated and sweep the mucus and trapped microbes up and away from the lungs. **Acid** secreted by cells in the epithelial lining of the stomach (see Figure 21.2) destroys many of the microbes that have been swallowed. However, some do survive the acid conditions and may gain further access to the body.

Inflammatory response

When the body suffers a physical injury such as a cut and/or invasion by microorganisms, it responds with a localised defence mechanism called the **inflammatory response** at the affected site (see Figure 21.3).

Mast cells and histamine

Mast cells are present in connective tissue throughout the body. They are closely related to (and arise from the same stem cells as) white blood cells. Mast cells possess many granules containing histamine. **Histamine** is a chemical that causes blood vessels to **dilate** (become wider) and capillaries to become **more permeable**.

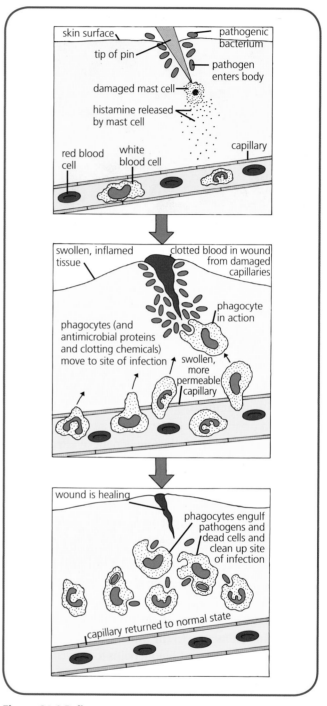

Figure 21.3 Inflammatory response

Following injury, mast cells become activated and release large quantities of histamine. This results in blood vessels in the injured area undergoing **vasodilation** and capillaries becoming swollen with blood. The additional supply of blood makes the injured area **red** and **inflamed**. It swells up because the stretched capillary walls become more permeable and leak fluid into neighbouring tissues.

313

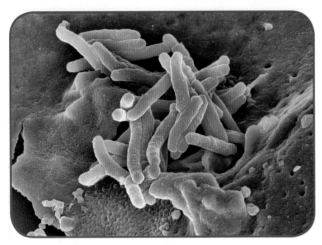

Figure 21.4 Phagocyte engulfing bacteria

Cytokines

Cytokines are cell-signalling protein molecules secreted by many types of cell, including white blood cells that have arrived at a site of injury or infection. Cytokines feature in both non-specific and specific defences.

During the inflammatory response, increased flow of blood and permeability of capillary walls at the site of injury bring about the following beneficial effects:

- **Enhanced migration of phagocytes to the damaged tissue, attracted by cytokines.** Within a short time, a variety of phagocytes have arrived at the scene and are engaged in engulfing pathogens by **phagocytosis** (see Figure 21.4). Some also clean up the injured site.

- **Speedy delivery of antimicrobial proteins to the infected site.** These proteins amplify the immune response.

- **Rapid delivery of blood-clotting chemicals (clotting elements) to the injured area.** Coagulation of blood stops loss of blood, helps to prevent further infection of the wound and marks the start of the tissue repair process.

Phagocytes and natural killer cells

Several types of specialised white blood cells protect the body against pathogenic microorganisms.

Phagocytosis

The process of **phagocytosis** is illustrated in Figure 21.5. A phagocyte is **motile**. When it detects chemicals released by a pathogen such as a bacterium, or antigens present on the surface of a pathogen (also see chapter 22), it moves towards the pathogen. It then engulfs the invader in an

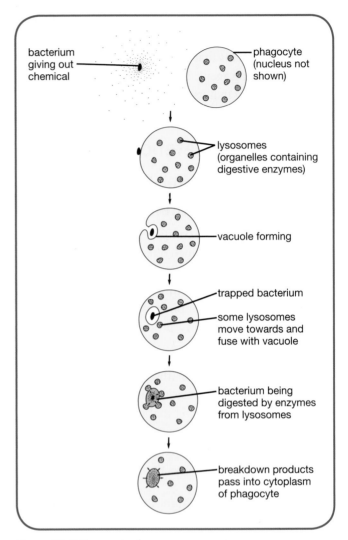

Figure 21.5 Phagocytosis

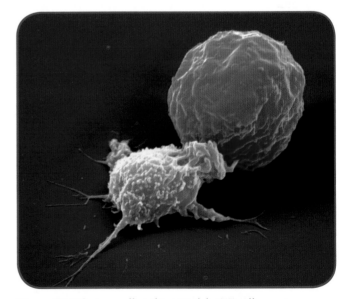

Figure 21.6 Cancer cell under attack by NK cell

infolding of the cell membrane which becomes pinched off to form a **vacuole** (sometimes called a phagocytic vesicle).

A phagocyte's cytoplasm contains a rich supply of **lysosomes** which contain digestive enzymes (such as lysozyme, proteases and nucleases). Some lysosomes fuse with the vacuole and release their enzymes into it. The bacterium becomes digested and the breakdown products are absorbed by the phagocyte.

Following digestion of the microorganism, the phagocyte releases **cytokines** which attract more phagocytes to the infected area to continue the battle against the pathogenic invasion. Dead bacteria and phagocytes may accumulate at an infected site as **pus**.

Natural killer cells

Natural killer (NK) cells (see Figure 1.4, page 4) also play a non-specific role in defence but they are not phagocytic. They mount an attack on virus-infected cells in general and cancer cells (see Figure 21.6) by the following means. An NK cell releases molecules of a protein which forms **pores** in the target cell's membrane. These allow a **'signal' molecule** from the NK cell to enter the target cell and trigger the genetically controlled series of events shown in Figure 21.7. The subsequent production by the target cell of **self-destructive enzymes** results in the cell's DNA and vital proteins being broken down into useless fragments. The cell then shrinks and dies. This process of programmed cell death is called **apoptosis**.

Cytokines

In addition to their role in non-specific immunity, natural killer cells and phagocytes release **cytokines** following contact with a pathogen. These molecules circulate in the bloodstream and stimulate the **specific** immune response by activating **lymphocytes** (see page 317).

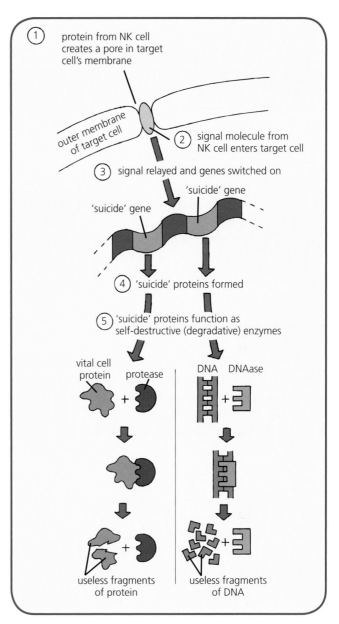

Figure 21.7 Events leading to apoptosis in an infected cell

Testing Your Knowledge

1 a) i) Identify the protective secretion produced by the epithelial lining of the trachea.
      ii) Name a different protective secretion made by the stomach lining.
      iii) Briefly describe how each of these secretions defends the body against attack by pathogens. (4)

   b) Against which other type of unwanted cell does the body defend itself? (1)

➜

315

2 a) What are *mast* cells? (1)

   b) i) Why does injured tissue at a cut quickly become red and swollen?

     ii) What name is given to this response to injury?

     iii) Briefly explain why it is of benefit to the body. (5)

3 a) Phagocytosis and antibody production are cellular responses to invasion by pathogens. Which of these is i) specific, ii) non-specific? (1)

   b) Briefly explain what is meant by the term *phagocytosis*. (2)

4 Decide whether each of the following statements is true or false and then use T or F to indicate your choice. Where a statement is false, give the word that should have been used in place of the word in bold print. (5)

   a) The process of programmed cell death of a pathogen, induced by a chemical from a natural killer cell, is called **phagocytosis**.

   b) If white blood cells detect tissue damage, they release **cytokines** which attract other white blood cells to the site of injury.

   c) A phagocyte's cytoplasm contains **ribosomes** full of digestive enzymes.

   d) Following injury, mast cells in connective tissue release **histamine** which causes vasodilation.

   e) **Decreased** capillary permeability occurs at an infected site showing inflammation.

22 Specific cellular defences

Lymphocytes

The third line of defence – the **specific** immune response – is brought about by **lymphocytes** derived from stem cells in bone marrow (see Figure 1.4, page 4). Some lymphocytes pass to the thymus (a gland in the chest cavity – see Figure 21.1) where they develop into **T lymphocytes** (T cells). Those that remain and mature in bone marrow become **B lymphocytes** (B cells).

Immune surveillance

A range of different types of white blood cell move round the body in the circulatory system and **continuously monitor** the state of the tissues. If damage or invasion by microorganisms is detected, some types of white blood cell release **cytokines** into the bloodstream. This results in large numbers of phagocytes (non-specific defence) and T cells (specific defence – see page 325) being attracted to, and accumulating at, the damaged or infected site.

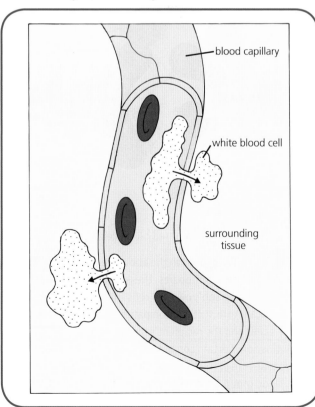

Figure 22.1 Exit of white blood cells from a capillary

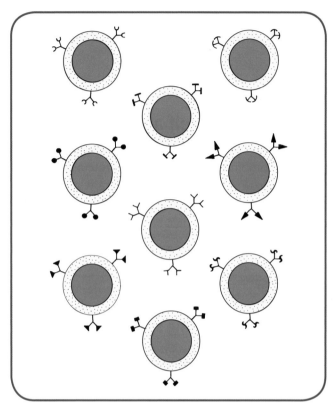

Figure 22.2 Pool of lymphocytes showing variety of antigen receptors

White blood cells often squeeze out through tiny spaces in the capillary wall in order to gain access to surrounding tissues, as shown in Figure 22.1.

Clonal selection theory

Any foreign molecule that is recognised by, and able to elicit a specific response from, a lymphocyte is called an **antigen**. Viruses, bacteria, bacterial toxins and molecules on the surfaces of transplanted cells and cancer cells can all act as antigens.

The body possesses an enormous number of different lymphocytes. Each lymphocyte has, on the surface of its cell membrane, several copies of a single type of **antigen receptor**. This antigen receptor is specific for one antigen and is different from that of any other type of lymphocyte (see Figure 22.2). Therefore, each lymphocyte is able to become attached to and be activated by only one type of antigen. When this occurs the lymphocyte is said to have been '**selected**' by the antigen. The lymphocyte then responds by dividing repeatedly to form a **clonal population** of identical lymphocytes. This process is called **clonal selection**.

Although each lymphocyte can only be activated by one type of antigen, when taken as a group, all of the body's lymphocytes possess such an enormous range and variety of cell surface antigen receptors that almost any antigen is recognised by one of them.

Recognition of self and non-self

Each person's body cells are different because they possess a combination of cell surface proteins (their 'antigen signature') that is unique to that person. It is of critical importance that a person's lymphocytes do not regard that person's own body cells' surface proteins as antigens and attack them. Normally this does not happen because, during the maturation of B cells and T cells, any lymphocyte bearing an antigen receptor that would fit a body cell surface protein is **weeded out** and rendered non-functional or destroyed by apoptosis.

T lymphocytes have specific surface proteins that enable them to distinguish between:

- 'self' molecules on the surfaces of the body's own cells (and therefore take no action)
- foreign molecules on the surfaces of cells not belonging to the body (and therefore initiate an immune response).

This ability to **recognise self and non-self** means that normally there are no lymphocytes acting against the proteins on the surfaces of 'self' body cells.

Related Topic

ABO blood-grouping system

Four types of blood group exist among human beings. These are A, B, AB and O.

Antigens

On the surfaces of their red blood cells, people with blood group A have A antigens, people with blood group B have B antigens, people with blood group AB have both A and B antigens and people with blood group O have neither A nor B antigens (see Figure 22.4)

Antibodies

In their plasma, people with blood group A have anti-B antibodies, people with blood group B have anti-A antibodies, people with blood group AB have neither anti-A nor anti-B antibodies and people with blood group O have both anti-A and anti-B antibodies.

Agglutination of blood

Certain combinations of different blood types are non-compatible. If, for example, a person with group A blood were to be given a transfusion of group B blood then anti-B antibodies in their plasma would combine with the B antigens on the surfaces of the donated red blood cells. This would result in clumping (**agglutination**) of red blood cells (see Figure 22.3), which would cause major problems by blocking small blood vessels.

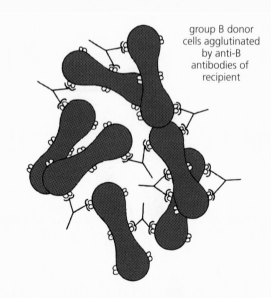

group B donor cells agglutinated by anti-B antibodies of recipient

Figure 22.3 Agglutination of red blood cells

Although the blood given in the above transfusion would also contain anti-A antibodies, these would be so diluted relative to the person's own red cells that they would have no significant effect. It is incompatibility between the donor's cells and the recipient's plasma which results in agglutination. Certain combinations of different blood group types are perfectly compatible. For example, a person with blood group B can receive blood of type O without risk of agglutination. Possible donors and recipients are given in Figure 22.4.

blood group	antigen(s) on red blood cell	naturally-occurring antibodies in plasma	can donate blood to groups	can receive blood from groups	percentage of Scottish population
A	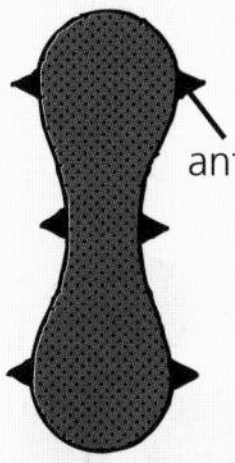 antigen A	 anti-B antibody	A and AB	A and O	35
B	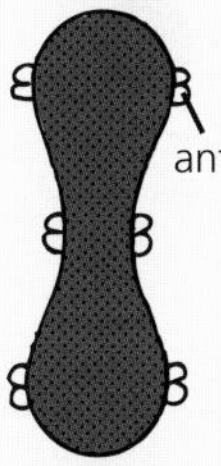 antigen B	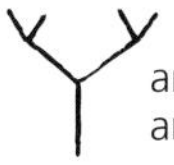 anti-A antibody	B and AB	B and O	11
AB	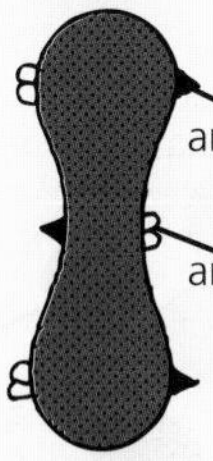 antigen A antigen B	neither anti-A nor anti-B antibodies present	AB	A, B, AB and O	3
O	 neither antigen A nor antigen B present	 both anti-A and anti-B antibodies present	A, B, AB and O	O	51

Figure 22.4 The ABO blood group system

Related Topic

Rh blood typing

Rhesus D-antigen

In addition to the ABO system of antigens, most people have a further antigen on the surface of their red blood cells. This is called **antigen D** and people who possess it are said to be **Rhesus positive (Rh+)**. A minority of people are described as being **Rhesus negative (Rh−)** because they lack antigen D but react to its presence by forming **anti-D antibodies**.

Incompatibility

Transfusion of Rh+ blood cells to a Rh− person must be avoided because the recipient's immune system would respond and produce anti-D antibodies, which would persist, leaving the person '**sensitised**'. Any subsequent transfusion of Rh+ red blood cells would be liable to cause the sensitised Rh− person to suffer severe or even fatal **agglutination**.

Antigen D is genetically determined by a dominant ➡

allele (D) and lack of antigen D by a recessive allele (d). Thus Rh+ individuals have genotype DD or Dd and Rh− individuals have genotype dd.

Risk factor

If a Rh− woman (dd) marries a Rh+ man (DD), each of their children will be Rh+. If a Rh− woman (dd) marries a Rh+ man (Dd), there is a 50% chance that each of their children will be Rh+ (Dd), as shown in Figure 22.5.

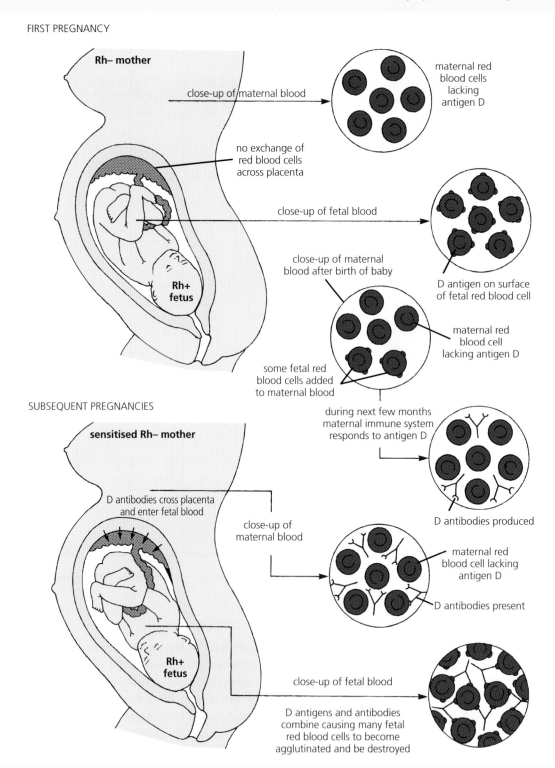

FIRST PREGNANCY

Rh− mother

close-up of maternal blood

maternal red blood cells lacking antigen D

no exchange of red blood cells across placenta

close-up of fetal blood

Rh+ fetus

D antigen on surface of fetal red blood cell

close-up of maternal blood after birth of baby

maternal red blood cell lacking antigen D

some fetal red blood cells added to maternal blood

during next few months maternal immune system responds to antigen D

SUBSEQUENT PREGNANCIES

sensitised Rh− mother

D antibodies produced

D antibodies cross placenta and enter fetal blood

close-up of maternal blood

maternal red blood cell lacking antigen D

D antibodies present

Rh+ fetus

close-up of fetal blood

D antigens and antibodies combine causing many fetal red blood cells to become agglutinated and be destroyed

Figure 22.6 Rhesus-incompatible pregnancies

Medical management

When a Rh– mother has a Rh+ fetus developing inside her body (see Figure 22.6) the baby has antigen D on its red blood cells which is regarded as foreign by the mother's immune system. Normally the placenta prevents maternal and fetal bloodstreams from coming into direct contact. As a result, only a very tiny number of fetal blood cells (if any) reach the maternal circulation. Therefore the mother's immune system normally remains 'unaware' of the presence of the 'foreign' fetus during the first pregnancy and does not attempt to reject it.

However, at the time of birth (or during a miscarriage) a small quantity of fetal blood does often become mixed with the mother's blood. The mother's immune system responds to this **sensitising event** by producing anti-D antibodies and she becomes **sensitised**. If, during subsequent pregnancies, the fetus is Rh+, antibodies against antigen D cross the placenta and attack and destroy fetal red blood cells. The resulting condition is called **haemolytic disease of the newborn**.

In the past the condition was treated by giving the baby transfusions of blood. Nowadays it is usually prevented by injecting the mother with anti-Rhesus antibodies soon after the birth of each Rh+ baby. These antibodies destroy any D antigens from the fetus before the mother's immune system has time to respond to them.

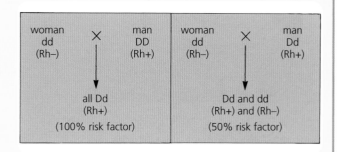

Figure 22.5 Inheritance of Rhesus D antigen

Autoimmunity

Sometimes the body no longer tolerates the antigens that make up the self message on cell surfaces. Such a failure in the regulation of the immune system leads to T lymphocytes launching an attack on the body's own cells. This process is called **autoimmunity** and it is the cause of **autoimmune diseases** (see Case Studies on rheumatoid arthritis, type 1 diabetes and multiple sclerosis).

Case Study | **Rheumatoid arthritis**

Rheumatoid arthritis is an autoimmune disease that causes chronic **inflammation** of the **synovial membranes** in joints. The membrane swells up and the cartilage and bone underneath are gradually destroyed (see Figure 22.7). They are replaced by fibrous tissue that, in advanced cases, joins the two bones together making the joint immovable.

Rheumatoid arthritis is a painful, disabling condition that can result in substantial loss of mobility. It is suffered by about 1% of the population (with women being three times more likely to be affected than men) and onset occurs most frequently between the ages of 40 and 50.

Role of cytokines
The cause of rheumatoid arthritis is unknown but cytokines are known to play a key role in its progression. **Cytokines** are chemical messengers that allow cells to communicate with one another. When a certain combination of cytokines occurs in a synovial joint of a potential sufferer, it promotes an **immune response** and white blood cells are stimulated to migrate to the joint. This creates a state of chronic inflammation that is followed by damage to bone and cartilage.

Treatment
Anti-inflammatory drugs and painkillers are used to suppress the symptoms. Immunosuppressant drugs may be used to inhibit the immune response in order to prevent irreversible, long-term damage to the joint's bone and cartilage.

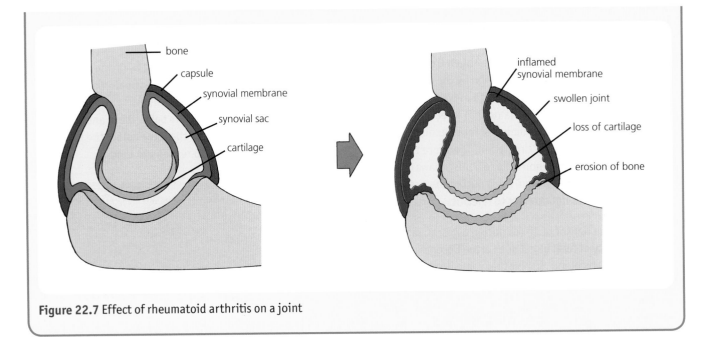

Figure 22.7 Effect of rheumatoid arthritis on a joint

Case Study   Type 1 diabetes

Type 1 diabetes (diabetes mellitus) is an autoimmune disease where **insulin-producing** beta cells in the pancreas are attacked and destroyed by **T cells** from the body's immune system.

A continuous supply of insulin is essential for regulation of blood glucose concentration (see page 190). Therefore, sufferers must inject insulin on a regular basis or risk diabetic shock and death. Poorly controlled diabetes increases the risk of further problems such as blindness and kidney disease.

The inheritance of particular cell surface proteins that make up a person's 'antigen signature' is associated with susceptibility to this autoimmune condition. However, studies of identical twins have shown that where one suffered type 1 diabetes, the other suffered it in only around 40% of the cases studied. This suggests that, in addition to inherited factors, environmental factors also play a part in this autoimmune disease.

Case Study   Multiple sclerosis

Multiple sclerosis is an autoimmune disease that affects the nervous system (see Table 19.1 on page 255). Each motor neuron possesses a long **axon** surrounded by a sheath of **myelin** which acts as an electrically insulating layer. The myelin sheath is interrupted by nodes which allow nerve impulses to pass rapidly through the nervous system.

In multiple sclerosis, some unknown trigger (perhaps a virus) activates T cells which then regard molecules on the myelin sheath as **antigens** and launch an attack on them. Clones of these T cells continue the

destruction of the myelin, resulting in the ability of the nerve cells to transmit impulses being seriously impaired.

Sufferers gradually develop symptoms such as numbness, walking difficulties and diminished vision as the ability to control muscles decreases. Although the trigger for multiple sclerosis is thought to be environmental, evidence from twin studies suggests that genetic factors also play a part in the development of this autoimmune disease.

Allergy

The immune system responds to a wide variety of agents that are molecularly foreign to it. This enables it to defend itself against pathogenic bacteria, fungi, viruses, worms etc. However, sometimes it **over-reacts** by B lymphocytes responding to **harmless** substances such as pollen, dust or feathers, or even a helpful substance such as the antibiotic penicillin. Such hypersensitivity in the form of an exaggerated (and sometimes damaging) immune response is called an **allergic reaction** (see Case Studies on hay fever, anaphylactic shock and allergic asthma).

Case Study   Hay fever

When airborne pollen grains (see Figure 22.8) enter the nose, throat and upper respiratory passages, they can act as an **allergen** by causing certain B cells to release antibodies. These become attached to mast cells in connective tissue causing the release of **histamine**. When stimulated in this way, the cells secrete excessive quantities of histamine, which produce the inflammatory response (see page 313). The sequence of events leading to this allergic reaction are summarised in Figure 22.9.

The symptoms typical of hay fever are:

- nasal congestion
- running nose
- red, itchy, watering eyes
- constriction of bronchioles.

They can be relieved by antihistamine drugs and other anti-inflammatory medication.

Figure 22.8 Pollen grains are essentially harmless but some of us react to them

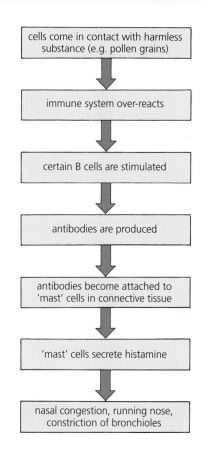

Figure 22.9 Events leading to an allergic reaction such as hay fever

Case Study   Anaphylactic shock

Anaphylactic shock is a life-threatening allergic response to an allergen that has been injected (for example, penicillin or bee venom) or consumed (for example, peanuts). The person is so allergic to the antigenic substance that many mast cells respond and secrete large quantities of **histamine** and other inflammatory agents. This triggers sudden dilation of peripheral blood vessels, loss of much circulatory fluid to surrounding tissues and a drop in blood pressure. Death can occur within minutes of exposure to the allergen.

People who know that they are hypersensitive to such allergens make every effort to avoid them and should always carry a **preloaded epinephrine (adrenaline) syringe** (see Figure 22.10) to counteract the allergic response and give symptomatic relief in an emergency.

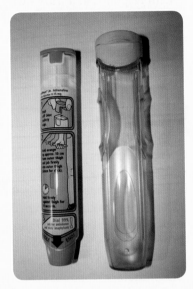

Figure 22.10 Pre-loaded epinephrine syringe

Case Study   Allergic asthma

Asthma is a respiratory condition in which the sufferer's air passages become narrower (see Figure 22.11). This makes it more difficult for the person to breathe. In addition to suffering shortness of breath, the person tends to **wheeze**, especially after exercise. In extreme situations their chest becomes so tight that they feel as if they are going to

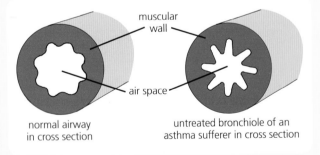

normal airway
in cross section

untreated bronchiole of an
asthma sufferer in cross section

muscular
wall

air space

Figure 22.11 Effect of asthma on a bronchiole

suffocate.

An asthmatic attack can be caused by an **allergic reaction** to dust mites, pollen, animal fur and certain foodstuffs such as peanuts and wheat. It can also be brought on by nervous tension. In the UK, boys below the age of 10 are twice as likely to be affected as girls of the same age. Asthma sufferers are able to control the condition if they follow a management plan.

Depending on the allergy, this might include:

- controlling dust mites
- avoiding foods to which they are allergic
- warming up gently before exercise
- using a bronchial dilator (see Figure 22.12) to inhale a drug that gives symptomatic relief by causing muscles in the bronchioles to relax and open the airways wider.

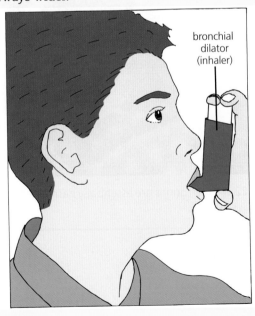

bronchial
dilator
(inhaler)

Figure 22.12 Bronchial dilator (inhaler)

Action of lymphocytes

T lymphocytes (T cells)

Two groups of T lymphocytes possessed by the human body are:

- **helper T cells** (T$_H$ **cells**), which secrete cytokines that activate phagocytes, T$_C$ cells and B cells
- **cytotoxic T cells** (T$_C$ **cells**), which destroy infected cells by several methods including the induction of apoptosis.

Antigen-presenting cell

Once a phagocyte has captured and destroyed an invading pathogen, it normally presents fragments of the pathogen's antigens at its surface, as shown in Figure 22.13. Such a phagocyte is described as an **antigen-presenting cell**. Other cells in the body infected with the pathogen also become antigen-presenting cells.

Action of helper T cells

Among the body's vast pool of helper T cells (T$_H$ cells) there will be a **type of** T$_H$ **cell** that bears antigen receptors able to recognise and bind with the foreign antigens on the surface of an antigen-presenting cell, such as the phagocyte shown in Figure 22.14. When this happens, the T$_H$ cell becomes activated, triggering the formation of a **clone of activated** T$_H$ **cells** and a **clone of memory** T$_H$ **cells**.

The activated T$_H$ cells secrete cytokines, which stimulate other cells, including T$_C$ cells and B cells. Proper functioning of the T$_C$ cells and B cells depends on this stimulation.

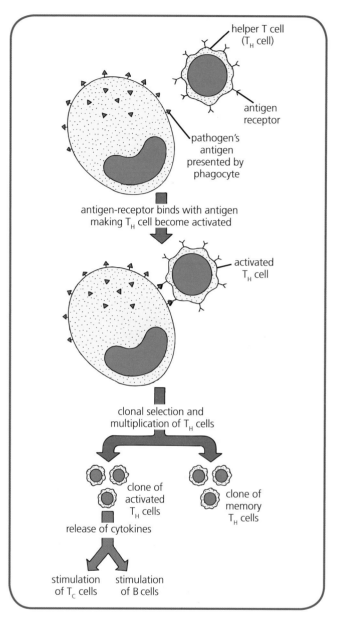

Figure 22.14 Activation of a T$_H$ cell

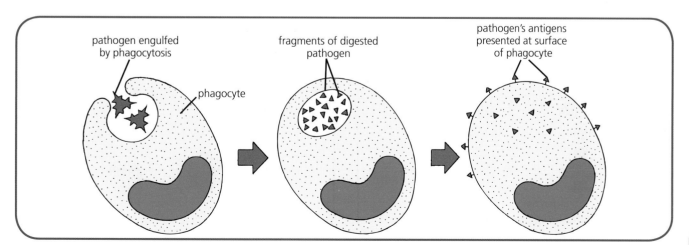

Figure 22.13 Phagocyte becoming an antigen-presenting cell

Action of cytotoxic T cells

Within the body's pool of cytotoxic T cells (T_C cells) there will be **a type of T_C cell** bearing copies of one type of antigen receptor on its surface that are **specific** to, and able to bind with, the type of foreign antigen on the surface of an antigen-presenting phagocyte, as shown in Figure 22.15. This binding process results in the T_C cell becoming **activated** by the antigen-presenting cell and then proliferating and differentiating. By this means it gives rise to a **clone of activated T_C cells** and a **clone of memory T_C cells**. The activated T_C cells move to the site of infection under the influence of cytokines (acting as chemical attractants) released by helper T cells and attack infected cells, as shown in Figure 22.16.

Some types of T_C cell bring about the death of infected, antigen-presenting cells at the site of infection by inducing them to undergo **apoptosis** (see page 315). Once a T_C cell has killed one target cell, it disengages from that cell and moves on to destroy another infected cell. An infected cell's membrane is not destroyed by apoptosis. Therefore its cell contents and pathogenic antigens remain enclosed and are not dispersed. Instead, the dead cell, which has shrunk, becomes engulfed and digested by a phagocyte.

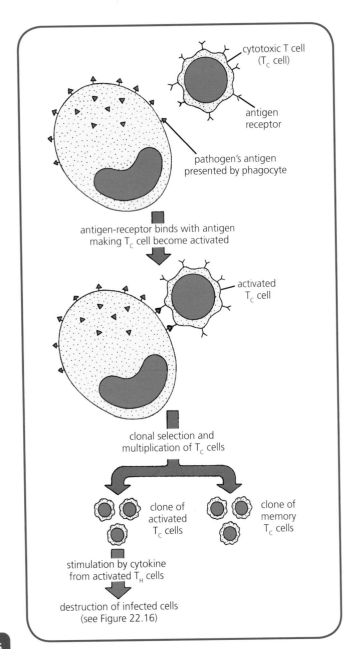

Figure 22.15 Activation of a T_C cell

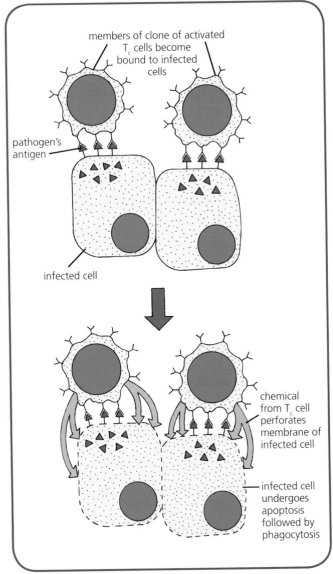

Figure 22.16 Destruction of infected cells by T_C cells

Other T_C cells recognise antigens on the surface of **cancer** cells and attack them. In Figure 22.17, the smaller T_C cell is releasing chemicals that bring about the **lysis** (bursting) of the larger cancer cell.

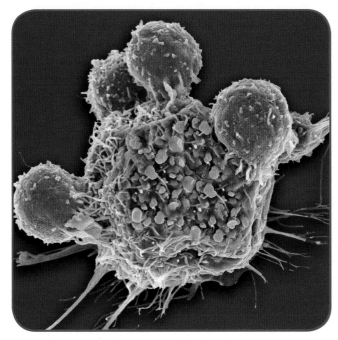

Figure 22.17 Lysis of a cancer cell by T_C cells

B lymphocytes (B cells)

Antigens and antibodies

An **antigen** is a complex molecule, such as a protein, that is recognised by the body as non-self and foreign. The antigen's presence triggers the production of antibodies by B lymphocytes (B cells). An **antibody** is a Y-shaped protein molecule, as shown in Figure 22.18. Each of its arms bears a **receptor** (binding site) that is specific to a particular antigen.

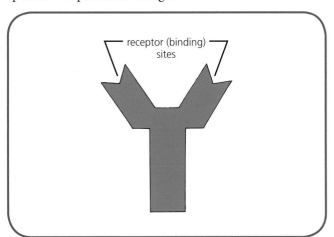

Figure 22.18 Antibody

Role of helper T cell

An antibody-producing response by a B cell specific to a foreign antigen can occur following direct contact between the B cell and the antigen but normally it only occurs with the help of a T_H **cell**. The B cell (see Figure 22.19) displays molecules of foreign antigen that it has taken in. These antigens are recognised by an activated T_H cell. It responds by releasing cytokines which stimulate the B cell to multiply and produce:

- **a clone of activated B cells** which make antibodies for immediate use
- **a clone of memory B cells** capable of making antibodies in the future if required.

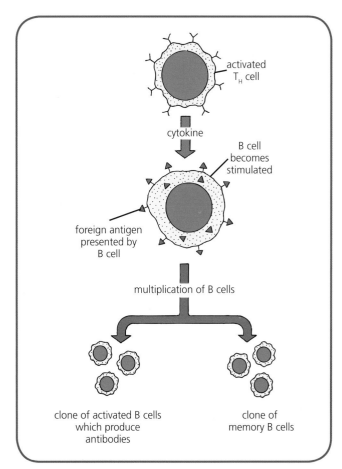

Figure 22.19 Cloning of B cells

Production of antibodies

Each clone of B cells produces one type of antibody molecule that is specific to one type of antigen surface molecule on a pathogenic cell (or toxin). A clone of B cells that produce antibodies is like a protein factory. Each B cell produces about 2000 antibody molecules per second during its 4–5-day life span. Once released into the blood and lymph systems, the antibodies are transported round the body and make their way to the infected area.

Action of antibodies

The antibodies recognise and combine with the antigens at the site of infection (see Figure 22.20). The binding of an antibody to an antigen does not in itself bring about destruction of the pathogen. However, the formation of an **antigen–antibody complex** inactivates the pathogen (or its toxin) and renders it more susceptible to phagocytosis. In some cases the antigen–antibody complex itself stimulates the activation of proteins that bring about lysis of the pathogen.

Immunological memory

Primary and secondary responses

When a person is infected by a disease-causing organism, the body responds by producing antibodies. This is called the **primary response** (see Figure 22.21).

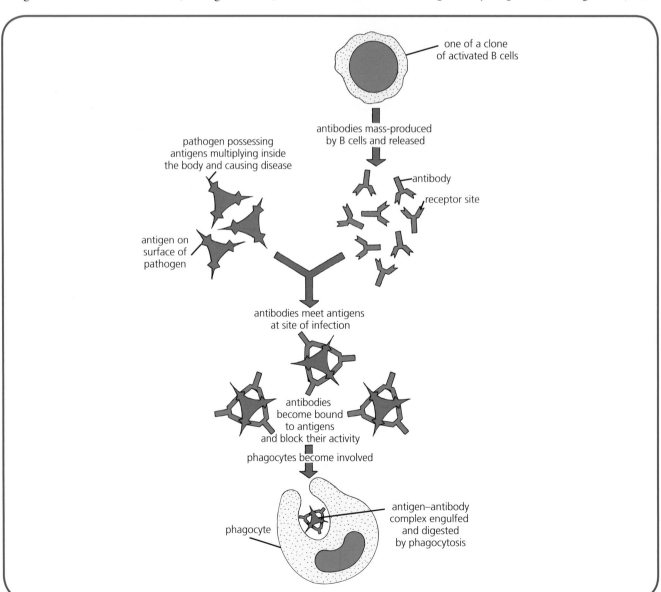

Figure 22.20 Action of antibodies

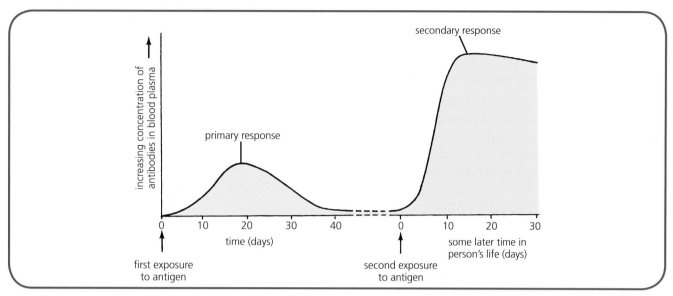

Figure 22.21 Primary and secondary responses

Due to a latent period elapsing before the appearance of the antibodies, this primary response is often unable to prevent the person from becoming ill.

If the person survives, exposure to the same antigen at a later date results in the **secondary response**. This time the disease is usually prevented because:

- antibody production is much **more rapid**
- the concentration of antibodies produced reaches a **higher level**
- the higher concentration of antibodies is maintained for a **longer time**.

Memory cells

The secondary response is made possible by the presence of **memory cells**. These are B and T lymphocytes specific to the antigen and produced in response to it by clonal selection following the body's first exposure to it. When the body becomes exposed to the disease-causing microorganism for a second time, the memory cells quickly proliferate and differentiate, producing clones of T cells and antibody-forming B cells.

A summary of some of the many interrelated components of the specific immune response is given in Figure 22.22.

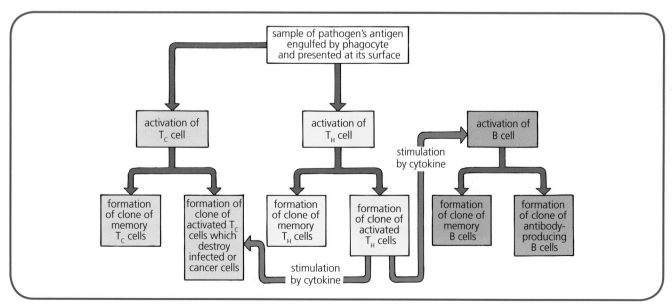

Figure 22.22 Summary of the specific immune response

Testing Your Knowledge

1 a) i) Name two types of lymphocyte.
      ii) Which of these is involved in the specific immune response to invasion by a pathogen? (2)
   b) Why do white blood cells at a damaged or infected site release cytokines? (1)
   c) How do white blood cells gain access to damaged tissue not directly in contact with blood capillaries? (1)

2 a) i) What is meant by the term *antigen*?
      ii) Give TWO examples of antigens. (3)
   b) If each lymphocyte can only recognise one specific antigen, how is it possible that lymphocytes offer effective protection against a vast variety of pathogens? (2)
   c) What name is given to the process by which a lymphocyte, having been selected by an antigen, divides repeatedly to form a population of identical lymphocytes? (1)

3 a) From each of the following pairs, choose the words that apply to the term *allergic reaction* and then use all three to explain what the term means. (3)
      underreacts/overreacts, hypersensitive response/hyposensitive response, normally harmful substance/normally harmless substance
   b) Explain what is meant by the term *autoimmunity*. (2)

4 a) Outline the series of events that takes place when a T_C cell binds to an antigen-presenting cell. (4)
   b) Outline the series of events that occurs when cytokine from an activated T_H cell stimulates a B cell displaying foreign antigens. (2)

What You Should Know Chapters 21–22

activated	complexes	lymphocytes
allergic	cytokines	lysis
antibody	dilate	memory
antigen	display	non-specific
antigen-presenting	distinguish	own
antimicrobial	epithelial	pathogens
apoptosis	faster	permeable
autoimmune	harmless	phagocytosis
bloodstream	helper	receptor
capture	histamine	selection
cells	immune	specific
cell-signalling	immunological	susceptible
chemical	infected	tolerate
clonal	inflammatory	
	killer	

Table 22.1 Word bank for chapters 21–22

1 The human body uses its _____ system to protect itself against _____, some toxins and cancer cells.

2 The body surface and cavity linings are covered with _____ cells that provide physical defence against infection and produce secretions that give _____ defence.

3 The body responds to an injury by making an _____ response. _____ released by mast cells causes blood vessels to _____ and capillaries to become more _____. Enhanced migration of phagocytes to the damaged site is accompanied by the rapid delivery of clotting elements and _____ proteins.

4 Certain specialised white blood cells display _____ responses to pathogens. Phagocytes destroy them by _____. Natural _____ cells make the pathogen destroy itself by apoptosis.

5 Phagocytes and NK cells release _____ molecules called cytokines. These stimulate the _____ immune response, which is brought about by T _____ (T cells) and B lymphocytes (B _____).

6 A foreign molecule able to elicit a specific response from a lymphocyte is called an _____. When an antigen binds to a _____ on the type of lymphocyte to which it is specific, the lymphocyte becomes _____ and divides, forming a _____ population.

7 Specific proteins present on T cells enable them to _____ between the body's _____ cells and cells bearing foreign antigens.

8 Failure to _____ the body's own antigens (that make up the 'self message' on cell surfaces) leads to an _____ disease. Over-reaction by the immune system to a _____ substance results in an _____ reaction.

9 Following infection of the body by a pathogen, some phagocytes _____ the microbe and _____ its antigen on their surface. These _____ cells activate T lymphocytes.

10 On becoming activated, one group of T lymphocytes destroys _____ cells by inducing them to undergo _____.

11 A second group of T lymphocytes (_____ T cells) releases _____ that activate B lymphocytes.

12 Each clone of B cells produces one type of _____ specific to one type of antigen surface molecule on a pathogen. These antibodies are secreted into the _____ and transported to the site of infection.

13 The subsequent formation of antigen–antibody _____ inactivates the pathogen and makes it more _____ to phagocytosis or leads to its destruction following cell _____.

14 Some of the T and B cells formed by clonal _____ in response to antigens persist as _____ cells. If the body is exposed to the same antigen, these cells produce new clones, which give a _____ and greater _____ response than during the first exposure.

23 Transmission and control of infectious diseases

An **infectious disease** is one that is capable of being transmitted from one person to another by direct or indirect contact.

Pathogens

Infectious diseases are caused by many types of **pathogen** including:

- viruses (e.g. measles – see Figure 23.1a)
- bacteria (e.g. cholera – see Figure 23.1b)
- fungi (e.g. athlete's foot – see Figure 23.1c)
- protozoa (e.g. malaria – see Figure 23.1d)
- multicellular parasites (e.g. hookworm – see Figure 23.2).

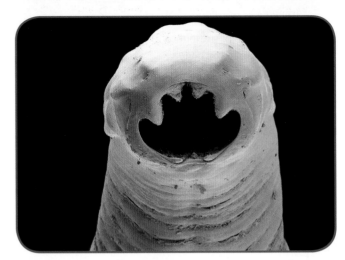

Figure 23.2 Head of a hookworm

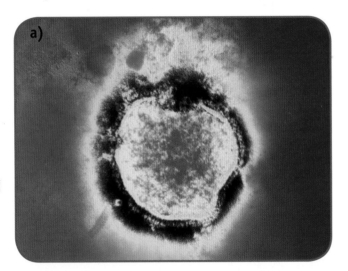

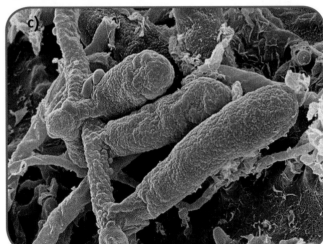

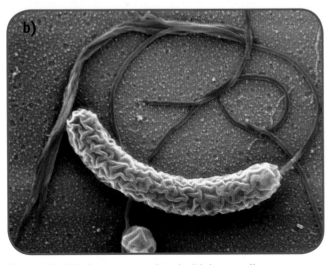

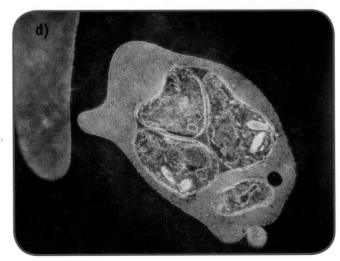

Figure 23.1 Pathogens associated with human diseases

Transmission

Infectious diseases are **transmitted** by many methods. Some of these are given in the following list:

- direct physical contact (e.g. by shaking hands – see Figure 23.3)

- inhaled air (e.g. by breathing in microbes or droplets contaminated with microbes released by an infected person during coughing or sneezing – see Figure 23.4)

Figure 23.3 Methods of disease transmission

Figure 23.4 Transmission by sneezing

- indirect physical contact (e.g. by sharing contaminated items such as cups or syringe needles)
- body fluids (e.g. by exchanging saliva during kissing or seminal fluids during sexual contact)
- faecal–oral route (e.g. by consuming food or drink contaminated with microorganisms from an infected person)
- vector organisms (e.g. by being bitten by a mosquito infected with malaria or yellow fever pathogens).

Case Study | Comparison of transmission methods

Different pathogens are transmitted by different means.

Measles
The virus responsible for this infectious disease is **air-borne**. It is transmitted in droplets released during coughing and also directly into the air during laboured breathing when the infected person's nasal passages are heavily congested.

HIV
HIV (human immunodeficiency virus) responsible for AIDS (acquired immune deficiency syndrome – see chapter 24, page 346) is transmitted in **body fluids** such as blood (on a syringe needle shared by drug addicts) and semen (during unprotected sexual intercourse).

Cholera
This infectious disease is caused by a bacterium that is **water-borne**. In regions of the world where sanitation conditions are poor or where a natural disaster (e.g. severe flooding) has occurred, the local water supply may become polluted with sewage contaminated with cholera bacteria. If the water is used for drinking without first being boiled, the pathogen gains access to the local population and a disease epidemic results. The epidemic may gain further momentum by the pathogen being spread by **direct contact** and by the consumption of food contaminated by insect **vectors** (e.g. flies) that have previously been in contact with sewage containing cholera microbes.

Control of transmission

Quarantine

This is a period of **compulsory isolation** of a person suffering a serious communicable disease or of those who have been in contact with an infected person. It is carried out in order to prevent the spread of the disease. Usually the length of the quarantine period is planned to match the length of the maximum known incubation period of the disease.

Antisepsis

Asepsis is the state of being completely free of live microorganisms. **Antisepsis** is the inhibition or destruction of microorganisms that cause disease or decay (in an attempt to approach a state of asepsis). The growth and multiplication of pathogens is arrested by adopting procedures such as the sterilisation of equipment, the application of antiseptics (see Figure 23.5) to wounds and the wearing of sterile gloves and surgical masks.

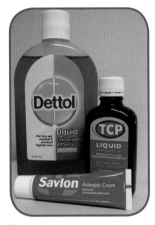

Figure 23.5 Familiar antiseptics

Individual responsibility

Good hygiene

Hygienic practices such as hand-washing, teeth-brushing and daily showering help to control populations of microbes and reduce the chance of a pathogen causing an infection and then being transmitted to another person.

Care in sexual health

Use of condoms gives protection against **sexually transmitted diseases** (STDs) such as gonorrhoea and HIV/AIDS.

Appropriate handling and storage of food

Adoption of **good practices** such as handling foods with clean hands, storing uncooked meats separately from cooked meats and sterilising knives and work surfaces in butchers' shops and abattoirs help to prevent the transmission of microbes.

Community responsibility

Quality water supply

In the UK, transmission of pathogens from wild animals to humans in drinking water is prevented by a series of treatments including **filtration** and **chlorination** to ensure that the water is free of pathogens.

Safe food webs

Food manufacturers in the UK are obliged to adopt **good manufacturing practice** to ensure that harmful microorganisms do not enter the food chain. Milk, for example, is pasteurised (heated to 72°C for 15 seconds) to kill most microorganisms including those that would cause tuberculosis.

In general, systems of control including **inspection**, **risk analysis** and **traceability** of food sources are employed to make the entire process from 'stable to table' safe. However, on rare occasions the system does break down. Toxins produced by algae, for example, sometimes manage to enter the food chain via shellfish. Consumption of these, in extreme cases, can lead to severe conditions such as paralysis and even death.

Appropriate waste disposal mechanisms

Dry refuse is collected on a regular basis and **recycled**, **incinerated** or **buried** under approximately 250 mm of soil.

Control of vectors

Many methods are employed to eradicate animals that transmit microbial pathogens.

The rat flea (see Figure 23.6) transmits **bubonic plague** (the 'Black Death') from rats to humans. When an infected rat is bitten by a flea, the plague bacterium (see Figure 23.7) passes from the rat's blood to the flea. If that flea then bites a human, the pathogenic bacterium is transmitted to the person in the flea bite.

One pandemic in the fourteenth century wiped out about 25% of the European population. Nowadays the control of rats (and their fleas) has reduced the incidence of bubonic plague to a very low level.

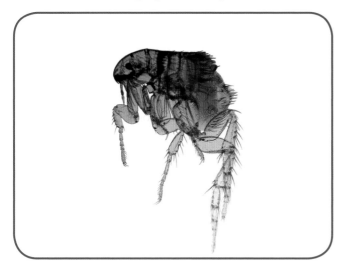

Figure 23.6 Rat flea

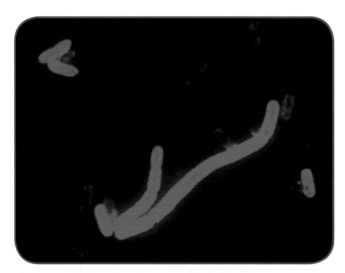

Figure 23.7 Bacterium that causes bubonic plague

Millions of people die every year of **malaria** caused by a protozoan carried by mosquitoes (see Figure 23.8) acting as vectors. Control of mosquitoes is attempted by:

- draining stagnant water to remove breeding sites
- introducing sterile male mosquitoes to reduce the breeding rate
- using chemicals such as insecticides and larvicides.

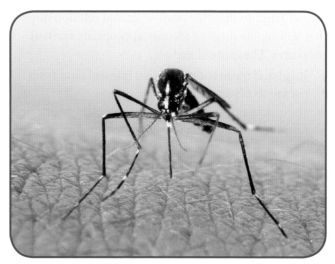

Figure 23.8 Mosquitoes are vectors for a number of diseases

Epidemiology of infectious diseases

The **epidemiology** of an infectious disease is the study of its characteristics such as:

- the location associated with its initial outbreak
- its pattern and speed of spread
- its geographical distribution.

This work is done by epidemiologists who are able to determine the factors that affect the spread of the disease. The spread patterns of infectious diseases are classified as shown in Table 23.1.

Spread pattern of disease	Description
Sporadic	Occurs in scattered or isolated instances with no connection between them
Endemic	Recurs as a regular number of cases in a particular area
Epidemic	Simultaneously affects an unusually large number of people in a particular area
Pandemic	Occurs as a series of epidemics that spreads across whole continents or even throughout the world

Table 23.1 Spread patterns of infectious diseases

Epidemiologists conduct a thorough surveillance (close observation) of the disease and collect data that are subjected to statistical analysis. From this work they attempt to discover possible **causal relationships** that will enable them to identify appropriate **control measures**. These could include:

- suggested means of preventing transmission from person to person and region to region
- appropriate drug therapy for people already infected
- immunisation for people not yet infected.

Active immunisation and vaccination

Immunisation is the process by which a person develops immunity to a disease-causing organism. **Active** immunity refers to the protection gained as a result of the person's body producing its own antibodies.

Naturally acquired active immunity

If a person survives infection by a pathogen, subsequent exposure to the same antigen at a later date results in a secondary response (see page 329) which prevents the disease from recurring. The person has acquired active immunity as an immunological memory by **natural** means.

Artificially acquired active immunity

Vaccination is the method of immunisation by which a weakened or altered form of the pathogen or its toxin (see Table 23.2) is deliberately introduced into the body by injection, ingestion or nasal spray in order to act as an **antigen** and initiate the **immune response**.

Form of antigen in vaccine	Examples of diseases to which active immunity is acquired
Dead pathogens	Hepatitis A and poliomyelitis
Parts of pathogens	Hepatitis B and HPV (human papilloma virus)
Weakened pathogens	Rubella, mumps and measles
Inactivated bacterial toxin	Diphtheria and tetanus

Table 23.2 Antigens in vaccines

Normally the antigen is mixed with an **adjuvant**. This is a chemical substance that promotes the activity of the antigen and enhances the immune response. In each case the antigen induces the production of B and T cells and the formation of antibodies but does not cause the disease. Some B and T cells persist in the body as **memory cells**. These initiate the secondary response

if the person is exposed to the normal disease-causing antigen at a later date. The person has acquired **active immunity** as an immunological memory by **artificial** means.

Vaccine clinical trials

Like other new pharmaceutical medicines, vaccines must be subjected to **clinical trials** on humans to establish that they are **safe** and **efficacious** (capable of producing the intended result). Only then can they be licensed for use.

Before a clinical trial is carried out, the potential treatment undergoes extensive testing on cells and on animals in the laboratory. If the new treatment works on them, approval for the next stage is sought from the regulatory authority. This group checks that the clinical trial's proposed design involving humans matches European **protocol** (a procedural method whose design and implementation meet certain agreed standards).

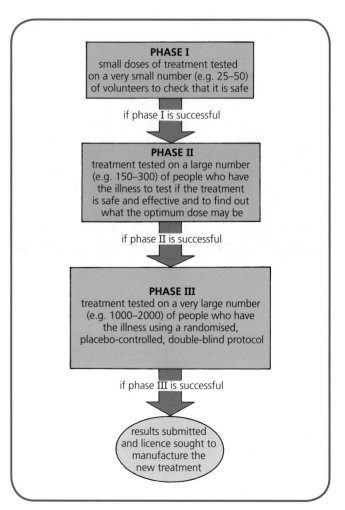

Figure 23.9 Phases in a clinical trial

Phases in the clinical trial

Once the protocol has been approved, testing on humans can take place. The three phases that make up a clinical trial are shown in Figure 23.9.

Design of phase III

In phase III of the trial, the target population is split into two groups – those in the **test group** who will receive the treatment and those in the **control group** who will not. The protocol employed at this stage is:

- placebo-controlled
- double-blind
- randomised.

Placebo effect

Instead of the treatment, the members of the control group are given a **placebo**. This is a 'sham' treatment that takes the same form as the real treatment except that it lacks the active ingredient being tested. This procedure is carried out to assess the **placebo effect**, that is the effect from receiving the treatment that does *not* depend on the active ingredient in the real treatment. For example, some patients receiving the placebo may show an improvement in their condition. This could be a result of the psychological effect of:

- thinking that they were receiving the real treatment
- receiving expert attention from health care staff
- expecting the treatment to be efficacious.

The use of the placebo allows a **valid comparison** to be made between the test group and the control group to assess the effect of the new treatment. If the control group had not been given the placebo, then there would be no way of knowing how many members of the group receiving the real treatment showed improvement in their condition for one of the above 'placebo effect' reasons and not as a result of the active ingredient present in the new treatment itself.

Double-blind trial

A **blind trial** is one in which the human subjects do not know whether they are receiving the active treatment or the placebo. A **double-blind trial** is one in which neither the subjects nor the doctors know who is receiving what.

A double-blind trial is used at phase III to **eliminate bias** (an irrational preference or prejudice). For example, if the trial was not double-blind, then a doctor's belief in the value of the new treatment could, consciously or subconsciously, affect their behaviour towards a patient if they knew who was receiving which 'treatment'.

Randomisation

Normally the gender, age and other relevant details of each subject taking part in phase III of the trial are entered into a computer. This then puts each person into one or other of the two groups **at random**. This procedure further **eliminates bias**.

In its absence a doctor might subconsciously avoid putting more seriously ill patients into the group receiving the new treatment. Therefore, at the end of the trial the new treatment would appear to be more effective than it really was because the members of the test group would have begun in better health than those in the control group.

Experimental error

The computer also ensures that the composition of the two groups is as **similar** as possible. Figure 23.10 shows this process in a very simple way. In the test population about to be split up, there are more females than males. The computer ensures that this difference is reflected in the two groups formed. Similarly, among males there are more older than younger subjects. Again this difference is maintained in the two groups formed and so on. This process reduces **experimental error** to a minimum.

If the process were not adopted, one group might receive, for example, an atypically large number of older subjects who are heavier in weight and more seriously ill. This could invalidate the results. Experimental error is also reduced by using a very **large sample** population. (The patient in Figure 23.11 has completely misunderstood the reasons for the protocol adopted in a clinical trial.)

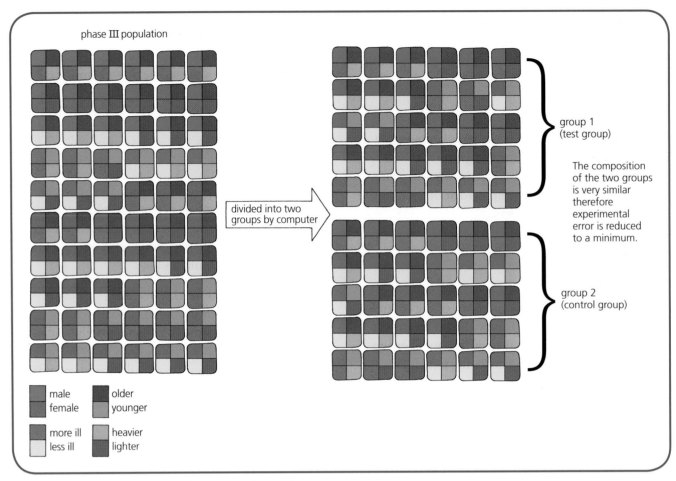

Figure 23.10 Using a computer to form test and control groups

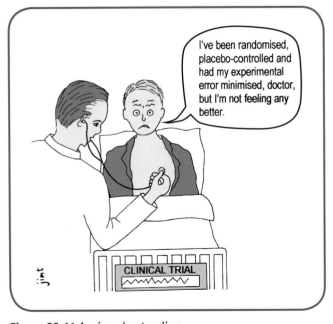

Figure 23.11 A misunderstanding

Statistical analysis

It is also important to use a very large population of patients at phase III so that the results obtained can be subjected to **statistical analysis** with confidence. The results for the two groups can then be compared to find out if significant differences exist between them that indicate that the new treatment is efficacious. If so, the researchers would then seek a licence to manufacture it.

Testing Your Knowledge

1 a) Identify THREE types of pathogenic microorganism. (3)

   b) Name the vector that transmits the malarial parasite and briefly describe a method used to try to control it. (2)

2 a) What is meant by each of the following terms: *antisepsis, quarantine, epidemiology*? (3)

   b) Distinguish between the following patterns of disease spread:

      i)   sporadic and endemic

      ii) epidemic and pandemic. (4)

3 a) Briefly describe how active immunisation can be acquired by artificial means. (2)

   b) Copy and complete Table 23.3. (3)

Feature of design of clinical trial	Reason for inclusion of design feature
Randomised	
Double-blind	
Placebo-controlled	

Table 23.3

Active immunisation and vaccination and the evasion of specific immune response by pathogens

Herd immunity

When most of the members of a population have been immunised by vaccination against a pathogen, the probability of the few remaining non-immune individuals coming into contact with an infected individual becomes very low. Under these circumstances, non-immune individuals are protected because the normal chain of infection has been disrupted. This form of protection, given indirectly to the non-immune minority by the immune majority, is called **herd immunity** (see Figure 24.1).

The greater the percentage of individuals in the population who are immune, the lower the chance that a non-immune person will come into contact with an infected person. Herd (community) immunity provides protection for **vulnerable sub-groups** of the population. These include people who must not be vaccinated because of a medical condition such as an immune disorder.

Mass vaccination

Herd immunity resulting from mass-vaccination programmes has successfully reduced the spread of diseases and even eradicated some (see the Case Studies on tuberculosis, poliomyelitis and smallpox).

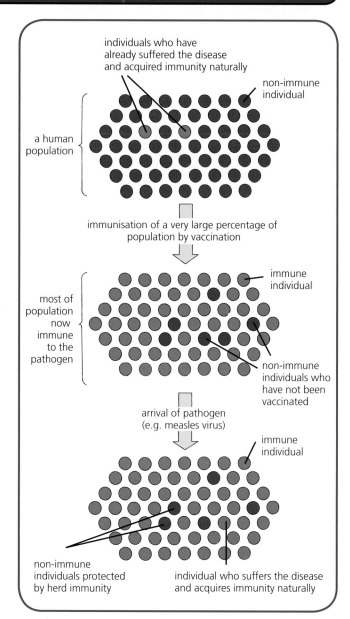

Figure 24.1 Herd immunity

Case Study | Tuberculosis mass vaccination programme

Tuberculosis (TB) is caused by a bacterium that is inhaled in droplets released by an infected person during coughing. The disease most commonly affects the **lungs** but many other organs can also become infected. In its later stages, sputum containing blood is coughed up. This stage of the disease was formerly called **consumption**. Tuberculosis used to be a common cause of death in the UK. A close correlation existed between incidence of the disease and poor social conditions. Poverty-stricken people eating a poor diet and living in overcrowded conditions with poor sanitation were much more likely to develop the disease than wealthier members of society. As social conditions gradually improved with time, the death rate decreased, as shown in Figure 24.2.

Vaccination

A programme of **mass vaccination** (accompanied by **antibiotic treatment** for infected people) was begun in the UK in the 1950s. The vaccine was routinely administered to school children aged 10–13 years. It has been so successful in achieving **herd immunity** over the years that it is no longer delivered to everyone. Instead, it has been replaced by **targeted vaccination** for those infants, school children, health care workers and older people considered to be at greatest risk.

In recent times the number of cases of TB has begun to rise again as it gains a foothold among problem drug and alcohol users and homeless people with an extremely low quality of life and very poor general health.

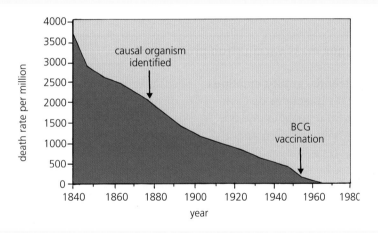

Figure 24.2 Decline in death rate from tuberculosis

Case Study | Poliomyelitis mass vaccination programme

Poliomyelitis (polio for short) is a disease caused by a virus. Its principal mode of transmission is the faecal–oral route, but it can be passed on in exhaled droplets. In severe cases of the disease, the virus attacks **motor neurons** in the spinal cord and hindbrain. This often leads to **paralysis** of limbs. In very severe cases, the disease can be fatal.

A summer epidemic of the disease developed in the UK in 1947 and then recurred in subsequent years. Therefore concerted efforts were made to develop a **vaccine**. The first type containing dead virus was

introduced in 1956. It reduced the incidence of the disease among vaccinated children to about 25% of that of non-vaccinated children. In 1962 an improved vaccine containing **attenuated** (weakened) virus was introduced. It was administered orally rather than by injection and has proved to be so effective that **herd immunity** has now been established and polio has been almost completely eradicated in developed countries such as the UK (see Figure 24.3). Experts have calculated that an infant immunisation rate of 80–86% must be maintained to keep polio in check. It is for this reason that polio vaccine is given to babies aged 2 months. →

It must be kept in mind, however, that this successful outcome in the battle against polio was not due solely to the mass vaccination programme. In the UK, over the years, ever-improving standards of living and hygiene, including effective means of sewage disposal, have also played their parts in wiping out the disease. Poliomyelitis is still common in many developing countries that lack mass vaccination programmes and effective sanitation.

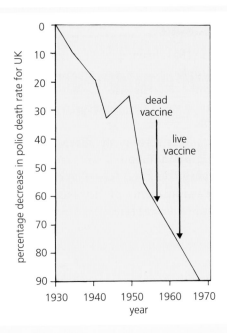

Figure 24.3 Effect of vaccination against polio

Case Study | Smallpox mass vaccination programme

Up until the end of the eighteenth century, **smallpox** was widespread in the UK. This infectious disease caused severe fever and was fatal in about one out of every five cases. Survivors were left permanently scarred. It was known at that time that smallpox could sometimes be prevented by deliberately inoculating people with pus from a pustule of a person suffering a mild form of the disease. However, this method of immunisation was not reliable and often produced the fatal form of the disease.

In 1796, a British doctor called Edward Jenner decided to act on observations that milkmaids who had

suffered **cowpox** (a similar, but milder, non-fatal disease) were immune to smallpox. He inoculated a healthy boy with cowpox. Once the boy had recovered, Jenner inoculated him with what would normally have been a deadly strain of smallpox virus. Fortunately for everyone concerned, the boy did not contract smallpox, showing that he was immune. The science of artificially acquired, active immunity had begun.

Eradication of smallpox

Free vaccination against smallpox became available in the UK in the 1840s and was made compulsory for babies 10 years later. In Britain and other developed countries, the death rate gradually decreased to a low level. However, it was not until many years later that one of the greatest triumphs in medical history was achieved. This was the **complete eradication worldwide** of smallpox which was brought about by a World Health Organization (WHO) programme begun in 1967. It involved vaccinating as many people as possible shortly after birth and quickly homing in on fresh outbreaks of the disease and then vaccinating all known and suspected contacts. This **surveillance-containment campaign** was so successful that the last recorded case of smallpox occurred in Somalia in 1977 (see Figure 24.4).

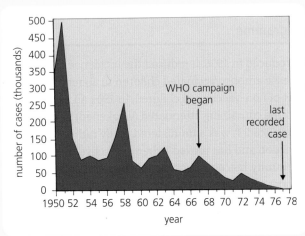

Figure 24.4 Worldwide eradication of smallpox

Herd immunity threshold

For herd immunity to be effective, only a minority of the population can be left unvaccinated. The percentage of immune individuals in a population above which a disease no longer manages to persist is called the **herd immunity threshold**. Its value varies from disease to disease (see Table 24.1) and depends on factors such as:

- the extent of the pathogen's **virulence** (its capacity for causing disease)
- the vaccine's **efficacy** (its effectiveness)
- the population's **contact parameters** (e.g. degree of population density that affects the pathogen's ability to spread).

Public health medicine

In many countries, the **public health policy** for combating a number of common diseases is to use mass vaccination programmes that create herd immunity to them. In the UK, for example, a person's vaccination schedule normally begins around the age of 2 months (when they are vaccinated against diphtheria, tetanus, poliomyelitis, influenza and whooping cough) and continues for many years.

Absence of herd immunity

In some **developing** countries where the majority of the population are **impoverished** and **malnourished**, it may not be possible to introduce a programme of widespread mass vaccination. Under these circumstances herd immunity cannot be established.

In a **developed** country, herd immunity for a vaccine-preventable disease (e.g. measles) may be compromised if parents believe **adverse publicity** about the vaccine and refuse to have their children vaccinated. Under these circumstances the level of herd immunity can slip below its threshold value. As a result, the incidence of the disease among non-vaccinated individuals will increase rapidly.

Evasion of specific immune responses

Over the course of evolution, many pathogens have evolved mechanisms that enable them to evade specific immune responses made to them by the human body. These often make it possible for a new version of the pathogen to appear that is 'one step ahead' of the current vaccination programme intended to give protection against it.

Change in genotype

Pathogenic microorganisms have vast powers of reproduction. Given suitable conditions, they increase rapidly in number. Within a population of a pathogen, new strains arise continuously as a result of changes occurring in the microbe's genotype. These changes are brought about, for example, by **mutations** and **genetic**

Related Topic

Comparison of herd immunity thresholds

Table 24.1 shows herd immunity thresholds for several diseases.

Disease	Mode of transmission	Herd immunity threshold (%)
Diphtheria	Airborne droplets of saliva	85
Measles	Airborne	83–94
Mumps	Saliva and airborne droplets	75–86
Poliomyelitis	Faecal–oral route	80–86
Rubella	Airborne droplets	80–85
Whooping cough	Airborne droplets	92–94

Table 24.1 Estimated herd immunity thresholds for vaccine-preventable diseases

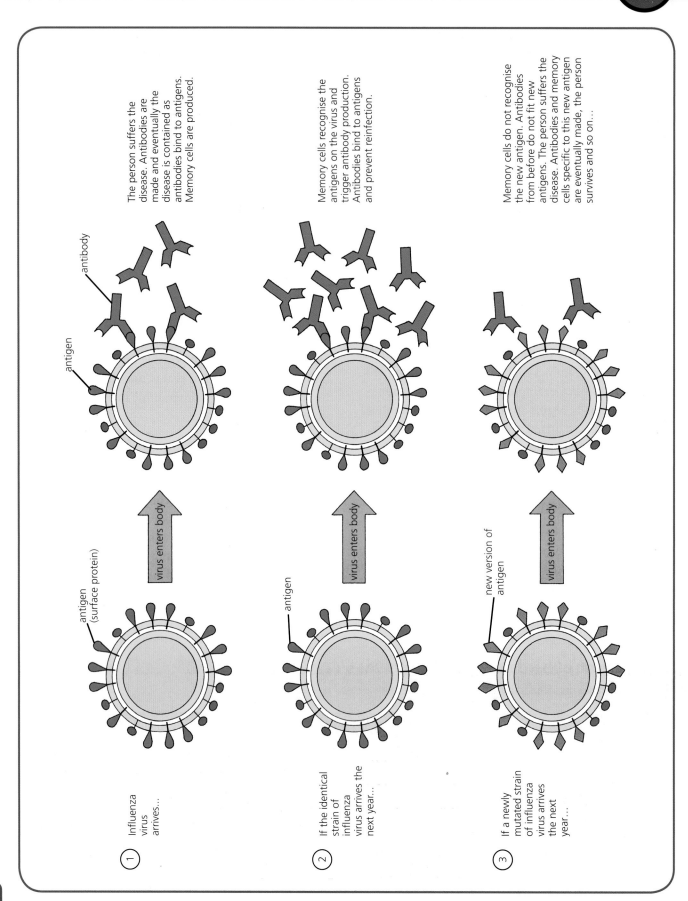

1. Influenza virus arrives…

   virus enters body

   antigen (surface protein)

   antigen

   antibody

   The person suffers the disease. Antibodies are made and eventually the disease is contained as antibodies bind to antigens. Memory cells are produced.

2. If the identical strain of influenza virus arrives the next year…

   virus enters body

   antigen

   Memory cells recognise the antigens on the virus and trigger antibody production. Antibodies bind to antigens and prevent reinfection.

3. If a newly mutated strain of influenza virus arrives the next year…

   virus enters body

   new version of antigen

   Memory cells do not recognise the new antigen. Antibodies from before do not fit new antigens. The person suffers the disease. Antibodies and memory cells specific to this new antigen are eventually made, the person survives and so on…

Figure 24.5 Effect of antigenic variation in influenza virus

recombination (such as a combination of genetic material from two different strains).

Antigenic variation

The new strains of a pathogen are described as demonstrating **antigenic variation** if they have antigens on their surface that are different from those of the original strain. A new strain of the pathogen showing antigenic variation is **genetically and immunologically distinct** from its parent strain(s) and succeeds because it enjoys a selective advantage.

Influenza virus

Production of new antigens enable the influenza virus, for example, to avoid the effects of the human body's immunological memory. This allows it to re-infect the person because its new antigens are not recognised by their memory cells (see Figure 24.5). It is for this reason that influenza remains a major public health problem. At-risk individuals need to be vaccinated every year with a new version of the vaccine to give protection (see Figure 24.6).

Pathogenic protozoa

Antigenic variation is not restricted to viruses and bacteria. It also occurs in **pathogenic protozoa** (unicellular animals), as in the following two examples.

Trypanosomiasis

Trypanosoma brucei (see Figure 24.7) is a protozoan that causes a fatal neurological disease called trypanosomiasis ('sleeping sickness' – see Figure 24.8) when it gains access to the bloodstream of humans and some other mammals. The pathogen is surrounded by a coat of **glycoprotein** molecules that varies in chemical composition depending on which of the many hundreds of genes that code for **variants** of the glycoprotein are switched on.

Figure 24.7 *Trypanosoma brucei* and red blood cells

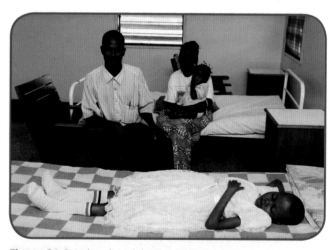

Figure 24.8 A sleeping sickness sufferer

The infected host responds to the pathogen by making antibodies against the antigen (the glycoprotein in the pathogen's coat). This process results in about 99% of the protozoa being neutralised and killed. However, 1% of them manage to shed their coat, switch on a different gene that codes for a variant of glycoprotein and produce a new, **antigenically distinct** coat.

This **antigenic variation** enables the remaining 1% of the pathogens to evade detection and give rise to a new

Figure 24.6 Epidemic or pandemic?

345

population where each individual is surrounded by a coat of the glycoprotein variant. The host's immune system responds by producing a new set of antibodies that again deal successfully with about 99% of the pathogens. But, as before, 1% of the pathogens switch to a different variant of glycoprotein and survive. This cycle of events continues until, eventually, in the absence of treatment, the host dies.

Malaria

Plasmodium falciparum (see Figure 23.1d on page 332) is a protozoan that causes **malaria**. It has a complex life cycle involving both humans and mosquitoes. For most of the time that the pathogen is in a human host, it is found inside red blood cells.

Great antigenic variation exists amongst the members of the population of the pathogen. This **genetic variability** enables the parasite to evade the host's immune response. In addition, individual pathogenic cells produce a protein that is transported to an infected red blood cell's surface. The protein makes the red blood cell adhere to the lining of the blood vessel and prevents it from being removed and destroyed by the body. The parasite is able to switch between the many genes that code for variants of this protein, making it impossible for the host to produce appropriate antibodies within the limited time available.

Antigenic variation in *Plasmodium falciparum* has so far prevented scientists from producing an effective vaccine. Malaria continues to kill millions of people annually.

Direct attack on the immune system

Microorganisms that invade the body soon come into contact with phagocytes. These might be the phagocytes present in the bloodstream or larger phagocytic cells called **macrophages** derived from blood cells but resident in connective tissues.

Bacteria that succeed as pathogens have often evolved a means of mounting a direct attack on the host's immune system. For example, by interfering with the host cell's phagocytic response, the pathogen manages to block an essential step in the process and bring it to a halt.

Tuberculosis

The above strategy is employed by *Mycobacterium tuberculosis*, the bacterium that causes **tuberculosis** (see the Case Study on page 341). *Mycobacterium tuberculosis* is described as an **intracellular** pathogen because it is able to survive inside phagocytes. When a macrophage engulfs a tuberculosis bacterium into a phagocytic vesicle, the microbe prevents **lysosomes** (containing digestive enzymes – see page 314) from fusing with the vesicle. The bacterial cell wall contains certain molecules that, on release, are thought to modify lysosomal membranes in a way that normally inhibits fusion.

Even if fusion does occur, *M. tuberculosis* is not easily attacked by lysosomal enzymes because it is protected by a waxy cell wall. Therefore the pathogen remains alive inside the macrophage and **avoids immune detection** (see Figure 24.9). Other macrophages often surround and wall off the infected macrophage to try to keep it contained.

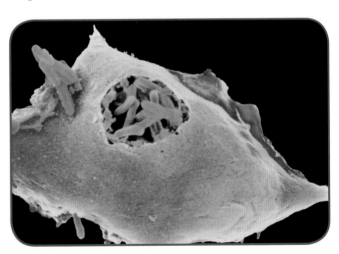

Figure 24.9 TB bacteria can thrive inside phagocytes

Immunodeficiency disease

An immunodeficiency disease results from the absence or failure of some component of the immune system which leaves the person susceptible to infection.

AIDS and HIV

AIDS (acquired immune deficiency syndrome) is a deficiency disease caused by **HIV** (human immunodeficiency virus) – see Figure 24.10.

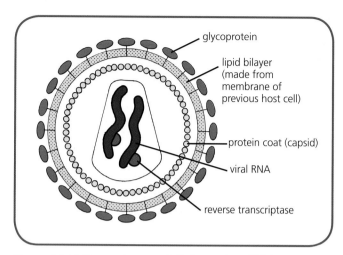

Figure 24.10 Human immunodeficiency virus (HIV)

HIV infection

HIV attacks **helper T lymphocytes**. It becomes attached by **glycoprotein** on its surface to specific receptors on the helper T cell surface (see Figure 24.11). The envelope surrounding the HIV particle fuses with the membrane of the helper T cell and the virus enters the host cell.

HIV is described as a **retrovirus** because it contains RNA. Since it lacks DNA to transcribe into mRNA, it adopts a different strategy. Along with its RNA, it introduces an enzyme called **reverse transcriptase** into the host cell. Using the RNA as a template, this enzyme brings about a reverse version of normal transcription and produces viral DNA from viral RNA.

Viral DNA becomes incorporated into the host cell's DNA where it can remain dormant for many years before directing the synthesis of new viral particles inside the host cells. These escape from the infected

helper T cell by **budding** (see Figure 24.12) and move off to infect other cells. This can cause destruction of the cell or it may undergo apoptosis (see page 315) following an attack by a T_C cell that recognises it as an infected cell.

B cells do make **antibodies** in response to HIV but these are ineffective against viral particles 'hiding' inside the helper T cells. Helper T cells are of critical importance to the immune system since they activate B cells and cytotoxic T cells. As the number of helper T cells gradually drops, the body's immunological activity decreases, leaving the person susceptible to serious **opportunistic infections** such as pneumonia and rare forms of cancer. At this point, after several years of infection by HIV, the person is suffering from AIDS.

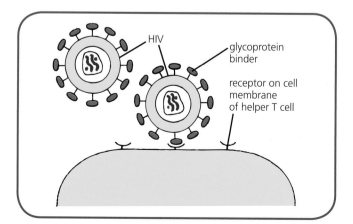

Figure 24.11 Attachment of HIV to a helper T cell

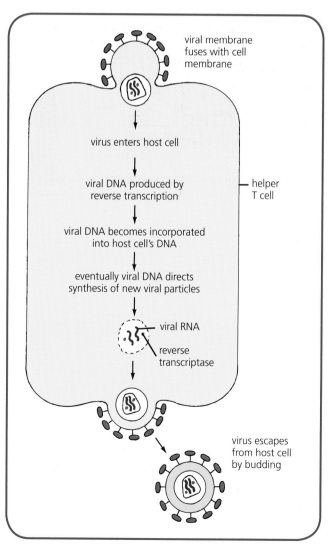

Figure 24.12 Life cycle of HIV

Case Study | HIV

Public health measures

An opportunity for HIV to be transmitted arises when a body fluid such as blood, semen or breast milk containing HIV from one person comes into contact with the mucous membranes or bloodstream of another person. Shared use of non-sterile needles by intravenous drug users and unprotected sex account for most cases of AIDS.

Public health measures to control the spread of AIDS set out to:

- raise public awareness of the problem (see Figure 24.13)

Figure 24.13 Raising awareness

- educate people about how the virus is spread
- supply drug addicts with sterile needles while trying to persuade them to seek treatment for their addiction
- promote the practice of safe sex and the use of condoms.

In the UK, screening of blood and blood products for HIV has eliminated, almost completely, the chance of the virus being transmitted via blood transfusions and blood products.

Drug therapies

At present there is no cure for AIDS. Some drugs do slow down the onset of AIDS but they are expensive and not available to many HIV-positive people, especially those living in developing countries.

Drugs developed to disrupt the action of HIV are called **anti-retrovirals**. They are designed to target different stages of the HIV life cycle by acting as:

- reverse transcriptase inhibitors
- DNA-synthesis inhibitors
- protease inhibitors (which prevent key steps in the synthesis of HIV proteins from occurring).

Combinations of these drugs, accompanied by medicines to treat opportunistic infections, prolong the life of an AIDS sufferer but do not offer a cure. Production of a successful vaccine has so far eluded scientists because the genetic material in HIV **mutates** frequently, forming many **new variants** with different antigenic properties.

Testing Your Knowledge

1 a) Explain how a minority of people in a population can be protected from an infectious disease even if they have not been immunised against it. (2)
   b) Identify TWO factors that affect the herd immunity threshold of an infectious disease. (2)
   c) Give TWO examples of situations where herd immunity to an infectious disease cannot be established, and explain why in each case. (4)

2 a) Name TWO ways by which antigenic variation can arise in a population of a pathogenic microorganism. (2)
   b) Explain why at-risk individuals need to be vaccinated against influenza every year. (2)
   c) Why does the development of an effective vaccine against malaria continue to elude scientists? (1)
   d) By what means does the bacterium responsible for tuberculosis avoid immune detection? (1)

3 Decide whether each of the following statements is true or false and then use T or F to indicate your choice. Where a statement is false, give the word that should have been used in place of the word in bold print. (5)
   a) The more **virulent** a pathogen, the higher the percentage of the population that needs to be vaccinated to establish herd immunity.
   b) The creation of herd immunity is particularly important in **sparsely** populated environments.
   c) Some pathogens can change their **antibodies** and evade the effect of the host's immunological memory.
   d) HIV, which attacks **phagocytes**, causes acquired immunodeficiency syndrome.
   e) **Antigenic** variation occurs in the 'sleeping sickness' pathogen, enabling its variants to avoid the effect of the host's antibodies.

What You Should Know Chapters 23–24

antigens	helper T cells	phagocytes
attack	herd	poor
bias	host	quarantine
clinical	hygiene	randomised
comparison	immunity	rejected
detection	influenza	sporadic
efficacy	inhaled	threshold
endemic	malaria	toxin
epidemic	non-immune	vaccination
epidemiology	pandemic	variation
evade	parameters	vector
experimental	pathogens	virulence

Table 24.2 Word bank for chapters 23–24

1 Infectious diseases caused by _____ such as bacteria and viruses are transmitted by many means including _____ air, body fluids and _____ organisms.

2 _____, antisepsis, good _____, clean water and removal of vectors are some of the ways by which transmission of infectious diseases can be controlled.

3 The study of an infectious disease's outbreak and the factors that affect its spread is called _____. The spread of a disease that occurs occasionally is described as _____, one that occurs as a regular number of cases in a particular area as _____, one that occurs as an unusually high number of cases in an area as _____ and one that occurs as a global epidemic as _____.

4 Weakened or altered forms of an infectious pathogen or its _____ are used as _____. These are administered to people by _____ so that they will develop active _____ to the disease but not suffer the disease.

5 The safety and _____ of new vaccines must be established by _____ trials before they can be licensed for use.

6 A clinical trial is _____ and double-blind to eliminate _____ and placebo-controlled to allow a valid _____ to be made. _____ error is reduced by using a very large number of people.

7 Within a population where most individuals are immune to an infectious disease, vulnerable _____ individuals are protected from it by _____ immunity.

8 The percentage of immune individuals in a population above which a disease no longer persists is called the

→

herd immunity _____. It depends on the pathogen's _____, the vaccine's efficacy and the population's contact _____.

9 Herd immunity can be difficult to establish if the country is too _____ to afford mass vaccination or if the vaccine is _____ by a large percentage of the population.

10 By altering their antigens, some pathogens are able to _____ the immune responses made by their _____.

11 Extensive antigenic _____ in the pathogens that cause _____ and trypanosomiasis has so far prevented scientists from developing a successful vaccine against them. Antigenic variation in the _____ virus makes it necessary to keep developing new vaccines.

12 Some pathogens take advantage of some part of the human immune system and launch a direct _____ on it. HIV attacks _____ and causes AIDS. Tuberculosis bacteria avoid immune _____ by surviving inside _____.

Applying Your Knowledge and Skills

Chapters 21–24

1 Figure 24.14 shows a simplified version of some of the events associated with the inflammatory response.

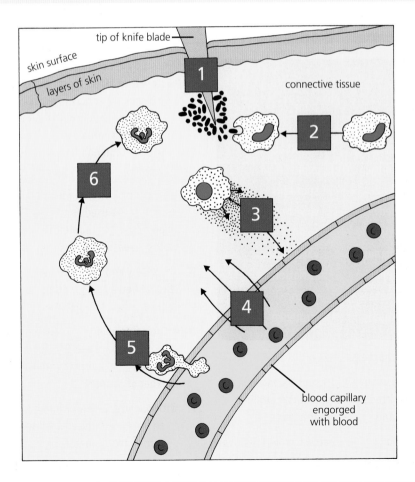

Figure 24.14

a) Match the following answers with blank boxes 1–6 in the diagram. (5)
  A release of histamine by mast cell
  B exit of phagocyte from blood capillary
  C entry of microorganisms at cut
  D attraction of newly arrived phagocyte to infected area
  E action by phagocyte already present in connective tissue
  F increased capillary permeability allowing increased flow of fluid out of capillary

b) *Pain, redness* and *swelling* are features of an inflamed area. Copy and complete Table 24.3 using these three words. (2)

2 Table 24.4 refers to antibody proteins called immunoglobulins found in human blood. The graph in Figure 24.15 refers to the sequence of events that occurs in response to two separate injections of a type of antigen into a small mammal.

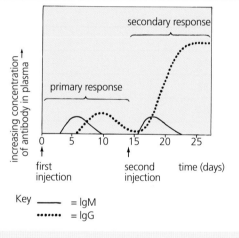

Figure 24.15

a) i) Which immunoglobulin in the table would be found in the blood of an unborn baby?
   ii) Suggest why these antibodies are only needed by the baby for a few months after birth. (2)
b) Of the five types of immunoglobulin molecule, which is the i) largest, ii) rarest? (2)
c) State the normal serum concentration of IgA in mg/ml. (1)
d) With reference to IgM and IgG, state ONE feature common to both the primary and secondary response shown in the graph. (1)
e) With reference to IgG, state THREE differences between the primary and the secondary response. (3)
f) Antibodies such as IgG are now known to be produced by the activity of long-lived lymphocytes. With reference to the graph, suggest why the latter are called *memory cells*. (1)

3 Give an account of the role played by T lymphocytes in the defence of the body. (9)

4 Table 24.5 shows the results for phase II of a clinical trial on an influenza vaccine.
  a) Present the data as a bar chart, including error bars. (See Appendix 3 for help.) (4)
  b) i) Based on these results, should this clinical trial proceed to phase III?
     ii) Explain your answer. (2)

Feature of an inflamed area	Reason
	An increased blood supply is sent to the affected area
	Fluid is forced out of blood vessels into the tissues at the site of injury
	Swollen tissues press against nerve receptors and nerves

Table 24.3

	Immunoglobulin (Ig)				
	IgA	IgD	IgE	IgG	IgM
Molecular weight	170 000	184 000	188 100	150 000	960 000
Normal serum concentration	1.4–4.0 g/l	0.1–0.4 g/l	0.1–1.3 mg/l	8.0–16.0 g/l	0.5–2.0 g/l
Ability to cross placenta	No	No	No	Yes	No

Table 24.4

351

Time from vaccination (days)	Relative viral load in blood plasma (units)	
	Vaccine group	Control group
1	39 ± 6	51 ± 8
2	38 ± 6	48 ± 7
3	23 ± 5	31 ± 5
4	12 ± 3	18 ± 4

Table 24.5

Year	Number of notifications of disease	Uptake of vaccine (%)
1965	34 000	No vaccine available
1967	31 000	No vaccine available
1969	32 000	80
1971	28 000	80
1973	29 000	80
1975	30 000	40
1977	48 000	32
1979	35 000	44
1981	28 000	54
1983	25 000	58
1985	22 000	64
1987	15 000	74
1989	31 000	78
1991	13 000	84
1993	4 000	86
1995	1 000	90

Table 24.6

5   The data in Table 24.6 refer to the disease pertussis (whooping cough) in a European country.
   a) Plot the data as two line graphs on the same sheet of graph paper with one x-axis and two y-axes. (4)
   b) i)   During which years were there 22 000 notifications of the disease?
      ii) What was the percentage uptake of the vaccine in 1982? (2)
   c) i)   Describe the pattern of uptake of vaccine during years 1973–77.
      ii) What effect did this have on incidence of the disease? (2)

d) i)   Describe the pattern of uptake of vaccine during years 1977–92.
   ii) In general, what effect did this have on incidence of the disease? (2)
e) i)   Which year required the introduction of a new vaccine for pertussis?
   ii) Suggest why. (2)

6 The graph in Figure 24.16 refers to a person infected with HIV.

a) i) What happens to the relative concentration of HIV between 6 months and 1 year after infection?

   ii) Suggest why. (2)

b) i) What relationship exists between the relative HIV concentration and relative helper T cell concentration from 2 years after infection onwards?

   ii) Explain why. (2)

c) i) Which type of white blood cell makes antibodies?

   ii) Why does the relative antibody concentration remain constant between years 1–9 after infection yet the relative HIV concentration increases? (2)

d) i) At which of the following times after infection would this person be most likely to be suffering AIDS?

   A 6 months   **B** 1 year, 3 months
   C 4 years, 6 months   **D** 10 years

   ii) Explain your choice of answer. (2)

e) Why has it proved impossible so far to make people immune to HIV by vaccination? (1)

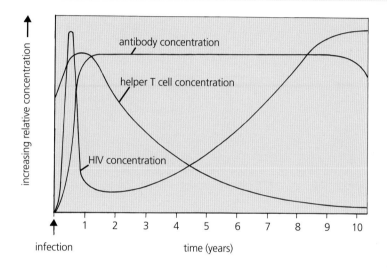

Figure 24.16

Note: Since this group of questions does not include examples of every type of question found in SQA exams, it is recommended that students also make use of past exam papers to aid learning and revision.

Appendix 1

The genetic code

		Second letter of triplet					
		A	**G**	**T**	**C**		
First letter of triplet	**A**	AAA	AGA	ATA	ACA	A	**Third letter of triplet**
		AAG	AGG	ATG	ACG	G	
		AAT	AGT	ATT	ACT	T	
		AAC	AGC	ATC	ACC	C	
	G	GAA	GGA	GTA	GCA	A	
		GAG	GGG	GTG	GCG	G	
		GAT	GGT	GTT	GCT	T	
		GAC	GGC	GTC	GCC	C	
	T	TAA	TGA	TTA	TCA	A	
		TAG	TGG	TTG	TCG	G	
		TAT	TGT	TTT	TCT	T	
		TAC	TGC	TTC	TCC	C	
	C	CAA	CGA	CTA	CCA	A	
		CAG	CGG	CTG	CCG	G	
		CAT	CGT	CTT	CCT	T	
		CAC	CGC	CTC	CCC	C	

Table Ap 1.1 The DNA bases grouped into 64 (4^3) triplets
(A = adenine, G = guanine, T = thymine, C = cytosine)

Abbreviation	Amino acid
ala	alanine
arg	arginine
asp	aspartic acid
asn	asparagine
cys	cysteine
glu	glutamic acid
gln	glutamine
gly	glycine
his	histidine
ile	isoleucine
leu	leucine
lys	lysine
met	methionine
phe	phenylalanine
pro	proline
ser	serine
thr	threonine
trp	tryptophan
tyr	tyrosine
val	valine

Table Ap 1.2 Key to amino acids

Appendix 2

Box plots

The data in Table Ap 2.1 refer to the birth weights of babies in three different groups. Each group contains 15 babies selected at random from a large number of individuals. The mothers of group A were non-smokers, those of group B smoked 20 cigarettes per day while pregnant and those of group C smoked 40 cigarettes per day while pregnant. It is difficult to compare the variability in birth weight between the three groups from the data table alone.

Baby number	Birth weight of baby (kg)		
	Group A	Group B	Group C
15	3.98	3.71	3.42
14	3.75	3.53	3.30
13	3.72	3.49	3.29
12	3.69	3.45	3.24
11	3.62	3.32	3.20
10	3.55	3.27	3.18
9	3.46	3.18	3.15
8	3.38	3.10	3.01
7	3.24	3.06	2.96
6	3.15	2.97	2.92
5	3.12	2.90	2.88
4	3.04	2.86	2.84
3	3.00	2.82	2.78
2	2.96	2.77	2.66
1	2.74	2.53	2.38

Table Ap 2.1

A **box plot** is a way of presenting information that allows differences between groups, sets, populations, etc. to be compared easily. Each box plot shows the **median**, which is the **central value** in the series of values when they are arranged in order. A box plot also displays the **upper quartile** (in this case the value 25% above the median) and the **lower quartile** (the value 25% below the median). The maximum and minimum values are called **upper and lower whiskers**. Figure Ap 2.1 shows how the data for group A are converted into a box plot.

Figure Ap 2.2 shows group A's box plot drawn alongside those for groups B and C. The box plots give a clear, visual representation that allows the variability between the three groups to be compared more easily than by studying the table of data alone.

In this case, the box plots show, at a glance, the depressant effect of smoking on the birth weight of babies.

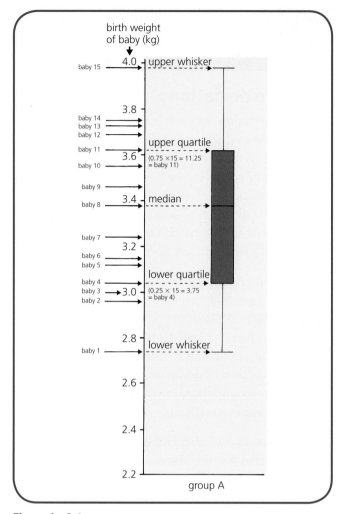

Figure Ap 2.1

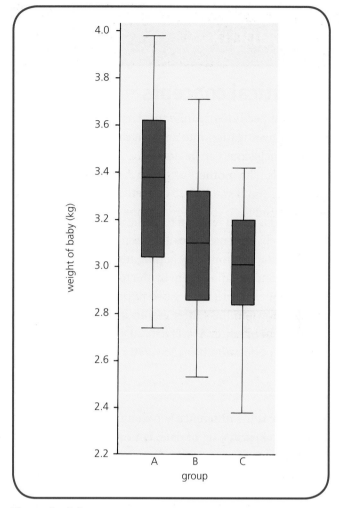

Figure Ap 2.2

Appendix 3

Statistical concepts

A scientist needs to organise the data collected as results from an investigation into a manageable form from which conclusions can be drawn.

Mean

The **mean** is often referred to as the average. It is the most widely used measure of the **central tendency** of a set of data. It is found by adding up all the values obtained and dividing them by the total number of values. For example, for the two populations of British miners shown in the scatter graphs in Figure Ap 3.1, the mean for population A = 11 830/70 = 169 cm and the mean for population B = 12 530/70 = 179 cm.

Range

The **range** is the difference between the two most extreme values in a set of data. For example, for population A the range = 180 − 158 = 22 cm and for population B the range = 192 − 164 = 28 cm.

Standard deviation

Standard deviation is a measure of the spread of individual data values around their mean and shows how much variation from the mean exists. A normal distribution of results can be divided into intervals of standard deviation as shown in Figure Ap 3.2. 68% of the values fall within plus or minus one standard deviation from the mean; 95% of the values fall within plus or minus two standard deviations from the mean.

The standard deviation of a set of data is calculated using a mathematical formula (often with the aid of an appropriate calculator or computer software). The deviation (as two standard deviations above or below the mean) for population A in Figure Ap 3.1 equals 9 cm. This low level of deviation reflects the clustering of the values around the mean, with 95% of values lying within the range 160–178 cm.

The deviation (as two standard deviations above or below the mean) for population B in Figure Ap 3.1 equals 11 cm. This higher level of deviation reflects the

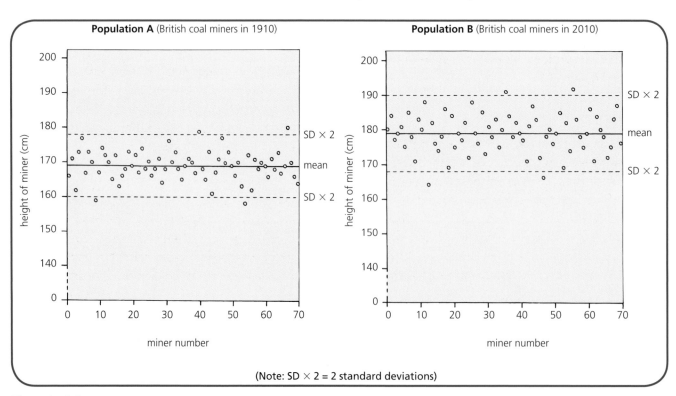

(Note: SD × 2 = 2 standard deviations)

Figure Ap 3.1

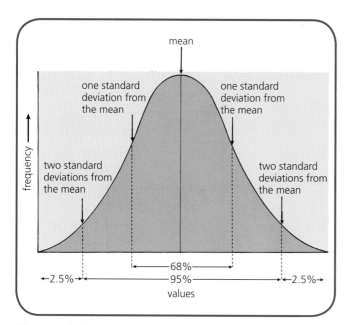

Figure Ap 3.2

Figure Ap 3.3

wider spread of the values around the mean, with 95% of values lying within the range 168–190 cm.

Quality of data

In a properly designed scientific investigation, several **replicates** of each treatment are set up to allow for experimental error. These replicates produce results with a **central tendency** around the **mean**. A set of results that are **clustered** around the mean indicates data of **high quality**. A comparable set of results (from a replicate of the same treatment) that are widespread are of lower quality.

Significant difference

In biology, an experiment is carried out to test a hypothesis. Once results have been obtained, the scientist needs to know whether these data (which rarely conform exactly to the expected outcome) support the hypothesis or not. A **significance test** (a type of statistical analysis) can be used to find out if the observed differences between two sets of data are **statistically significant** or simply the result of **chance**.

In an investigation into perceptual set (see page 235), three groups of 50 volunteers were used. Before seeing and being asked to identify the ambiguous image shown in Figure Ap 3.3, the members of group A were shown images of birds, those of group B images of small mammals and those of group C no images. Table Ap 3.1 shows the results.

A **plus sign** after a result indicates that the significance test shows the value to be **significantly higher** than would be expected by chance alone; a **minus sign** after a result indicates a value to be **significantly lower** than would be expected by chance alone.

The results from groups A and B (each affected by a different form of perceptual set) both differ significantly from what would be expected by chance alone. The results from group C (not affected by perceptual set) show no such significant difference.

Error bars

When a bar chart of mean values of data is drawn, it is often important to be able to show variability on the chart. This can be done using **error bars**. These are lines that extend outside and inside each bar and indicate how far from the mean value the true error-free value

Number of people in group A who said:		Number of people in group B who said:		Number of people in group C who said:	
Bird	Rabbit	Bird	Rabbit	Bird	Rabbit
47 (+)	3 (−)	5 (−)	45 (+)	24	26

Table Ap 3.1

is likely to be. Error bars can be based on aspects of variability such as 95% level of confidence and standard deviation.

Figure Ap 3.4 shows a bar chart of the results from a survey carried out on several thousand young people in a country to estimate the incidence of asthma. Each bar represents a mean value with a 95% level of confidence whose range is indicated by error bars. Based on the information in the bar chart, health care experts could be 95% confident that the percentage number of asthma cases for the whole population would be between 13% and 21% for 2–4-year-old males, between 7 and 14% for 9–15-year-old females, etc.

Error bars also allow a comparison to be made between two means to determine if they are **significantly different** from one another. If their error bars (based on 95% level of confidence) do not overlap, the difference between the two means is regarded as being significant.

Figure Ap 3.5 shows a bar chart of the results from Table 12.1 on page 145 of mean concentration of urea in the blood plasma of non-pregnant women (NP), pregnant women (P) and pregnant women suffering pre-eclampsia (PE). Each bar represents a mean value with its variability indicated by error bars. The mean for PE is seen to be significantly different from those for both NP and P, which are not significantly different from one another.

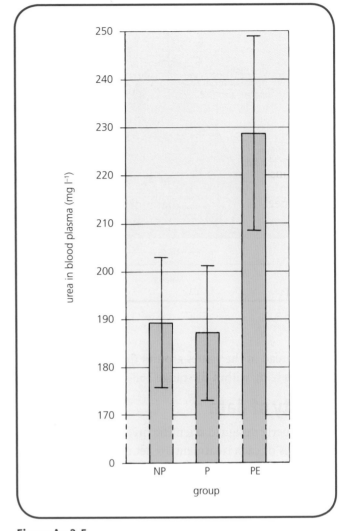

Figure Ap 3.5

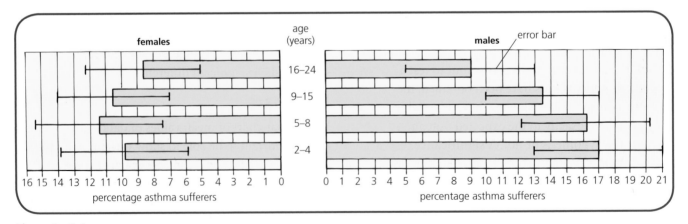

Figure Ap 3.4

Appendix 4

False positives and negatives

A **false positive** is a result that indicates that the outcome of an investigative procedure is a positive result when in reality the outcome for the set of conditions being tested is negative.

A **false negative** is a result that indicates that the outcome of an investigative procedure is a negative result when in reality the outcome for the set of conditions being tested is positive.

Table Ap 4.1 shows data from an investigation into the effectiveness of ultrasound imaging on the antenatal detection of inherited malformations in babies born in a region of the UK.

The results in row 3 of the table are examples of **false positives** because the procedure (screening for malformations) has produced results indicating that these fetuses had inherited malformations when in fact the babies were found to be normal at birth.

The results in row 4 are examples of **false negatives** because the procedure (screening for malformations) has produced results indicating that these fetuses had not inherited malformations when in fact the babies were found at birth to possess them.

Row	Category	Year		
		2001	**2005**	**2009**
1	Total number of births	36 433	33 292	29 779
2	Number of babies born with malformations	790	785	762
3	Number of babies reported antenatally as having malformations but normal at birth	151	119	52
4	Number of babies reported antenatally as being normal but born with malformations	487	464	398

Table Ap 4.1

Index